An Overview of Twistor, String & Quantum Field Theory

" Mapping Minkowski Space to Twistor Space "

Edited by Paul F. Kisak

Contents

Chapter 1

Twistor theory

In theoretical and mathematical physics, **twistor theory** maps the geometric objects of conventional 3+1 space-time (Minkowski space) into geometric objects in a 4-dimensional space endowed with a Hermitian form of signature (2,2). This space is called twistor space, and its complex valued coordinates are called "twistors."

Twistor theory was first proposed by Roger Penrose in 1967,[1] as a possible path to a theory of quantum gravity. The twistor approach is especially natural for solving the equations of motion of massless fields of arbitrary spin.

In 2003, Edward Witten[2] proposed uniting twistor and string theory by embedding the topological B model of string theory in twistor space. His objective was to model certain Yang–Mills amplitudes. The resulting model has come to be known as twistor string theory (read below). Simone Speziale and collaborators have also applied it to loop quantum gravity.[3]

1.1　Details

Twistor theory is unique to 4D Minkowski space and the (2,2) signature, and does not generalize to other dimensions or signatures. At the heart of twistor theory lies the isomorphism between the conformal group Spin(4,2) and SU(2,2), which is the group of unitary transformations of determinant 1 over a four-dimensional complex vector space that leave invariant a Hermitian form of signature (2,2), see classical group.

- $\mathbb{R}^6$ is the real 6D vector space corresponding to the vector representation of Spin(4,2).

- $\mathbf{RP}^5$ is the real 5D projective representation corresponding to the equivalence class of nonzero points in $\mathbb{R}^6$ under scalar multiplication.

- M^c corresponds to the subspace of $\mathbf{RP}^5$ corresponding to vectors of zero norm. This is conformally compactified Minkowski space.

- $\mathbb{T}$ is the 4D complex Weyl spinor representation, called twistor space. It has an invariant Hermitian sesquilinear norm of signature (2,2).

- $\mathbb{PT}$ is a 3D complex manifold corresponding to projective twistor space.

- $\mathbb{PT}^+$ is the subspace of $\mathbb{PT}$ corresponding to projective twistors with positive norm (the sign of the norm, but not its absolute value is projectively invariant). This is a 3D complex manifold.

- $\mathbb{PN}$ is the subspace of $\mathbb{PT}$ consisting of null projective twistors (zero norm). This is a real-complex manifold (i.e., it has 5 real dimensions, with four of the real dimensions having a complex structure making them two complex dimensions).

- $\mathbb{PT}^-$ is the subspace of $\mathbb{PT}$ of projective twistors with negative norm.

$\mathbb{M}^c$, $\mathbb{PT}^+$, $\mathbb{PN}$ and $\mathbb{PT}^-$ are all homogeneous spaces of the conformal group.

$\mathbb{M}^c$ admits a conformal metric (i.e., an equivalence class of metric tensors under Weyl rescalings) with signature (+++−). Straight null rays map to straight null rays under a conformal transformation and there is a unique canonical isomorphism between null rays in $\mathbb{M}^c$ and points in $\mathbb{PN}$ respecting the conformal group.

In $\mathbb{M}^c$, it is the case that positive and negative frequency solutions cannot be locally separated. However, this is possible in twistor space.

$$\mathbb{PT}^+ \simeq SU(2,2)/\left[SU(2,1) \times U(1)\right]$$

1.2   Twistor string theory

Main article: Twistor string theory

For many years after Penrose's foundational 1967 paper, twistor theory progressed slowly, in part because of mathematical challenges. Twistor theory also seemed unrelated to ideas in mainstream physics. While twistor theory appeared to say something about quantum gravity, its potential contributions to understanding the other fundamental interactions and particle physics were less obvious.

Witten (2003) proposed a connection between string theory and twistor geometry, called *twistor string theory*. Witten (2004)[2] built on this insight to propose a way to do string theory in twistor space, whose dimensionality is necessarily the same as that of 3+1 Minkowski spacetime. Although Witten has said that "I think twistor string theory is something that only partly works," his work has given new life to the twistor research program. For example, twistor string theory may simplify calculating scattering amplitudes from Feynman diagrams by using a geometric structure called an amplituhedron.

1.3   Supertwistors

Witten's twistor string theory is defined on the supertwistor space $\mathbb{CP}^{3|4}$. Supertwistors are a supersymmetric extension of twistors introduced by Alan Ferber in 1978.[4] Along with the standard twistor degrees of freedom, a supertwistor contains N fermionic scalars, where N is the number of supersymmetries. The superconformal algebra can be realized on supertwistor space.

1.4   See also

- Penrose transform
- Twistor space
- Invariance mechanics

1.5   Notes

[1] Penrose, R. (1967) "Twistor algebra," *J. Math. Phys.* 8: 345.

[2] Witten, E. (2004) "Perturbative gauge theory as a string theory in twistor space," *Commun. Math. Phys.* 252: 189–258.

[3] http://arxiv.org/abs/1006.0199

[4] Ferber, A (1978), "Supertwistors and conformal supersymmetry", *Nuclear Physics B* **132**: 55–64, Bibcode:1978NuPhB.132...55F, doi:10.1016/0550-3213(78)90257-2.

1.6 Further reading

- Baird, Paul "An Introduction To Twistors"

- Penrose,Roger(1967),"Twistor algebra",*Journal of Mathematical Physics* **8**(2): 345–366,Bibcode:1967JMP.....8.. 345P, doi:10.1063/1.1705200, MR 0216828

- Penrose, Roger (1968), "Twistor quantisation and curved space-time", *International Journal of Theoretical Physics* (Springer Netherlands) **1**: 61–99, Bibcode:1968IJTP....1...61P, doi:10.1007/BF00668831

- Penrose, Roger (1969), "Solutions of the Zero-Rest-Mass Equations", *Journal of Mathematical Physics* **10** (1): 38–39, Bibcode:1969JMP....10...38P, doi:10.1063/1.1664756

- Penrose,Roger(1977), "The twistor programme",*Reports on Mathematical Physics* **12**(1): 65–76,Bibcode:1977 RpMP...12...65P, doi:10.1016/0034-4877(77)90047-7, MR 0465032

- Penrose, Roger (1987) "On the Origins of Twistor Theory" in *Gravitation and Geometry*, a volume in honour of I. Robinson. Naples: Bibliopolis.

- Penrose, Roger (1999) "The Central Programme of Twistor Theory," *Chaos, Solitons and Fractals* 10: 581-611.

- Arkani-Hamed,Nima;Cachazo,Freddy;Cheung,Clifford;Kaplan,Jared(2009) "The S-Matrix in Twistor Space."

- Witten, Edward (2003), "Perturbative Gauge Theory As A String Theory In Twistor Space", *Communications in Mathematical Physics* **252**: 189–258,arXiv:hep-th/0312171,Bibcode:2004CMaPh.252..189W,doi:10.1007/s00220-004-1187-3

1.7 External links

- Penrose, Roger (1999) "Einstein's Equation and Twistor Theory: Recent Developments"

- Penrose, Roger; Hadrovich, Fedja. "Twistor Theory."

- Dunajski, Maciej, "Twistor Theory and Differential Equations."

- Hadrovich, Fedja, "Twistor primer."

- Andrew Hodges, "Twistor Theory and the Twistor Programme." Includes many links.

- Huggett, Stephen (2005) "The Elements of Twistor Theory."

- Richard Jozsa (1976) "Applications of Sheaf Cohomology in Twistor Theory."

- Mason, L. J., "The twistor programme and twistor strings:From twistor strings to quantum gravity?"

- Sämann, Christian (2006) "Aspects of Twistor Geometry and Supersymmetric Field Theories within Superstring Theory."

- Sparling, George (1999) "On Time Asymmetry."

- Spradlin, Marcus (2006), "Progress And Prospects In Twistor String Theory."

Chapter 2

Twistor space

In mathematics, **twistor space** is the complex vector space of solutions of the twistor equation $\nabla^{(A}_{A'}\Omega^{B)} = 0$. It was described in the 1960s by Roger Penrose and MacCallum.[1] According to Andrew Hodges, twistor space is useful for conceptualizing the way photons travel through space, using four complex numbers. He also posits that twistor space may aid in understanding the asymmetry of the weak nuclear force.[2]

For Minkowski space, denoted $\mathbb{M}$, the solutions to the twistor equation are of the form

$$\Omega^A(x) = \omega^A - ix^{AA'}\pi_{A'}$$

where ω^A and $\pi_{A'}$ are two constant Weyl spinors and $x^{AA'} = \sigma^{AA'}_\mu x^\mu$ is a point in Minkowski space. This twistor space is a four-dimensional complex vector space, whose points are denoted by $Z^\alpha = (\omega^A, \pi_{A'})$, and with a hermitian form

$$\Sigma(Z) = \omega^A\bar{\pi}_A + \bar{\omega}^{A'}\pi_{A'}$$

which is invariant under the group SU(2,2) which is a quadruple cover of the conformal group C(1,3) of compactified Minkowski spacetime.

Points in Minkowski space are related to subspaces of twistor space through the incidence relation

$$\omega^A = ix^{AA'}\pi_{A'}.$$

This incidence relation is preserved under an overall re-scaling of the twistor, so usually one works in projective twistor space, denoted **PT**, which is isomorphic as a complex manifold to $\mathbb{CP}^3$.

Given a point $x \in M$ it is related to a line in projective twistor space where we can see the incidence relation as giving the linear embedding of a $\mathbb{CP}^1$ parametrized by $\pi_{A'}$.

The geometric relation between projective twistor space and complexified compactified Minkowski space is the same as the relation between lines and two-planes in twistor space; more precisely, twistor space is

T := $\mathbf{C}^4$. It has associated to it the double fibration of flag manifolds **P** $\leftarrow_\mu$ **F** $_\nu\rightarrow$ **M**, where

    projective twistor space

        P := $\mathbf{F}_1(\mathbf{T}) = \mathbf{P}_3(\mathbf{C}) = \mathbf{P}(\mathbf{C}^4)$

    compactified complexified Minkowski space

        M := $\mathbf{F}_2(\mathbf{T}) = \mathbf{G}_2(\mathbf{C}^4) = \mathbf{G}_{2,4}(\mathbf{C})$

the correspondence space between **P** and **M**

$$F := F_{1,2}(T)$$

In the above, **P** stands for projective space, **G** a Grassmannian, and **F** a flag manifold. The double fibration gives rise to two correspondences, $c := \nu \cdot \mu^{-1}$ and $c^{-1} := \mu \cdot \nu^{-1}$.

M is embedded in $\mathbf{P}_5 \simeq \mathbf{P}(\Lambda^2 T)$ by the Plücker embedding and the image is the Klein quadric.

2.1 Rationale

In the (translated) words of Jacques Hadamard: "the shortest path between two truths in the real domain passes through the complex domain." Therefore when studying $\mathbf{R}^4$ it might be valuable to identify it with $\mathbf{C}^2$. However, since there is no canonical way of doing so, instead all isomorphisms respecting orientation and metric between the two are considered. It turns out that complex projective 3-space $\mathbf{P}_3(\mathbf{C})$ parametrizes such isomorphisms together with complex coordinates. Thus one complex coordinate describes the identification and the other two describe a point in $\mathbf{R}^4$. It turns out that vector bundles with self-dual connections on $\mathbf{R}^4$(instantons) correspond bijectively to holomorphic bundles on complex projective 3-space $\mathbf{P}_3(\mathbf{C})$.

2.2 See also

- Roger Penrose
- Twistor theory

2.3 References

[1] R. Penrose and M. A. H. MacCallum, Twistor theory: An approach to the quantisation of fields and space-time. doi:10.1016/0370-1573(73)90008-2

[2] Hodges, Andrew, "One to Nine" 2009

- Ward, R.S. and Wells, Raymond O. Jr., *Twistor Geometry and Field Theory*, Cambridge University Press (1991). ISBN 0-521-42268-X.

- Huggett, S. A. and Tod, K. P., *An introduction to twistor theory*, Cambridge University Press (1994). ISBN 978-0-521-45689-0.

Chapter 3

Twistor string theory

Twistor string theory is an equivalence between N = 4 supersymmetric Yang–Mills theory and the peturbative topological B model string theory in twistor space.[1] It was initially proposed by Edward Witten in 2003.

3.1 Related links

- Amplituhedron
- BCFW recursion
- MHV amplitudes

3.2 References

[1] Witten, Edward (December 2003). "Perturbative Gauge Theory As A String Theory In Twistor Space:". *Communications in Mathematical Physics*. 1 **252** (1): 189. arXiv:hep-th/0312171. Bibcode:2004CMaPh.252..189W. doi:10.1007/s00220-004-1187-3.

Chapter 4

Vector space

This article is about linear (vector) spaces. For the structure in incidence geometry, see Linear space (geometry).

A **vector space** (also called a **linear space**) is a collection of objects called **vectors**, which may be added together and

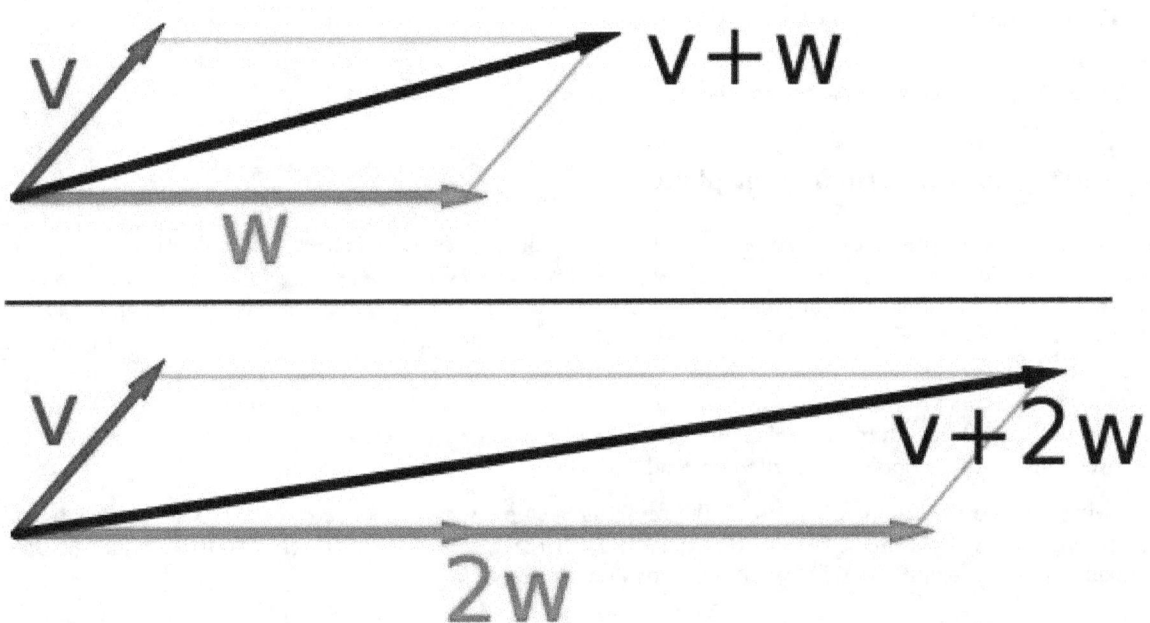

Vector addition and scalar multiplication: a vector v (blue) is added to another vector w (red, upper illustration). Below, w is stretched by a factor of 2, yielding the sum v + 2w.

multiplied ("scaled") by numbers, called *scalars* in this context. Scalars are often taken to be real numbers, but there are also vector spaces with scalar multiplication by complex numbers, rational numbers, or generally any field. The operations of vector addition and scalar multiplication must satisfy certain requirements, called *axioms*, listed below.

Euclidean vectors are an example of a vector space. They represent physical quantities such as forces: any two forces (of the same type) can be added to yield a third, and the multiplication of a force vector by a real multiplier is another force vector. In the same vein, but in a more geometric sense, vectors representing displacements in the plane or in three-dimensional space also form vector spaces. Vectors in vector spaces do not necessarily have to be arrow-like objects as they appear in the mentioned examples: vectors are regarded as abstract mathematical objects with particular properties, which in some cases can be visualized as arrows.

Vector spaces are the subject of linear algebra and are well understood from this point of view since vector spaces are characterized by their dimension, which, roughly speaking, specifies the number of independent directions in the space. A vector space may be endowed with additional structure, such as a norm or inner product. Such spaces arise naturally in mathematical analysis, mainly in the guise of infinite-dimensional function spaces whose vectors are functions. Analytical problems call for the ability to decide whether a sequence of vectors converges to a given vector. This is accomplished by considering vector spaces with additional structure, mostly spaces endowed with a suitable topology, thus allowing the consideration of proximity and continuity issues. These topological vector spaces, in particular Banach spaces and Hilbert spaces, have a richer theory.

Historically, the first ideas leading to vector spaces can be traced back as far as the 17th century's analytic geometry, matrices, systems of linear equations, and Euclidean vectors. The modern, more abstract treatment, first formulated by Giuseppe Peano in 1888, encompasses more general objects than Euclidean space, but much of the theory can be seen as an extension of classical geometric ideas like lines, planes and their higher-dimensional analogs.

Today, vector spaces are applied throughout mathematics, science and engineering. They are the appropriate linear-algebraic notion to deal with systems of linear equations; offer a framework for Fourier expansion, which is employed in image compression routines; or provide an environment that can be used for solution techniques for partial differential equations. Furthermore, vector spaces furnish an abstract, coordinate-free way of dealing with geometrical and physical objects such as tensors. This in turn allows the examination of local properties of manifolds by linearization techniques. Vector spaces may be generalized in several ways, leading to more advanced notions in geometry and abstract algebra.

4.1   Introduction and definition

The concept of vector space will first be explained by describing two particular examples:

4.1.1   First example: arrows in the plane

The first example of a vector space consists of arrows in a fixed plane, starting at one fixed point. This is used in physics to describe forces or velocities. Given any two such arrows, $\mathbf{v}$ and $\mathbf{w}$, the parallelogram spanned by these two arrows contains one diagonal arrow that starts at the origin, too. This new arrow is called the *sum* of the two arrows and is denoted $\mathbf{v} + \mathbf{w}$. In the special case of two arrows on the same line, their sum is the arrow on this line whose length is the sum or the difference of the lengths, depending on whether the arrows have the same direction. Another operation that can be done with arrows is scaling: given any positive real number a, the arrow that has the same direction as $\mathbf{v}$, but is dilated or shrunk by multiplying its length by a, is called *multiplication* of $\mathbf{v}$ by a. It is denoted $a\mathbf{v}$. When a is negative, $a\mathbf{v}$ is defined as the arrow pointing in the opposite direction, instead.

The following shows a few examples: if $a = 2$, the resulting vector $a\mathbf{w}$ has the same direction as $\mathbf{w}$, but is stretched to the double length of $\mathbf{w}$ (right image below). Equivalently $2\mathbf{w}$ is the sum $\mathbf{w} + \mathbf{w}$. Moreover, $(-1)\mathbf{v} = -\mathbf{v}$ has the opposite direction and the same length as $\mathbf{v}$ (blue vector pointing down in the right image).

4.1.2   Second example: ordered pairs of numbers

A second key example of a vector space is provided by pairs of real numbers x and y. (The order of the components x and y is significant, so such a pair is also called an ordered pair.) Such a pair is written as (x, y). The sum of two such pairs and multiplication of a pair with a number is defined as follows:

$$(x_1, y_1) + (x_2, y_2) = (x_1 + x_2, y_1 + y_2)$$

and

$$a\,(x, y) = (ax, ay).$$

The first example above reduces to this one if the arrows are represented by the pair of Cartesian coordinates of their end points.

4.1.3 Definition

A vector space over a field F is a set V together with two operations that satisfy the eight axioms listed below. Elements of V are commonly called *vectors*. Elements of F are commonly called *scalars*. The first operation, called *vector addition* or simply *addition*, takes any two vectors **v** and **w** and assigns to them a third vector which is commonly written as **v** + **w**, and called the sum of these two vectors. The second operation, called *scalar multiplication* takes any scalar a and any vector **v** and gives another vector a**v**.

In this article, vectors are distinguished from scalars by boldface.[nb 1] In the two examples above, the field is the field of the real numbers and the set of the vectors consists of the planar arrows with fixed starting point and of pairs of real numbers, respectively.

To qualify as a vector space, the set V and the operations of addition and multiplication must adhere to a number of requirements called axioms.[1] In the list below, let **u**, **v** and **w** be arbitrary vectors in V, and a and b scalars in F.

These axioms generalize properties of the vectors introduced in the above examples. Indeed, the result of addition of two ordered pairs (as in the second example above) does not depend on the order of the summands:

$$(x_v, y_v) + (x_w, y_w) = (x_w, y_w) + (x_v, y_v).$$

Likewise, in the geometric example of vectors as arrows, **v** + **w** = **w** + **v** since the parallelogram defining the sum of the vectors is independent of the order of the vectors. All other axioms can be checked in a similar manner in both examples. Thus, by disregarding the concrete nature of the particular type of vectors, the definition incorporates these two and many more examples in one notion of vector space.

Subtraction of two vectors and division by a (non-zero) scalar can be defined as

$$\mathbf{v} - \mathbf{w} = \mathbf{v} + (-\mathbf{w}),$$
$$\mathbf{v}/a = (1/a)\mathbf{v}.$$

When the scalar field F is the real numbers **R**, the vector space is called a *real vector space*. When the scalar field is the complex numbers, it is called a *complex vector space*. These two cases are the ones used most often in engineering. The general definition of a vector space allows scalars to be elements of any fixed field F. The notion is then known as an F-*vector spaces* or a *vector space over F*. A field is, essentially, a set of numbers possessing addition, subtraction, multiplication and division operations.[nb 3] For example, rational numbers also form a field.

In contrast to the intuition stemming from vectors in the plane and higher-dimensional cases, there is, in general vector spaces, no notion of nearness, angles or distances. To deal with such matters, particular types of vector spaces are introduced; see below.

4.1.4 Alternative formulations and elementary consequences

The requirement that vector addition and scalar multiplication be binary operations includes (by definition of binary operations) a property called closure: that **u** + **v** and a**v** are in V for all a in F, and **u**, **v** in V. Some older sources mention these properties as separate axioms.[2]

In the parlance of abstract algebra, the first four axioms can be subsumed by requiring the set of vectors to be an abelian group under addition. The remaining axioms give this group an F-module structure. In other words, there is a ring homomorphism f from the field F into the endomorphism ring of the group of vectors. Then scalar multiplication a**v** is defined as $(f(a))(\mathbf{v})$.[3]

There are a number of direct consequences of the vector space axioms. Some of them derive from elementary group theory, applied to the additive group of vectors: for example the zero vector **0** of V and the additive inverse $-$**v** of any vector **v** are unique. Other properties follow from the distributive law, for example a**v** equals **0** if and only if a equals 0 or **v** equals **0**.

4.2   History

Vector spaces stem from affine geometry via the introduction of coordinates in the plane or three-dimensional space. Around 1636, Descartes and Fermat founded analytic geometry by equating solutions to an equation of two variables with points on a plane curve.[4] In 1884, to achieve geometric solutions without using coordinates, Bolzano introduced certain operations on points, lines and planes, which are predecessors of vectors.[5] His work was then used in the conception of barycentric coordinates by Möbius in 1827.[6] The definition of vectors was founded on Bellavitis' notion of the bipoint, an oriented segment of which one end is the origin and the other a target, then further elaborated with the presentation of complex numbers by Argand and Hamilton and the introduction of quaternions and biquaternions by the latter.[7] They are elements in $\mathbf{R}^2$, $\mathbf{R}^4$, and $\mathbf{R}^8$; their treatment as linear combinations can be traced back to Laguerre in 1867, who also defined systems of linear equations.

In 1857, Cayley introduced matrix notation, which allows for a harmonization and simplification of linear maps. Around the same time, Grassmann studied the barycentric calculus initiated by Möbius. He envisaged sets of abstract objects endowed with operations.[8] In his work, the concepts of linear independence and dimension, as well as scalar products, are present. In fact, Grassmann's 1844 work exceeds the framework of vector spaces, since his consideration of multiplication led him to what are today called algebras. Peano was the first to give the modern definition of vector spaces and linear maps in 1888.[9]

An important development of vector spaces is due to the construction of function spaces by Lebesgue. This was later formalized by Banach and Hilbert, around 1920.[10] At that time, algebra and the new field of functional analysis began to interact, notably with key concepts such as spaces of p-integrable functions and Hilbert spaces.[11] Vector spaces, including infinite-dimensional ones, then became a firmly established notion, and many mathematical branches started making use of this concept.

4.3   Examples

Main article: Examples of vector spaces

4.3.1   Coordinate spaces

Main article: Coordinate space

The most simple example of a vector space over a field F is the field itself, equipped with its standard addition and multiplication. More generally, a vector space can be composed of n-tuples (sequences of length n) of elements of F, such as

$(a_1, a_2, ..., an)$, where each ai is an element of F.[12]

A vector space composed of all the n-tuples of a field F is known as a *coordinate space*, usually denoted F^n. The case $n = 1$ is the above-mentioned simplest example, in which the field F is also regarded as a vector space over itself. The case $F = \mathbf{R}$ and $n = 2$ was discussed in the introduction above.

4.3.2   Complex numbers and other field extensions

The set of complex numbers $\mathbf{C}$, i.e., numbers that can be written in the form $x + iy$ for real numbers x and y where i is the imaginary unit, form a vector space over the reals with the usual addition and multiplication: $(x + iy) + (a + ib) = (x + a) + i(y + b)$ and $c \cdot (x + iy) = (c \cdot x) + i(c \cdot y)$ for real numbers x, y, a, b and c. The various axioms of a vector space follow from the fact that the same rules hold for complex number arithmetic.

In fact, the example of complex numbers is essentially the same (i.e., it is *isomorphic*) to the vector space of ordered pairs of real numbers mentioned above: if we think of the complex number $x + iy$ as representing the ordered pair (x, y) in the complex plane then we see that the rules for sum and scalar product correspond exactly to those in the earlier example.

More generally, field extensions provide another class of examples of vector spaces, particularly in algebra and algebraic number theory: a field F containing a smaller field E is an E-vector space, by the given multiplication and addition operations of F.[13] For example, the complex numbers are a vector space over **R**, and the field extension $\mathbf{Q}(i\sqrt{5})$ is a vector space over **Q**.

4.3.3 Function spaces

Functions from any fixed set Ω to a field F also form vector spaces, by performing addition and scalar multiplication pointwise. That is, the sum of two functions f and g is the function $(f + g)$ given by

$$(f + g)(w) = f(w) + g(w),$$

and similarly for multiplication. Such function spaces occur in many geometric situations, when Ω is the real line or an interval, or other subsets of **R**. Many notions in topology and analysis, such as continuity, integrability or differentiability are well-behaved with respect to linearity: sums and scalar multiples of functions possessing such a property still have that property.[14] Therefore, the set of such functions are vector spaces. They are studied in greater detail using the methods of functional analysis, see below. Algebraic constraints also yield vector spaces: the vector space $F[x]$ is given by polynomial functions:

$$f(x) = r_0 + r_1 x + \dots + r_{n-1}x^{n-1} + r_n x^n, \text{ where the coefficients } r_0, \dots, r_n \text{ are in } F.[15]$$

4.3.4 Linear equations

Main articles: Linear equation, Linear differential equation and Systems of linear equations

Systems of homogeneous linear equations are closely tied to vector spaces.[16] For example, the solutions of

are given by triples with arbitrary a, $b = a/2$, and $c = -5a/2$. They form a vector space: sums and scalar multiples of such triples still satisfy the same ratios of the three variables; thus they are solutions, too. Matrices can be used to condense multiple linear equations as above into one vector equation, namely

$$A\mathbf{x} = \mathbf{0},$$

where $A = \begin{bmatrix} 1 & 3 & 1 \\ 4 & 2 & 2 \end{bmatrix}$ is the matrix containing the coefficients of the given equations, $\mathbf{x}$ is the vector (a, b, c), $A\mathbf{x}$ denotes the matrix product, and $\mathbf{0} = (0, 0)$ is the zero vector. In a similar vein, the solutions of homogeneous *linear differential equations* form vector spaces. For example,

$$f''(x) + 2f'(x) + f(x) = 0$$

yields $f(x) = a\,e^{-x} + bx\,e^{-x}$, where a and b are arbitrary constants, and e^x is the natural exponential function.

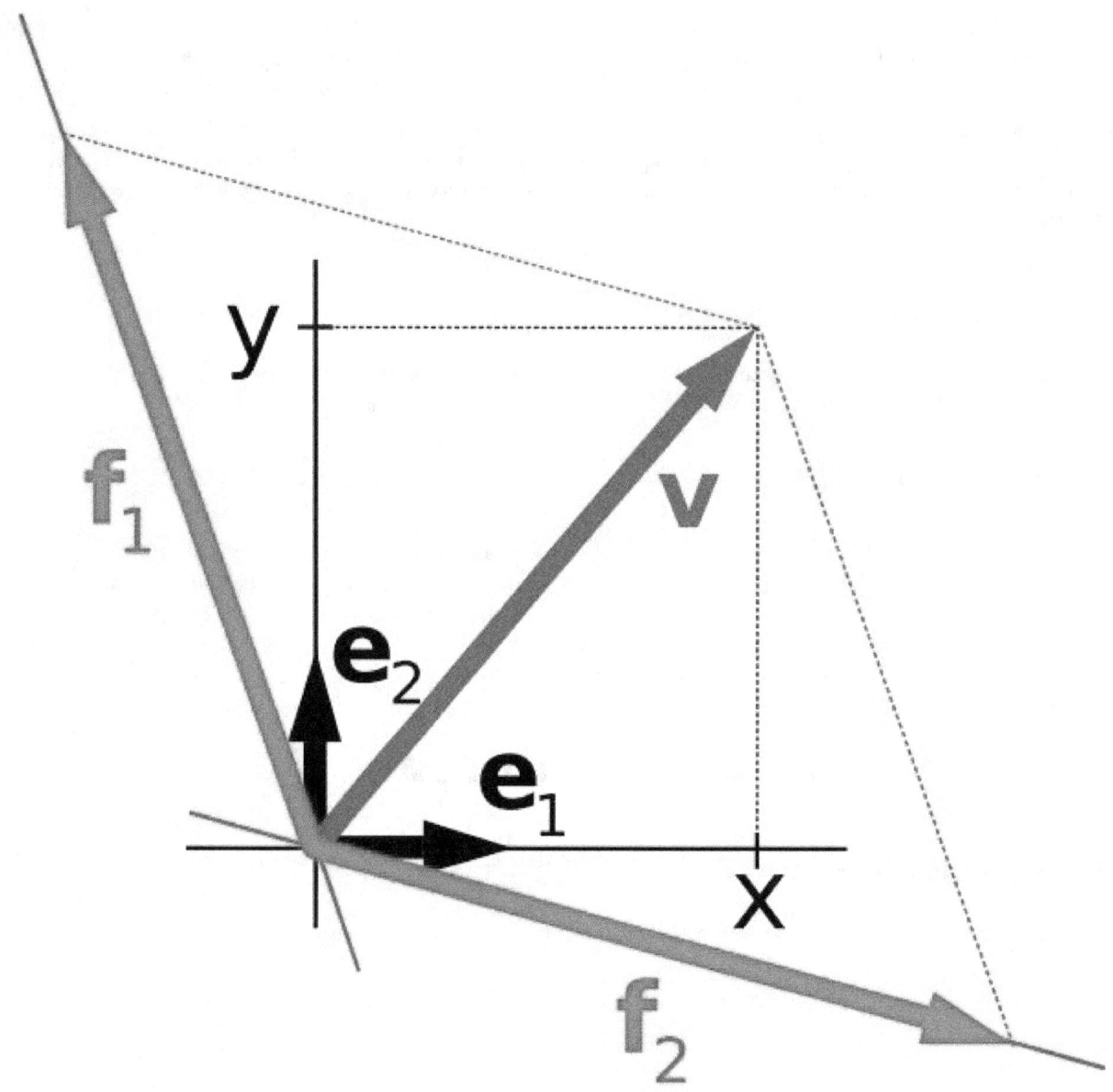

A vector v in $\mathbf{R}^2$ (blue) expressed in terms of different bases: using the standard basis of $\mathbf{R}^2$ $v = xe_1 + ye_2$ (black), and using a different, non-orthogonal basis: $v = f_1 + f_2$ (red).

4.4   Basis and dimension

Main articles: Basis and Dimension

Bases allow to represent vectors by a sequence of scalars called *coordinates* or *components*. A basis is a (finite or infinite) set $B = \{\mathbf{b}i\}i \in I$ of vectors $\mathbf{b}i$, for convenience often indexed by some index set I, that spans the whole space and is linearly independent. "Spanning the whole space" means that any vector $\mathbf{v}$ can be expressed as a finite sum (called a *linear combination*) of the basis elements:

where the ak are scalars, called the coordinates (or the components) of the vector $\mathbf{v}$ with respect to the basis B, and $\mathbf{b}ik$ ($k = 1, ..., n$) elements of B. Linear independence means that the coordinates ak are uniquely determined for any vector in the vector space.

For example, the coordinate vectors $e_1 = (1, 0, ..., 0)$, $e_2 = (0, 1, 0, ..., 0)$, to $e_n = (0, 0, ..., 0, 1)$, form a basis of F^n, called the standard basis, since any vector $(x_1, x_2, ..., x_n)$ can be uniquely expressed as a linear combination of these vectors:

$$(x_1, x_2, ..., x_n) = x_1(1, 0, ..., 0) + x_2(0, 1, 0, ..., 0) + ... + x_n(0, ..., 0, 1) = x_1 e_1 + x_2 e_2 + ... + x_n e_n.$$

The corresponding coordinates $x_1, x_2, ..., x_n$ are just the Cartesian coordinates of the vector.

Every vector space has a basis. This follows from Zorn's lemma, an equivalent formulation of the Axiom of Choice.[17] Given the other axioms of Zermelo–Fraenkel set theory, the existence of bases is equivalent to the axiom of choice.[18] The ultrafilter lemma, which is weaker than the axiom of choice, implies that all bases of a given vector space have the same number of elements, or cardinality (cf. *Dimension theorem for vector spaces*).[19] It is called the *dimension* of the vector space, denoted dim V. If the space is spanned by finitely many vectors, the above statements can be proven without such fundamental input from set theory.[20]

The dimension of the coordinate space F^n is n, by the basis exhibited above. The dimension of the polynomial ring $F[x]$ introduced above is countably infinite, a basis is given by $1, x, x^2, ...$ A fortiori, the dimension of more general function spaces, such as the space of functions on some (bounded or unbounded) interval, is infinite.[nb 4] Under suitable regularity assumptions on the coefficients involved, the dimension of the solution space of a homogeneous ordinary differential equation equals the degree of the equation.[21] For example, the solution space for the above equation is generated by e^{-x} and xe^{-x}. These two functions are linearly independent over **R**, so the dimension of this space is two, as is the degree of the equation.

A field extension over the rationals **Q** can be thought of as a vector space over **Q** (by defining vector addition as field addition, defining scalar multiplication as field multiplication by elements of **Q**, and otherwise ignoring the field multiplication). The dimension (or degree) of the field extension $\mathbf{Q}(\alpha)$ over **Q** depends on α. If α satisfies some polynomial equation

$$q_n \alpha^n + q_{n-1} \alpha^{n-1} + ... + q_0 = 0, \text{ with rational coefficients } q_n, ..., q_0.$$

("α is algebraic"), the dimension is finite. More precisely, it equals the degree of the minimal polynomial having α as a root.[22] For example, the complex numbers **C** are a two-dimensional real vector space, generated by 1 and the imaginary unit i. The latter satisfies $i^2 + 1 = 0$, an equation of degree two. Thus, **C** is a two-dimensional **R**-vector space (and, as any field, one-dimensional as a vector space over itself, **C**). If α is not algebraic, the dimension of $\mathbf{Q}(\alpha)$ over **Q** is infinite. For instance, for $\alpha = \pi$ there is no such equation, in other words π is transcendental.[23]

4.5 Linear maps and matrices

Main article: Linear map

The relation of two vector spaces can be expressed by *linear map* or *linear transformation*. They are functions that reflect the vector space structure—i.e., they preserve sums and scalar multiplication:

$$f(\mathbf{x} + \mathbf{y}) = f(\mathbf{x}) + f(\mathbf{y}) \text{ and } f(a \cdot \mathbf{x}) = a \cdot f(\mathbf{x}) \text{ for all } \mathbf{x} \text{ and } \mathbf{y} \text{ in } V, \text{ all } a \text{ in } F.[24]$$

An *isomorphism* is a linear map $f : V \to W$ such that there exists an inverse map $g : W \to V$, which is a map such that the two possible compositions $f \circ g : W \to W$ and $g \circ f : V \to V$ are identity maps. Equivalently, f is both one-to-one (injective) and onto (surjective).[25] If there exists an isomorphism between V and W, the two spaces are said to be *isomorphic*; they are then essentially identical as vector spaces, since all identities holding in V are, via f, transported to similar ones in W, and vice versa via g.

For example, the "arrows in the plane" and "ordered pairs of numbers" vector spaces in the introduction are isomorphic: a planar arrow **v** departing at the origin of some (fixed) coordinate system can be expressed as an ordered pair by considering the x- and y-component of the arrow, as shown in the image at the right. Conversely, given a pair (x, y), the arrow going by x to the right (or to the left, if x is negative), and y up (down, if y is negative) turns back the arrow **v**.

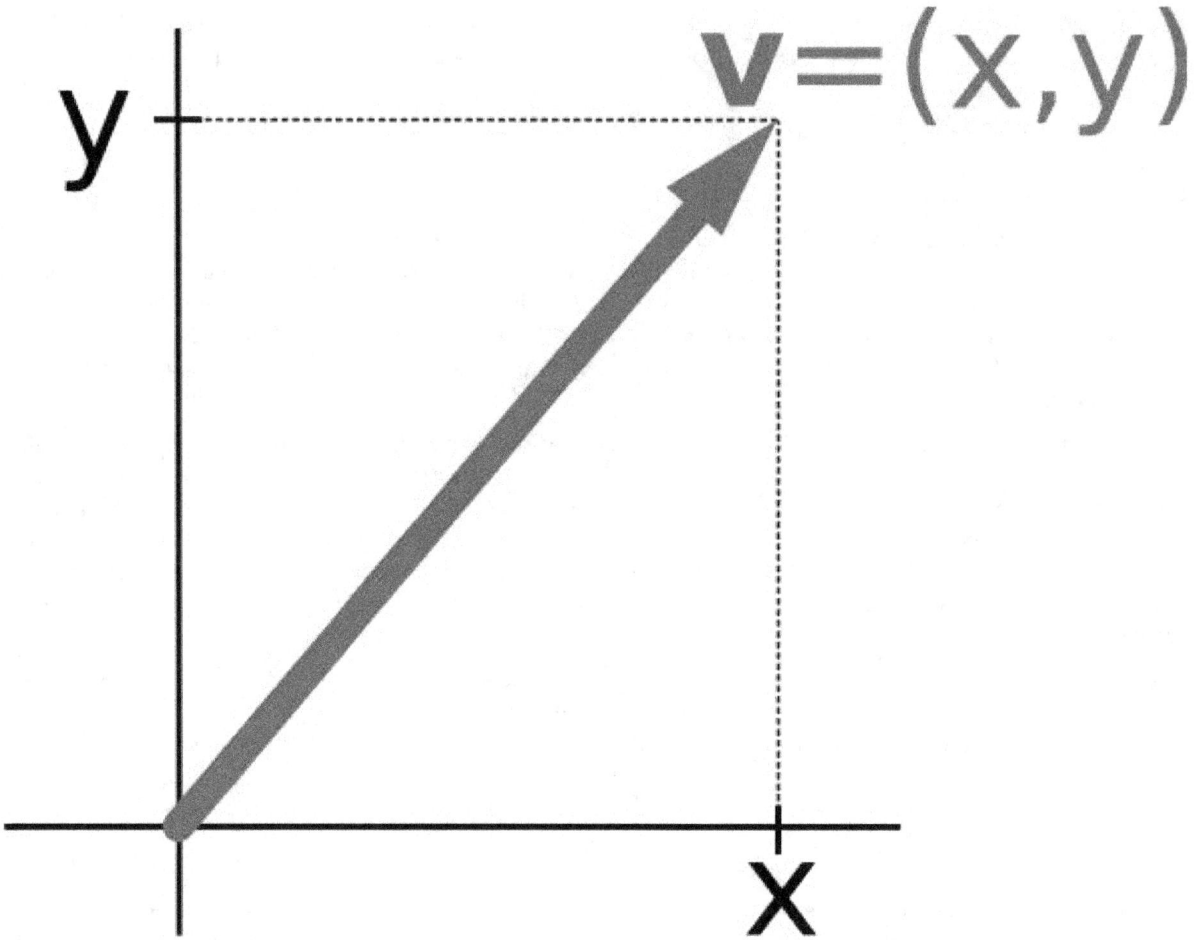

Describing an arrow vector v by its coordinates x and y yields an isomorphism of vector spaces.

Linear maps $V \to W$ between two vector spaces form a vector space $\mathrm{Hom}F(V, W)$, also denoted $L(V, W)$.[26] The space of linear maps from V to F is called the *dual vector space*, denoted V^*.[27] Via the injective natural map $V \to V^{**}$, any vector space can be embedded into its *bidual*; the map is an isomorphism if and only if the space is finite-dimensional.[28]

Once a basis of V is chosen, linear maps $f : V \to W$ are completely determined by specifying the images of the basis vectors, because any element of V is expressed uniquely as a linear combination of them.[29] If $\dim V = \dim W$, a 1-to-1 correspondence between fixed bases of V and W gives rise to a linear map that maps any basis element of V to the corresponding basis element of W. It is an isomorphism, by its very definition.[30] Therefore, two vector spaces are isomorphic if their dimensions agree and vice versa. Another way to express this is that any vector space is *completely classified* (up to isomorphism) by its dimension, a single number. In particular, any n-dimensional F-vector space V is isomorphic to F^n. There is, however, no "canonical" or preferred isomorphism; actually an isomorphism $\varphi : F^n \to V$ is equivalent to the choice of a basis of V, by mapping the standard basis of F^n to V, via φ. The freedom of choosing a convenient basis is particularly useful in the infinite-dimensional context, see below.

4.5.1 Matrices

Main articles: Matrix and Determinant
Matrices are a useful notion to encode linear maps.[31] They are written as a rectangular array of scalars as in the image at the right. Any m-by-n matrix A gives rise to a linear map from F^n to F^m, by the following

$$\mathbf{x} = (x_1, x_2, \cdots, x_n) \mapsto \left(\sum_{j=1}^n a_{1j}x_j, \sum_{j=1}^n a_{2j}x_j, \cdots, \sum_{j=1}^n a_{mj}x_j \right), \text{ where } \sum \text{ denotes summation,}$$

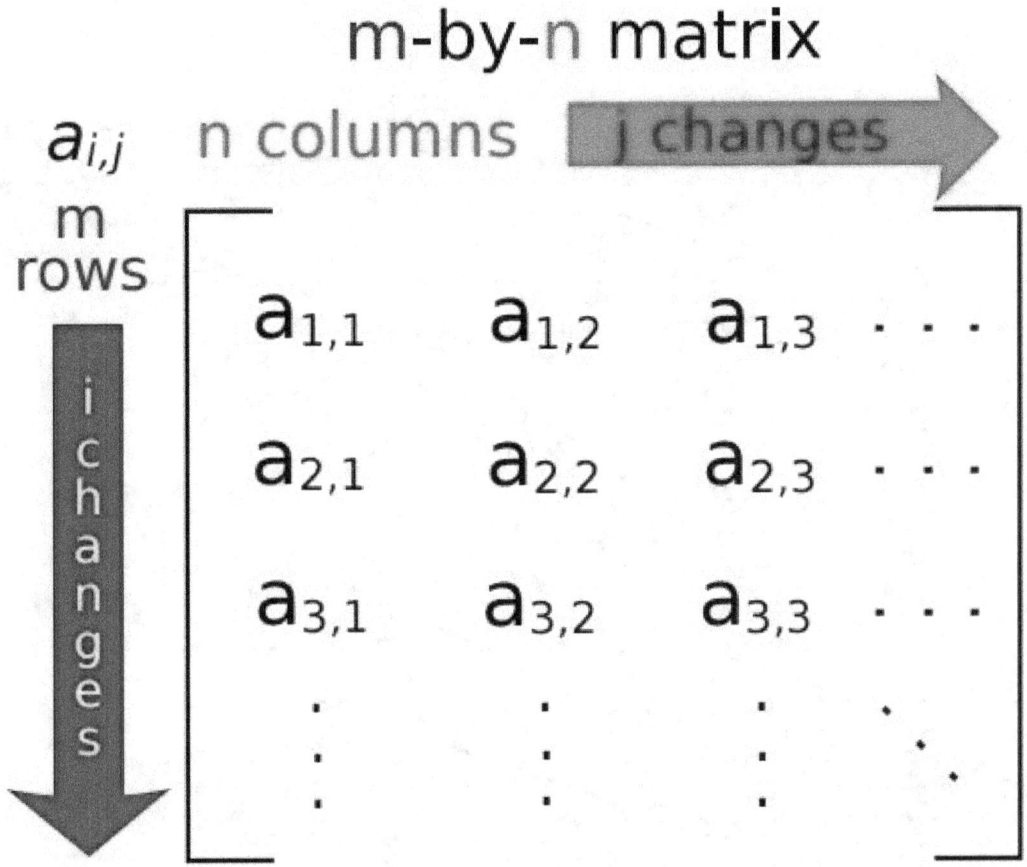

A typical matrix

or, using the matrix multiplication of the matrix A with the coordinate vector $\mathbf{x}$:

$$\mathbf{x} \mapsto A\mathbf{x}.$$

Moreover, after choosing bases of V and W, *any* linear map $f : V \to W$ is uniquely represented by a matrix via this assignment.[32]

The determinant det (A) of a square matrix A is a scalar that tells whether the associated map is an isomorphism or not: to be so it is sufficient and necessary that the determinant is nonzero.[33] The linear transformation of $\mathbf{R}^n$ corresponding to a real n-by-n matrix is orientation preserving if and only if its determinant is positive.

4.5.2 Eigenvalues and eigenvectors

Main article: Eigenvalues and eigenvectors

Endomorphisms, linear maps $f : V \to V$, are particularly important since in this case vectors $\mathbf{v}$ can be compared with their image under f, $f(\mathbf{v})$. Any nonzero vector $\mathbf{v}$ satisfying $\lambda\mathbf{v} = f(\mathbf{v})$, where λ is a scalar, is called an *eigenvector* of f with *eigenvalue* λ.[nb 5][34] Equivalently, $\mathbf{v}$ is an element of the kernel of the difference $f - \lambda \cdot$ Id (where Id is the identity map $V \to V$). If V is finite-dimensional, this can be rephrased using determinants: f having eigenvalue λ is equivalent to

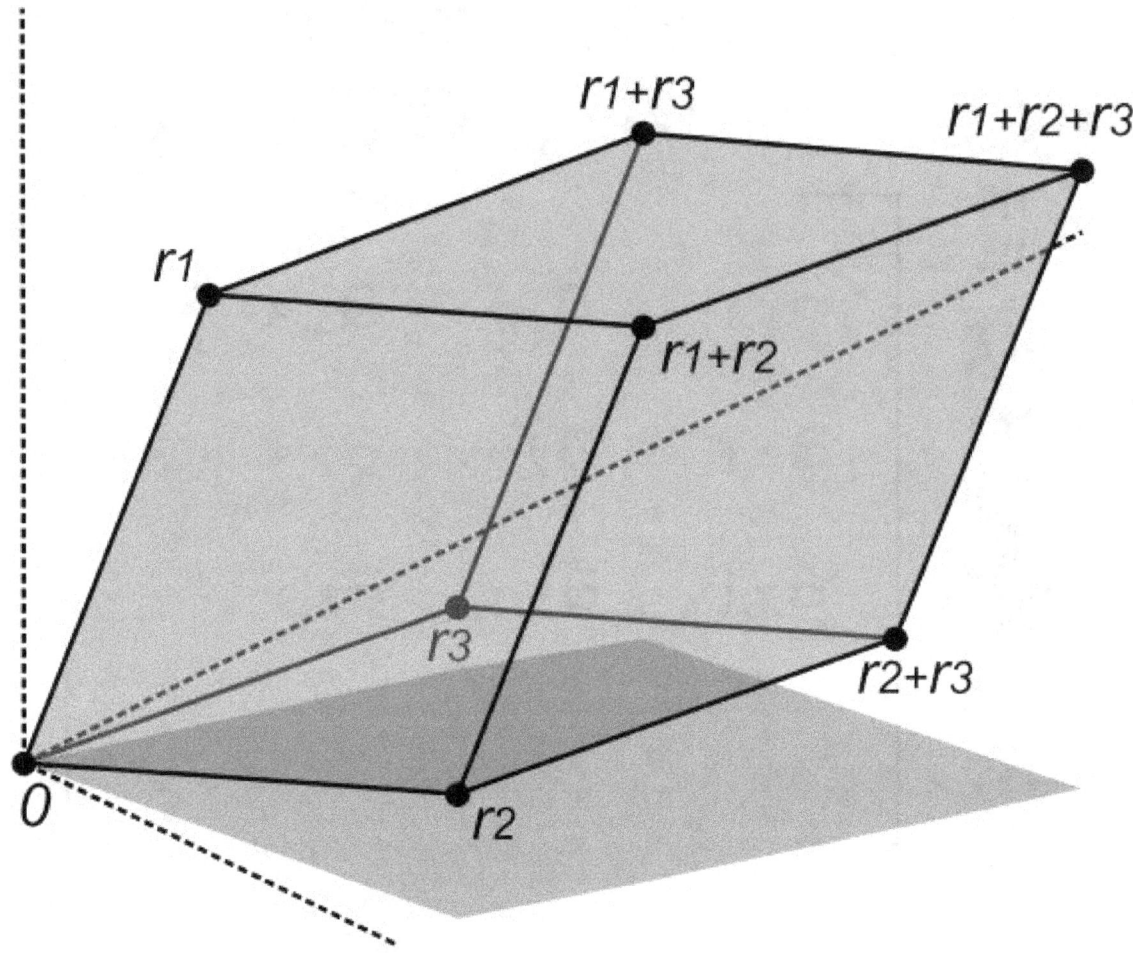

The volume of this parallelepiped is the absolute value of the determinant of the 3-by-3 matrix formed by the vectors r_1, r_2, and r_3.

$$\det(f - \lambda \cdot \mathrm{Id}) = 0.$$

By spelling out the definition of the determinant, the expression on the left hand side can be seen to be a polynomial function in λ, called the characteristic polynomial of f.[35] If the field F is large enough to contain a zero of this polynomial (which automatically happens for F algebraically closed, such as $F = \mathbf{C}$) any linear map has at least one eigenvector. The vector space V may or may not possess an eigenbasis, a basis consisting of eigenvectors. This phenomenon is governed by the Jordan canonical form of the map.[nb 6] The set of all eigenvectors corresponding to a particular eigenvalue of f forms a vector space known as the *eigenspace* corresponding to the eigenvalue (and f) in question. To achieve the spectral theorem, the corresponding statement in the infinite-dimensional case, the machinery of functional analysis is needed, see below.

4.6   Basic constructions

In addition to the above concrete examples, there are a number of standard linear algebraic constructions that yield vector spaces related to given ones. In addition to the definitions given below, they are also characterized by universal properties, which determine an object X by specifying the linear maps from X to any other vector space.

4.6.1 Subspaces and quotient spaces

Main articles: Linear subspace and Quotient vector space

A nonempty subset W of a vector space V that is closed under addition and scalar multiplication (and therefore contains

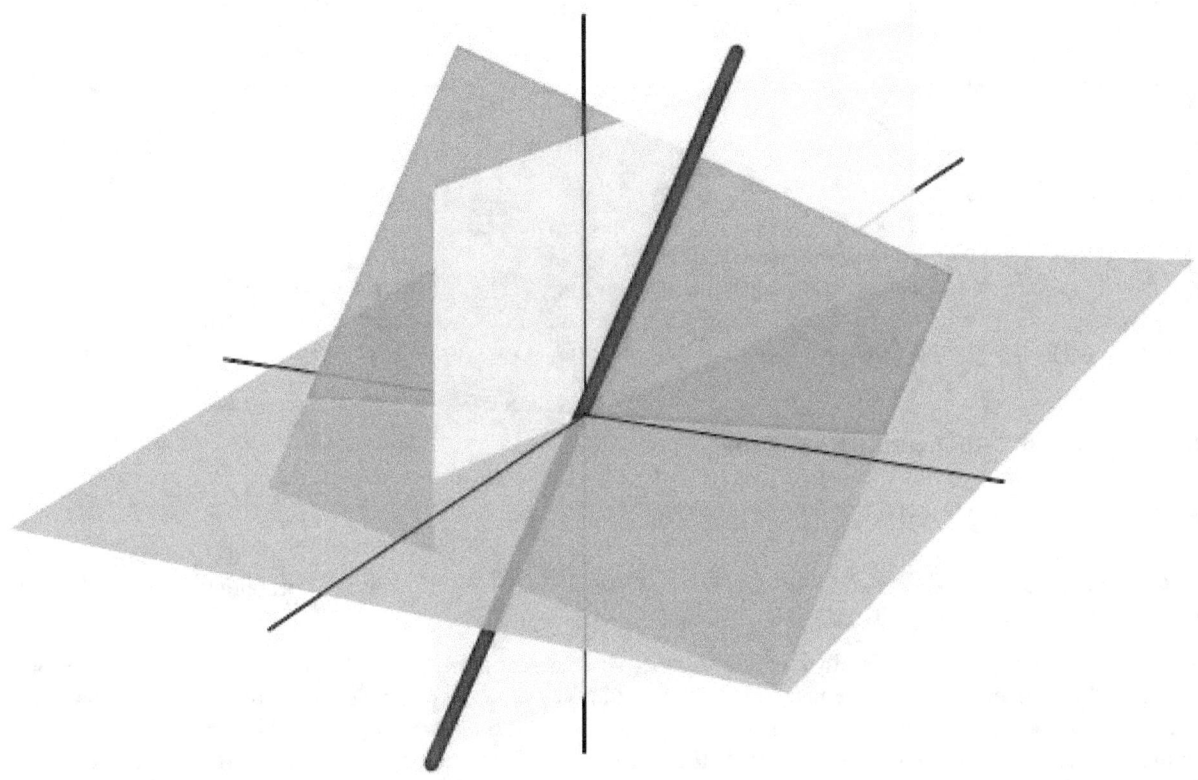

A line passing through the origin (blue, thick) in $\mathbf{R}^3$ is a linear subspace. It is the intersection of two planes (green and yellow).

the **0**-vector of V) is called a *subspace* of V.[36] Subspaces of V are vector spaces (over the same field) in their own right. The intersection of all subspaces containing a given set S of vectors is called its span, and it is the smallest subspace of V containing the set S. Expressed in terms of elements, the span is the subspace consisting of all the linear combinations of elements of S.[37]

The counterpart to subspaces are *quotient vector spaces*.[38] Given any subspace $W \subset V$, the quotient space V/W ("V modulo W") is defined as follows: as a set, it consists of $\mathbf{v} + W = \{\mathbf{v} + \mathbf{w} : \mathbf{w} \in W\}$, where $\mathbf{v}$ is an arbitrary vector in V. The sum of two such elements $\mathbf{v}_1 + W$ and $\mathbf{v}_2 + W$ is $(\mathbf{v}_1 + \mathbf{v}_2) + W$, and scalar multiplication is given by $a \cdot (\mathbf{v} + W) = (a \cdot \mathbf{v}) + W$. The key point in this definition is that $\mathbf{v}_1 + W = \mathbf{v}_2 + W$ if and only if the difference of $\mathbf{v}_1$ and $\mathbf{v}_2$ lies in W.[nb 7] This way, the quotient space "forgets" information that is contained in the subspace W.

The kernel $\ker(f)$ of a linear map $f : V \to W$ consists of vectors $\mathbf{v}$ that are mapped to **0** in W.[39] Both kernel and image $\mathrm{im}(f) = \{f(\mathbf{v}) : \mathbf{v} \in V\}$ are subspaces of V and W, respectively.[40] The existence of kernels and images is part of the statement that the category of vector spaces (over a fixed field F) is an abelian category, i.e. a corpus of mathematical objects and structure-preserving maps between them (a category) that behaves much like the category of abelian groups.[41] Because of this, many statements such as the first isomorphism theorem (also called rank–nullity theorem in matrix-related terms)

$$V / \ker(f) \equiv \mathrm{im}(f).$$

and the second and third isomorphism theorem can be formulated and proven in a way very similar to the corresponding statements for groups.

An important example is the kernel of a linear map $x \mapsto Ax$ for some fixed matrix A, as above. The kernel of this map is the subspace of vectors x such that $Ax = 0$, which is precisely the set of solutions to the system of homogeneous linear equations belonging to A. This concept also extends to linear differential equations

$$a_0 f + a_1 \frac{df}{dx} + a_2 \frac{d^2 f}{dx^2} + \cdots + a_n \frac{d^n f}{dx^n} = 0 \,,$$ where the coefficients a_i are functions in x, too.

In the corresponding map

$$f \mapsto D(f) = \sum_{i=0}^{n} a_i \frac{d^i f}{dx^i}$$

the derivatives of the function f appear linearly (as opposed to $f''(x)^2$, for example). Since differentiation is a linear procedure (i.e., $(f + g)' = f' + g'$ and $(c \cdot f)' = c \cdot f'$ for a constant c) this assignment is linear, called a linear differential operator. In particular, the solutions to the differential equation $D(f) = 0$ form a vector space (over $\mathbf{R}$ or $\mathbf{C}$).

4.6.2   Direct product and direct sum

Main articles: Direct product and Direct sum of modules

The *direct product* of vector spaces and the *direct sum* of vector spaces are two ways of combining an indexed family of vector spaces into a new vector space.

The *direct product* $\prod_{i \in I} V_i$ of a family of vector spaces V_i consists of the set of all tuples $(v_i)_{i \in I}$, which specify for each index i in some index set I an element v_i of V_i.[42] Addition and scalar multiplication is performed componentwise. A variant of this construction is the *direct sum* $\oplus_{i \in I} V_i$ (also called coproduct and denoted $\coprod_{i \in I} V_i$), where only tuples with finitely many nonzero vectors are allowed. If the index set I is finite, the two constructions agree, but in general they are different.

4.6.3   Tensor product

Main article: Tensor product of vector spaces

The *tensor product* $V \otimes_F W$, or simply $V \otimes W$, of two vector spaces V and W is one of the central notions of multilinear algebra which deals with extending notions such as linear maps to several variables. A map $g : V \times W \to X$ is called bilinear if g is linear in both variables v and w. That is to say, for fixed w the map $v \mapsto g(v, w)$ is linear in the sense above and likewise for fixed v.

The tensor product is a particular vector space that is a *universal* recipient of bilinear maps g, as follows. It is defined as the vector space consisting of finite (formal) sums of symbols called tensors

$$v_1 \otimes w_1 + v_2 \otimes w_2 + \ldots + v_n \otimes w_n,$$

subject to the rules

$$a \cdot (v \otimes w) = (a \cdot v) \otimes w = v \otimes (a \cdot w), \text{ where } a \text{ is a scalar,}$$

$$(v_1 + v_2) \otimes w = v_1 \otimes w + v_2 \otimes w, \text{ and}$$

$$v \otimes (w_1 + w_2) = v \otimes w_1 + v \otimes w_2.[43]$$

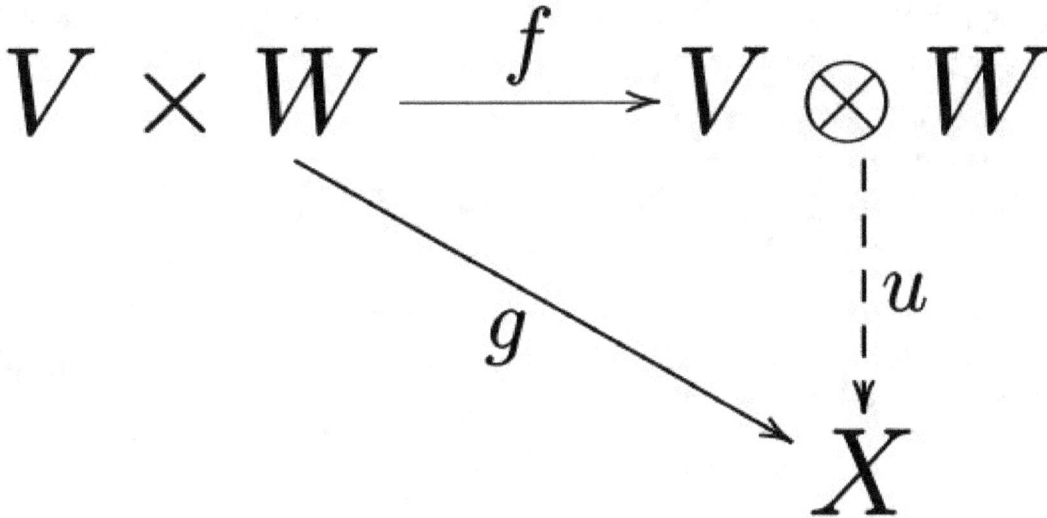

Commutative diagram depicting the universal property of the tensor product.

These rules ensure that the map f from the $V \times W$ to $V \otimes W$ that maps a tuple $(\mathbf{v}, \mathbf{w})$ to $\mathbf{v} \otimes \mathbf{w}$ is bilinear. The universality states that given *any* vector space X and *any* bilinear map $g : V \times W \to X$, there exists a unique map u, shown in the diagram with a dotted arrow, whose composition with f equals g: $u(\mathbf{v} \otimes \mathbf{w}) = g(\mathbf{v}, \mathbf{w})$.[44] This is called the universal property of the tensor product, an instance of the method—much used in advanced abstract algebra—to indirectly define objects by specifying maps from or to this object.

4.7 Vector spaces with additional structure

From the point of view of linear algebra, vector spaces are completely understood insofar as any vector space is characterized, up to isomorphism, by its dimension. However, vector spaces *per se* do not offer a framework to deal with the question—crucial to analysis—whether a sequence of functions converges to another function. Likewise, linear algebra is not adapted to deal with infinite series, since the addition operation allows only finitely many terms to be added. Therefore, the needs of functional analysis require considering additional structures.

A vector space may be given a partial order $\leq$, under which some vectors can be compared.[45] For example, n-dimensional real space $\mathbf{R}^n$ can be ordered by comparing its vectors componentwise. Ordered vector spaces, for example Riesz spaces, are fundamental to Lebesgue integration, which relies on the ability to express a function as a difference of two positive functions

$$f = f^+ - f^-,$$

where f^+ denotes the positive part of f and f^- the negative part.[46]

4.7.1 Normed vector spaces and inner product spaces

Main articles: Normed vector space and Inner product space

"Measuring" vectors is done by specifying a norm, a datum which measures lengths of vectors, or by an inner product, which measures angles between vectors. Norms and inner products are denoted $|\mathbf{v}|$ and $\langle \mathbf{v}, \mathbf{w} \rangle$, respectively. The datum of an inner product entails that lengths of vectors can be defined too, by defining the associated norm $|\mathbf{v}| := \sqrt{\langle \mathbf{v}, \mathbf{v} \rangle}$. Vector spaces endowed with such data are known as *normed vector spaces* and *inner product spaces*, respectively.[47]

Coordinate space F^n can be equipped with the standard dot product:

$$\langle \mathbf{x}, \mathbf{y} \rangle = \mathbf{x} \cdot \mathbf{y} = x_1 y_1 + \cdots + x_n y_n.$$

In $\mathbf{R}^2$, this reflects the common notion of the angle between two vectors $\mathbf{x}$ and $\mathbf{y}$, by the law of cosines:

$$\mathbf{x} \cdot \mathbf{y} = \cos\left(\angle(\mathbf{x}, \mathbf{y})\right) \cdot |\mathbf{x}| \cdot |\mathbf{y}|.$$

Because of this, two vectors satisfying $\langle \mathbf{x}, \mathbf{y} \rangle = 0$ are called orthogonal. An important variant of the standard dot product is used in Minkowski space: $\mathbf{R}^4$ endowed with the Lorentz product

$$\langle \mathbf{x} | \mathbf{y} \rangle = x_1 y_1 + x_2 y_2 + x_3 y_3 - x_4 y_4. \text{ [48]}$$

In contrast to the standard dot product, it is not positive definite: $\langle \mathbf{x} | \mathbf{x} \rangle$ also takes negative values, for example for $\mathbf{x} = (0, 0, 0, 1)$. Singling out the fourth coordinate—corresponding to time, as opposed to three space-dimensions—makes it useful for the mathematical treatment of special relativity.

4.7.2   Topological vector spaces

Main article: Topological vector space

Convergence questions are treated by considering vector spaces V carrying a compatible topology, a structure that allows one to talk about elements being close to each other.[49][50] Compatible here means that addition and scalar multiplication have to be continuous maps. Roughly, if $\mathbf{x}$ and $\mathbf{y}$ in V, and a in F vary by a bounded amount, then so do $\mathbf{x} + \mathbf{y}$ and $a\mathbf{x}$.[nb 8] To make sense of specifying the amount a scalar changes, the field F also has to carry a topology in this context; a common choice are the reals or the complex numbers.

In such *topological vector spaces* one can consider series of vectors. The infinite sum

$$\sum_{i=0}^{\infty} f_i$$

denotes the limit of the corresponding finite partial sums of the sequence $(f_i)_{i \in \mathbb{N}}$ of elements of V. For example, the f_i could be (real or complex) functions belonging to some function space V, in which case the series is a function series. The mode of convergence of the series depends on the topology imposed on the function space. In such cases, pointwise convergence and uniform convergence are two prominent examples.

A way to ensure the existence of limits of certain infinite series is to restrict attention to spaces where any Cauchy sequence has a limit; such a vector space is called complete. Roughly, a vector space is complete provided that it contains all necessary limits. For example, the vector space of polynomials on the unit interval [0,1], equipped with the topology of uniform convergence is not complete because any continuous function on [0,1] can be uniformly approximated by a sequence of polynomials, by the Weierstrass approximation theorem.[51] In contrast, the space of *all* continuous functions on [0,1] with the same topology is complete.[52] A norm gives rise to a topology by defining that a sequence of vectors $\mathbf{v}_n$ converges to $\mathbf{v}$ if and only if

$$\lim_{n \to \infty} |\mathbf{v}_n - \mathbf{v}| = 0.$$

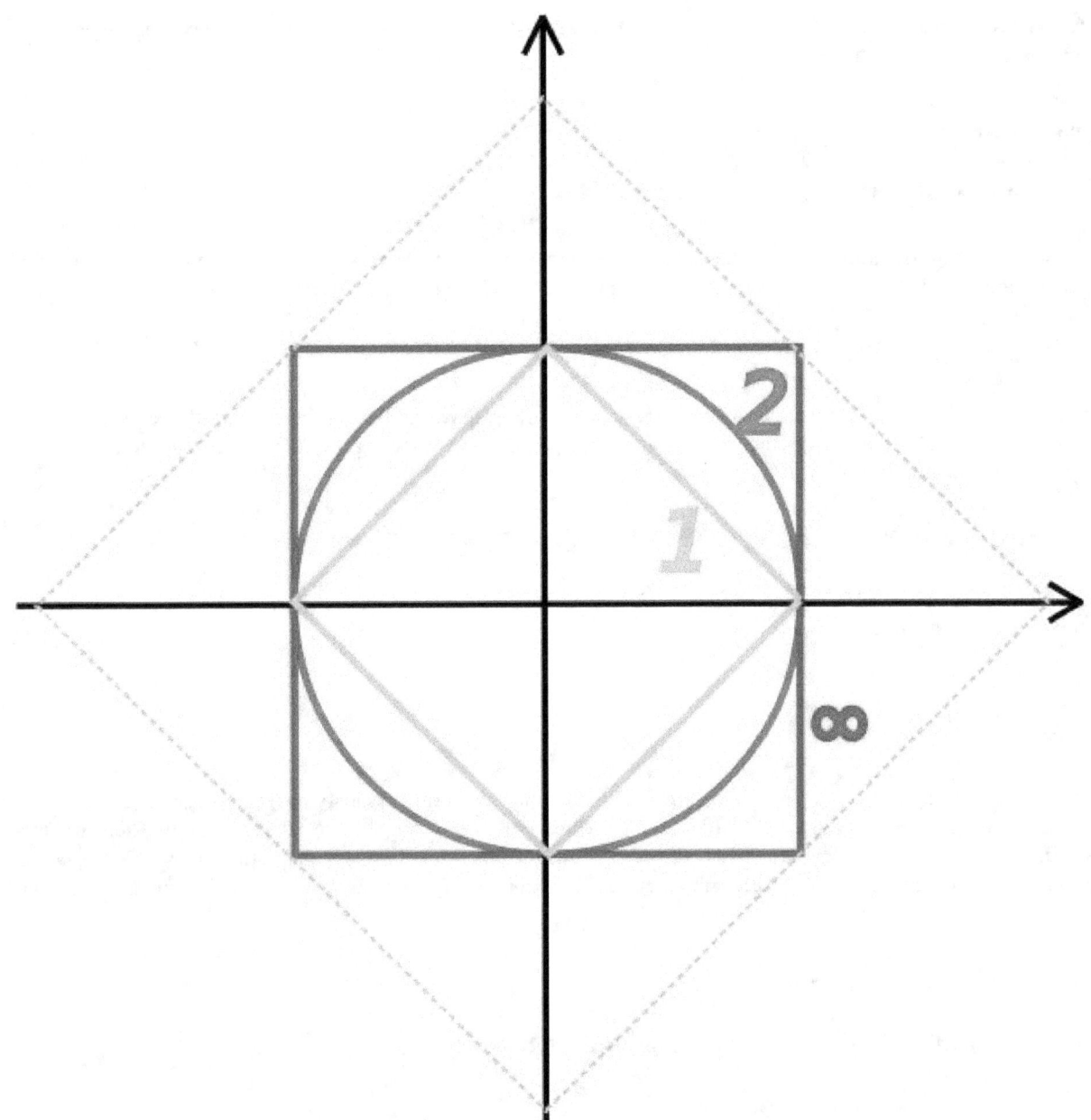

Unit "spheres" in $\mathbf{R}^2$ consist of plane vectors of norm 1. Depicted are the unit spheres in different p-norms, for p = 1, 2, and ∞. The bigger diamond depicts points of 1-norm equal to $\sqrt{2}$.

Banach and Hilbert spaces are complete topological vector spaces whose topologies are given, respectively, by a norm and an inner product. Their study—a key piece of functional analysis—focusses on infinite-dimensional vector spaces, since all norms on finite-dimensional topological vector spaces give rise to the same notion of convergence.[53] The image at the right shows the equivalence of the 1-norm and ∞-norm on $\mathbf{R}^2$: as the unit "balls" enclose each other, a sequence converges to zero in one norm if and only if it so does in the other norm. In the infinite-dimensional case, however, there will generally be inequivalent topologies, which makes the study of topological vector spaces richer than that of vector spaces without additional data.

From a conceptual point of view, all notions related to topological vector spaces should match the topology. For example, instead of considering all linear maps (also called functionals) $V \to W$, maps between topological vector spaces are required to be continuous.[54] In particular, the (topological) dual space V^* consists of continuous functionals $V \to \mathbf{R}$ (or to $\mathbf{C}$). The fundamental Hahn–Banach theorem is concerned with separating subspaces of appropriate topological vector spaces by continuous functionals.[55]

Banach spaces

Main article: Banach space

Banach spaces, introduced by Stefan Banach, are complete normed vector spaces.[56] A first example is the vector space ℓ^p consisting of infinite vectors with real entries $\mathbf{x} = (x_1, x_2, ...)$ whose p-norm ($1 \le p \le \infty$) given by

$$|\mathbf{x}|_p := \left(\sum_i |x_i|^p\right)^{1/p} \text{ for } p < \infty \text{ and } |\mathbf{x}|_\infty := \sup_i |x_i|$$

is finite. The topologies on the infinite-dimensional space ℓ^p are inequivalent for different p. E.g. the sequence of vectors $\mathbf{x}n = (2^{-n}, 2^{-n}, ..., 2^{-n}, 0, 0, ...)$, i.e. the first 2^n components are 2^{-n}, the following ones are 0, converges to the zero vector for $p = \infty$, but does not for $p = 1$:

$$|x_n|_\infty = \sup(2^{-n}, 0) = 2^{-n} \to 0 \text{ , but } |x_n|_1 = \sum_{i=1}^{2^n} 2^{-n} = 2^n \cdot 2^{-n} = 1.$$

More generally than sequences of real numbers, functions $f: \Omega \to \mathbf{R}$ are endowed with a norm that replaces the above sum by the Lebesgue integral

$$|f|_p := \left(\int_\Omega |f(x)|^p \, dx\right)^{1/p}.$$

The space of integrable functions on a given domain Ω (for example an interval) satisfying $|f|p < \infty$, and equipped with this norm are called Lebesgue spaces, denoted $L^p(\Omega)$.[nb 9] These spaces are complete.[57] (If one uses the Riemann integral instead, the space is *not* complete, which may be seen as a justification for Lebesgue's integration theory.[nb 10]) Concretely this means that for any sequence of Lebesgue-integrable functions $f_1, f_2, ...$ with $|f_n|p < \infty$, satisfying the condition

$$\lim_{k,\, n\to\infty} \int_\Omega |f_k(x) - f_n(x)|^p \, dx = 0$$

there exists a function $f(x)$ belonging to the vector space $L^p(\Omega)$ such that

$$\lim_{k\to\infty} \int_\Omega |f(x) - f_k(x)|^p \, dx = 0.$$

Imposing boundedness conditions not only on the function, but also on its derivatives leads to Sobolev spaces.[58]

Hilbert spaces

Main article: Hilbert space

Complete inner product spaces are known as *Hilbert spaces*, in honor of David Hilbert.[59] The Hilbert space $L^2(\Omega)$, with inner product given by

$$\langle f, g \rangle = \int_\Omega f(x)\overline{g(x)} \, dx,$$

where $\overline{g(x)}$ denotes the complex conjugate of $g(x)$,[60][nb 11] is a key case.

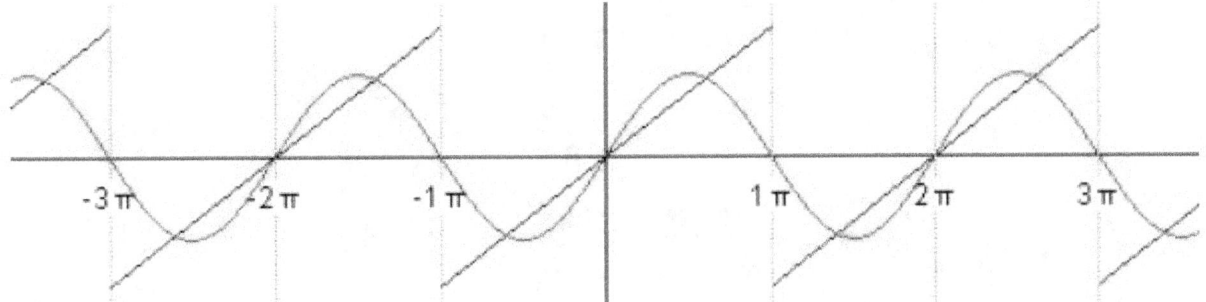

The succeeding snapshots show summation of 1 to 5 terms in approximating a periodic function (blue) by finite sum of sine functions (red).

By definition, in a Hilbert space any Cauchy sequence converges to a limit. Conversely, finding a sequence of functions f_n with desirable properties that approximates a given limit function, is equally crucial. Early analysis, in the guise of the Taylor approximation, established an approximation of differentiable functions f by polynomials.[61] By the Stone–Weierstrass theorem, every continuous function on $[a, b]$ can be approximated as closely as desired by a polynomial.[62] A similar approximation technique by trigonometric functions is commonly called Fourier expansion, and is much applied in engineering, see below. More generally, and more conceptually, the theorem yields a simple description of what "basic functions", or, in abstract Hilbert spaces, what basic vectors suffice to generate a Hilbert space H, in the sense that the closure of their span (i.e., finite linear combinations and limits of those) is the whole space. Such a set of functions is called a *basis* of H, its cardinality is known as the Hilbert space dimension.[nb 12] Not only does the theorem exhibit suitable basis functions as sufficient for approximation purposes, but together with the Gram-Schmidt process, it enables one to construct a basis of orthogonal vectors.[63] Such orthogonal bases are the Hilbert space generalization of the coordinate axes in finite-dimensional Euclidean space.

The solutions to various differential equations can be interpreted in terms of Hilbert spaces. For example, a great many fields in physics and engineering lead to such equations and frequently solutions with particular physical properties are used as basis functions, often orthogonal.[64] As an example from physics, the time-dependent Schrödinger equation in quantum mechanics describes the change of physical properties in time by means of a partial differential equation, whose solutions are called wavefunctions.[65] Definite values for physical properties such as energy, or momentum, correspond to eigenvalues of a certain (linear) differential operator and the associated wavefunctions are called eigenstates. The spectral theorem decomposes a linear compact operator acting on functions in terms of these eigenfunctions and their eigenvalues.[66]

4.7.3 Algebras over fields

Main articles: Algebra over a field and Lie algebra

General vector spaces do not possess a multiplication between vectors. A vector space equipped with an additional bilinear operator defining the multiplication of two vectors is an *algebra over a field*.[67] Many algebras stem from functions on some geometrical object: since functions with values in a given field can be multiplied pointwise, these entities form algebras. The Stone–Weierstrass theorem mentioned above, for example, relies on Banach algebras which are both Banach spaces and algebras.

Commutative algebra makes great use of rings of polynomials in one or several variables, introduced above. Their multiplication is both commutative and associative. These rings and their quotients form the basis of algebraic geometry, because they are rings of functions of algebraic geometric objects.[68]

Another crucial example are *Lie algebras*, which are neither commutative nor associative, but the failure to be so is limited by the constraints ($[x, y]$ denotes the product of x and y):

- $[x, y] = -[y, x]$ (anticommutativity), and

- $[x, [y, z]] + [y, [z, x]] + [z, [x, y]] = 0$ (Jacobi identity).[69]

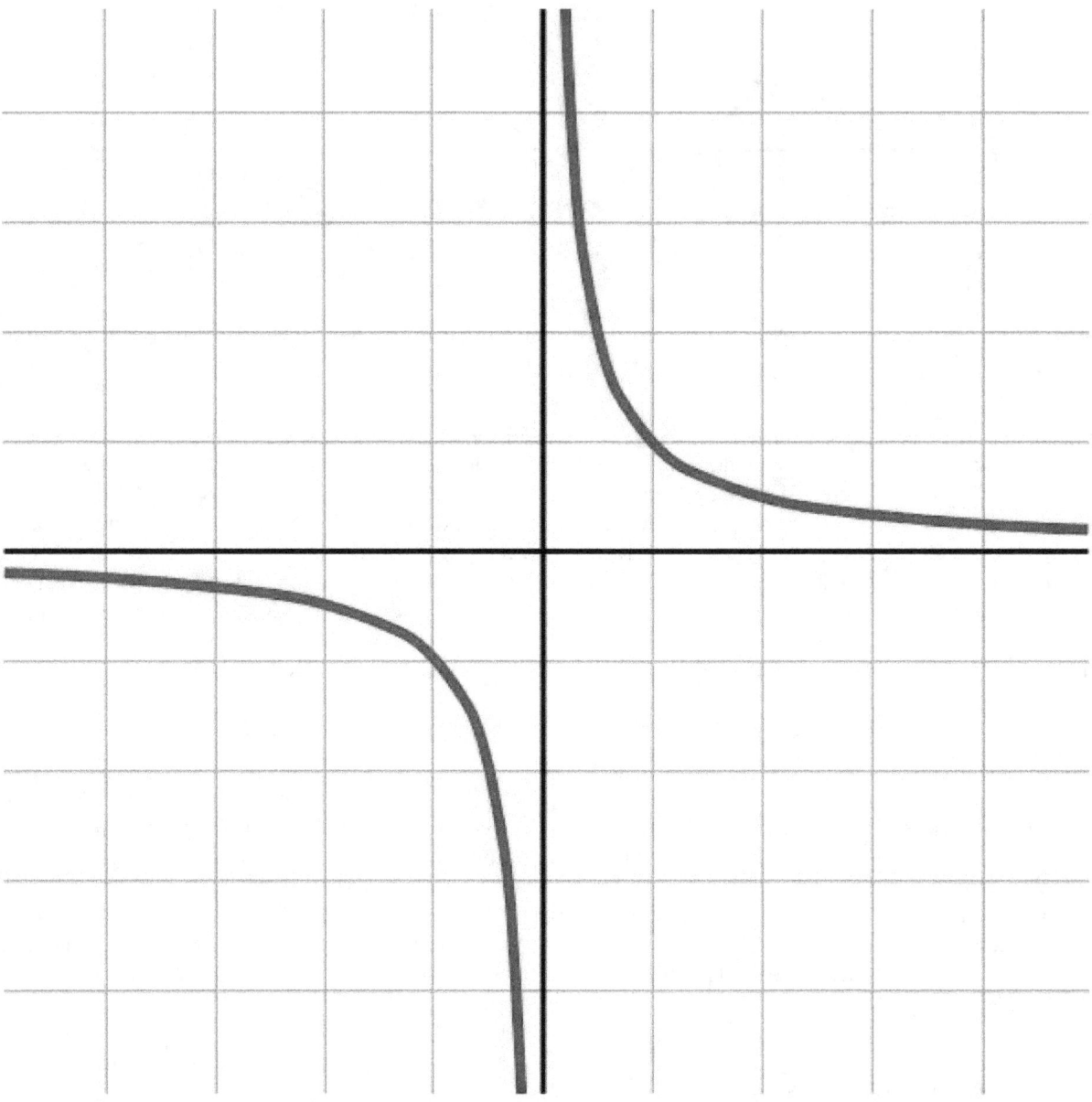

A hyperbola, given by the equation x ☒ y = 1. *The coordinate ring of functions on this hyperbola is given by* **R**[x, y] /(x · y − 1), *an infinite-dimensional vector space over* **R**.

Examples include the vector space of n-by-n matrices, with $[x, y] = xy − yx$, the commutator of two matrices, and $\mathbf{R}^3$, endowed with the cross product.

The tensor algebra T(V) is a formal way of adding products to any vector space V to obtain an algebra.[70] As a vector space, it is spanned by symbols, called simple tensors

$\mathbf{v}_1 \otimes \mathbf{v}_2 \otimes ... \otimes \mathbf{v}n$, where the degree n varies.

The multiplication is given by concatenating such symbols, imposing the distributive law under addition, and requiring that scalar multiplication commute with the tensor product $\otimes$, much the same way as with the tensor product of two vector spaces introduced above. In general, there are no relations between $\mathbf{v}_1 \otimes \mathbf{v}_2$ and $\mathbf{v}_2 \otimes \mathbf{v}_1$. Forcing two such elements to be equal leads to the symmetric algebra, whereas forcing $\mathbf{v}_1 \otimes \mathbf{v}_2 = − \mathbf{v}_2 \otimes \mathbf{v}_1$ yields the exterior algebra.[71]

When a field, F is explicitly stated, a common term used is F-algebra.

4.8 Applications

Vector spaces have manifold applications as they occur in many circumstances, namely wherever functions with values in some field are involved. They provide a framework to deal with analytical and geometrical problems, or are used in the Fourier transform. This list is not exhaustive: many more applications exist, for example in optimization. The minimax theorem of game theory stating the existence of a unique payoff when all players play optimally can be formulated and proven using vector spaces methods.[72] Representation theory fruitfully transfers the good understanding of linear algebra and vector spaces to other mathematical domains such as group theory.[73]

4.8.1 Distributions

Main article: Distribution

A *distribution* (or *generalized function*) is a linear map assigning a number to each "test" function, typically a smooth function with compact support, in a continuous way: in the above terminology the space of distributions is the (continuous) dual of the test function space.[74] The latter space is endowed with a topology that takes into account not only f itself, but also all its higher derivatives. A standard example is the result of integrating a test function f over some domain Ω:

$$I(f) = \int_\Omega f(x)\, dx.$$

When $\Omega = \{p\}$, the set consisting of a single point, this reduces to the Dirac distribution, denoted by δ, which associates to a test function f its value at the p: $\delta(f) = f(p)$. Distributions are a powerful instrument to solve differential equations. Since all standard analytic notions such as derivatives are linear, they extend naturally to the space of distributions. Therefore, the equation in question can be transferred to a distribution space, which is bigger than the underlying function space, so that more flexible methods are available for solving the equation. For example, Green's functions and fundamental solutions are usually distributions rather than proper functions, and can then be used to find solutions of the equation with prescribed boundary conditions. The found solution can then in some cases be proven to be actually a true function, and a solution to the original equation (e.g., using the Lax–Milgram theorem, a consequence of the Riesz representation theorem).[75]

4.8.2 Fourier analysis

Main article: Fourier analysis

Resolving a periodic function into a sum of trigonometric functions forms a *Fourier series*, a technique much used in physics and engineering.[nb 13][76] The underlying vector space is usually the Hilbert space $L^2(0, 2\pi)$, for which the functions $\sin mx$ and $\cos mx$ (m an integer) form an orthogonal basis.[77] The Fourier expansion of an L^2 function f is

$$\frac{a_0}{2} + \sum_{m=1}^{\infty} [a_m \cos(mx) + b_m \sin(mx)].$$

The coefficients a_m and b_m are called Fourier coefficients of f, and are calculated by the formulas[78]

$$a_m = \frac{1}{\pi} \int_0^{2\pi} f(t) \cos(mt)\, dt \, , \, b_m = \frac{1}{\pi} \int_0^{2\pi} f(t) \sin(mt)\, dt.$$

In physical terms the function is represented as a superposition of sine waves and the coefficients give information about the function's frequency spectrum.[79] A complex-number form of Fourier series is also commonly used.[78] The concrete formulae above are consequences of a more general mathematical duality called Pontryagin duality.[80] Applied to the group $\mathbf{R}$, it yields the classical Fourier transform; an application in physics are reciprocal lattices, where the underlying

The heat equation describes the dissipation of physical properties over time, such as the decline of the temperature of a hot body placed in a colder environment (yellow depicts colder regions than red).

group is a finite-dimensional real vector space endowed with the additional datum of a lattice encoding positions of atoms in crystals.[81]

Fourier series are used to solve boundary value problems in partial differential equations.[82] In 1822, Fourier first used this technique to solve the heat equation.[83] A discrete version of the Fourier series can be used in sampling applications where the function value is known only at a finite number of equally spaced points. In this case the Fourier series is finite and its value is equal to the sampled values at all points.[84] The set of coefficients is known as the discrete Fourier transform (DFT) of the given sample sequence. The DFT is one of the key tools of digital signal processing, a field whose applications include radar, speech encoding, image compression.[85] The JPEG image format is an application of the closely related discrete cosine transform.[86]

The fast Fourier transform is an algorithm for rapidly computing the discrete Fourier transform.[87] It is used not only for calculating the Fourier coefficients but, using the convolution theorem, also for computing the convolution of two finite sequences.[88] They in turn are applied in digital filters[89] and as a rapid multiplication algorithm for polynomials and large integers (Schönhage-Strassen algorithm).[90][91]

4.8.3 Differential geometry

Main article: Tangent space
 The tangent plane to a surface at a point is naturally a vector space whose origin is identified with the point of contact. The tangent plane is the best linear approximation, or linearization, of a surface at a point.[nb 14] Even in a three-dimensional Euclidean space, there is typically no natural way to prescribe a basis of the tangent plane, and so it is conceived of as an abstract vector space rather than a real coordinate space. The *tangent space* is the generalization to higher-dimensional differentiable manifolds.[92]

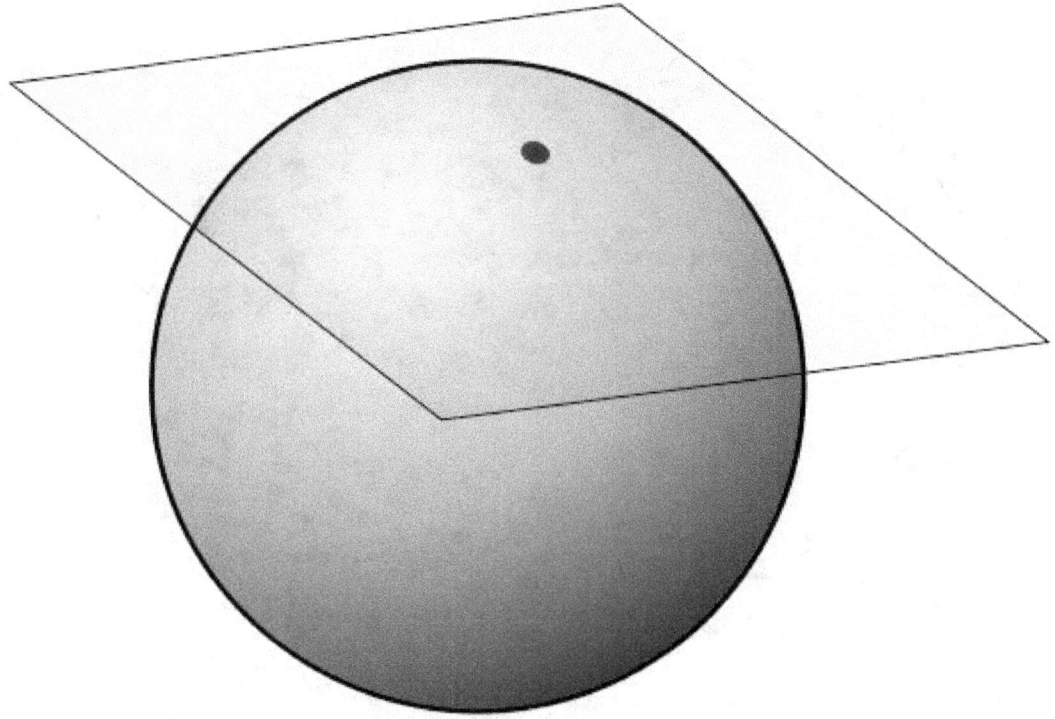

The tangent space to the 2-sphere at some point is the infinite plane touching the sphere in this point.

Riemannian manifolds are manifolds whose tangent spaces are endowed with a suitable inner product.[93] Derived there-from, the Riemann curvature tensor encodes all curvatures of a manifold in one object, which finds applications in general relativity, for example, where the Einstein curvature tensor describes the matter and energy content of space-time.[94][95] The tangent space of a Lie group can be given naturally the structure of a Lie algebra and can be used to classify compact Lie groups.[96]

4.9   Generalizations

4.9.1   Vector bundles

Main articles: Vector bundle and Tangent bundle

A *vector bundle* is a family of vector spaces parametrized continuously by a topological space X.[92] More precisely, a vector bundle over X is a topological space E equipped with a continuous map

$$\pi : E \to X$$

such that for every x in X, the fiber $\pi^{-1}(x)$ is a vector space. The case dim $V = 1$ is called a line bundle. For any vector space V, the projection $X \times V \to X$ makes the product $X \times V$ into a "trivial" vector bundle. Vector bundles over X are required to be locally a product of X and some (fixed) vector space V: for every x in X, there is a neighborhood U of x such that the restriction of π to $\pi^{-1}(U)$ is isomorphic[nb 15] to the trivial bundle $U \times V \to U$. Despite their locally trivial character, vector bundles may (depending on the shape of the underlying space X) be "twisted" in the large (i.e., the bundle need not be (globally isomorphic to) the trivial bundle $X \times V$). For example, the Möbius strip can be seen as a line bundle over the circle S^1 (by identifying open intervals with the real line). It is, however, different from the cylinder $S^1 \times$ **R**, because the latter is orientable whereas the former is not.[97]

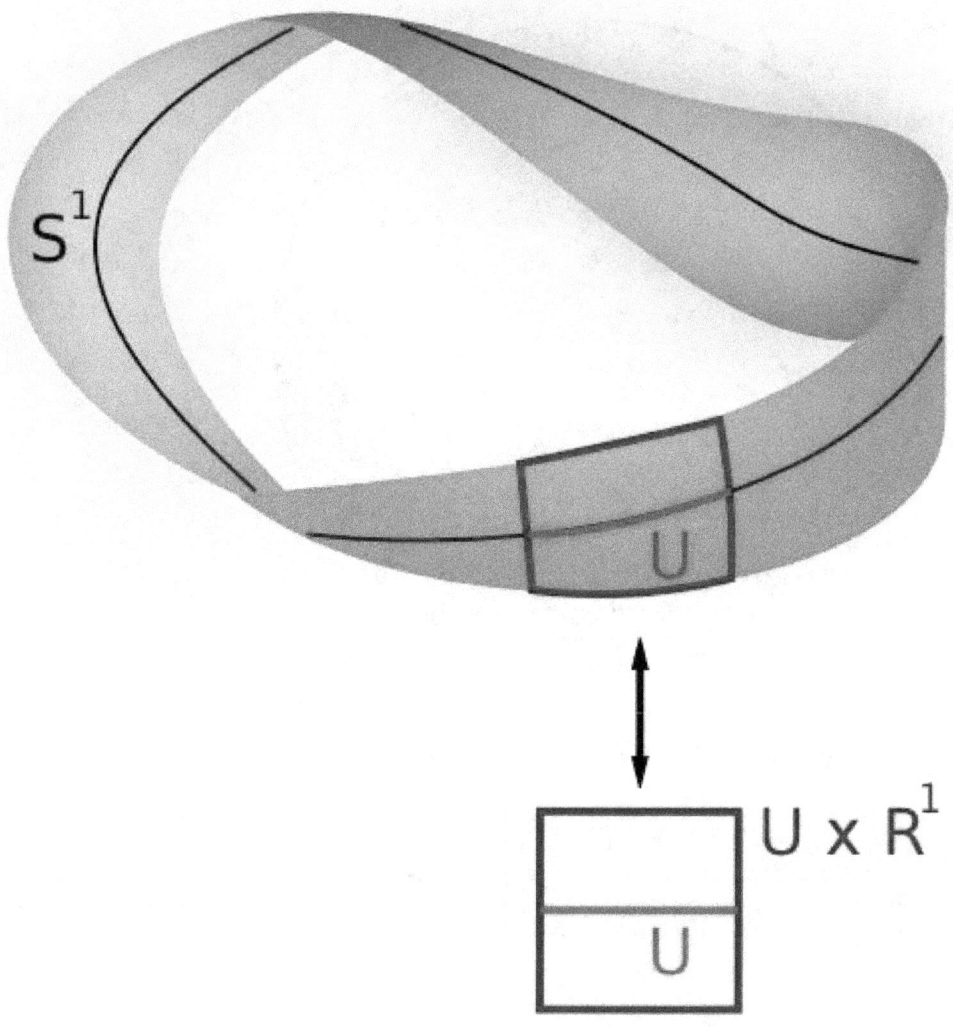

A Möbius strip. Locally, it looks like U × *R.*

Properties of certain vector bundles provide information about the underlying topological space. For example, the tangent bundle consists of the collection of tangent spaces parametrized by the points of a differentiable manifold. The tangent bundle of the circle S^1 is globally isomorphic to $S^1 \times \mathbf{R}$, since there is a global nonzero vector field on S^1.[nb 16] In contrast, by the hairy ball theorem, there is no (tangent) vector field on the 2-sphere S^2 which is everywhere nonzero.[98] K-theory studies the isomorphism classes of all vector bundles over some topological space.[99] In addition to deepening topological and geometrical insight, it has purely algebraic consequences, such as the classification of finite-dimensional real division algebras: $\mathbf{R}$, $\mathbf{C}$, the quaternions $\mathbf{H}$ and the octonions.

The cotangent bundle of a differentiable manifold consists, at every point of the manifold, of the dual of the tangent space, the cotangent space. Sections of that bundle are known as differential one-forms.

4.9.2 Modules

Main article: Module

Modules are to rings what vector spaces are to fields. The very same axioms, applied to a ring *R* instead of a field *F* yield modules.[100] The theory of modules, compared to that of vector spaces, is complicated by the presence of ring elements that do not have multiplicative inverses. For example, modules need not have bases, as the **Z**-module (i.e., abelian group) **Z**/2**Z** shows; those modules that do (including all vector spaces) are known as free modules. Nevertheless, a vector space can be compactly defined as a module over a ring which is a field with the elements being called vectors. Some authors use the term *vector space* to mean modules over a division ring.[101] The algebro-geometric interpretation of commutative rings via their spectrum allows the development of concepts such as locally free modules, the algebraic counterpart to vector bundles.

4.9.3 Affine and projective spaces

Main articles: Affine space and Projective space

Roughly, *affine spaces* are vector spaces whose origins are not specified.[102] More precisely, an affine space is a set with

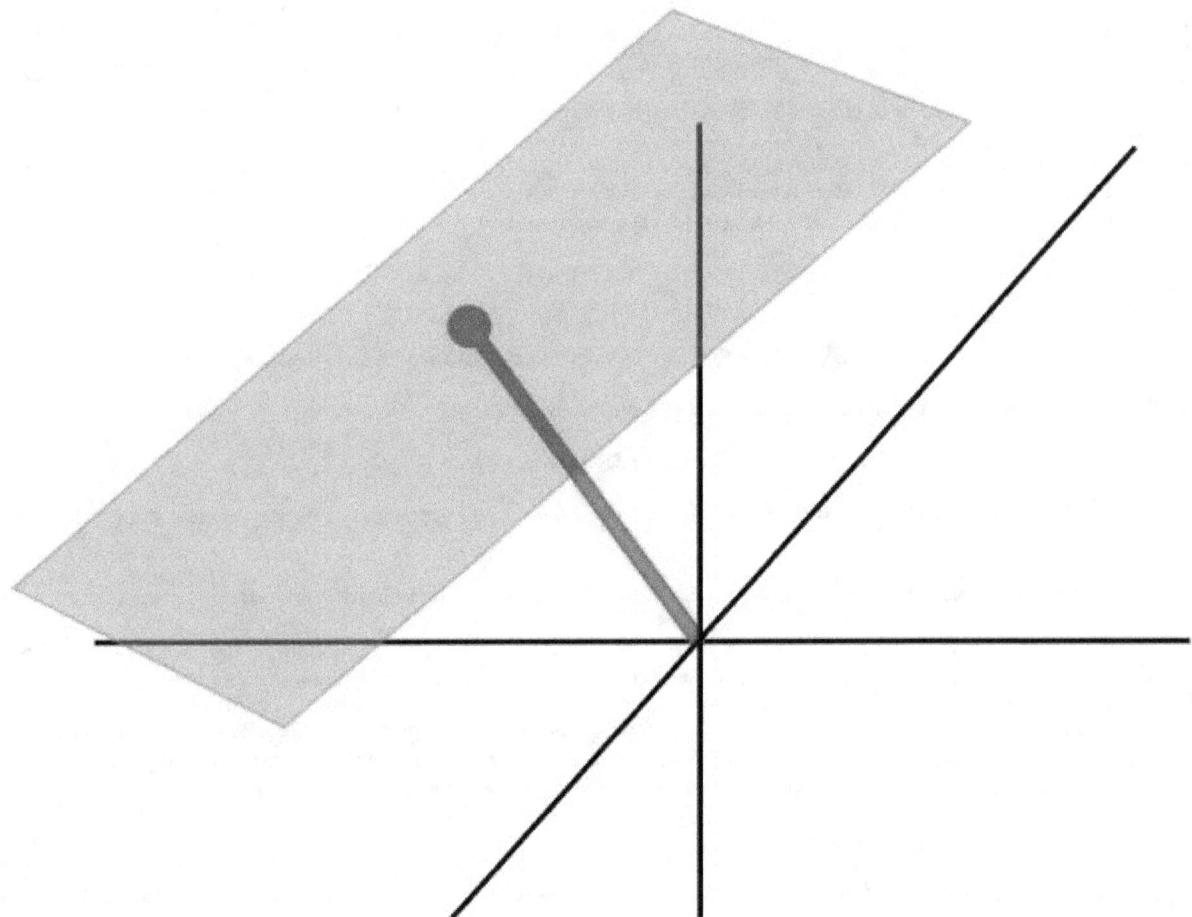

An affine plane (light blue) in $\mathbf{R}^3$. *It is a two-dimensional subspace shifted by a vector* x *(red).*

a free transitive vector space action. In particular, a vector space is an affine space over itself, by the map

$$V \times V \to V, (\mathbf{v}, \mathbf{a}) \mapsto \mathbf{a} + \mathbf{v}.$$

If *W* is a vector space, then an affine subspace is a subset of *W* obtained by translating a linear subspace *V* by a fixed vector $\mathbf{x} \in W$; this space is denoted by $\mathbf{x} + V$ (it is a coset of *V* in *W*) and consists of all vectors of the form $\mathbf{x} + \mathbf{v}$ for $\mathbf{v} \in V$. An important example is the space of solutions of a system of inhomogeneous linear equations

$$Ax = b$$

generalizing the homogeneous case $b = 0$ above.[103] The space of solutions is the affine subspace $x + V$ where x is a particular solution of the equation, and V is the space of solutions of the homogeneous equation (the nullspace of A).

The set of one-dimensional subspaces of a fixed finite-dimensional vector space V is known as *projective space*; it may be used to formalize the idea of parallel lines intersecting at infinity.[104] Grassmannians and flag manifolds generalize this by parametrizing linear subspaces of fixed dimension k and flags of subspaces, respectively.

4.10   See also

- Vector (mathematics and physics), for a list of various kinds of vectors

4.11   Notes

[1] It is also common, especially in physics, to denote vectors with an arrow on top: $\vec{v}$.

[2] This axiom refers to two different operations: scalar multiplication: bv; and field multiplication: ab. It does not assert the associativity of either operation. More formally, scalar multiplication is the *semigroup action* of the scalars on the vector space. Combined with the axiom of the identity element of scalar multiplication, it is a *monoid action*.

[3] Some authors (such as Brown 1991) restrict attention to the fields **R** or **C**, but most of the theory is unchanged for an arbitrary field.

[4] The indicator functions of intervals (of which there are infinitely many) are linearly independent, for example.

[5] The nomenclature derives from German "eigen", which means own or proper.

[6] Roman 2005, ch. 8, p. 140. See also Jordan–Chevalley decomposition.

[7] Some authors (such as Roman 2005) choose to start with this equivalence relation and derive the concrete shape of V/W from this.

[8] This requirement implies that the topology gives rise to a uniform structure, Bourbaki 1989, ch. II

[9] The triangle inequality for $|-|p$ is provided by the Minkowski inequality. For technical reasons, in the context of functions one has to identify functions that agree almost everywhere to get a norm, and not only a seminorm.

[10] "Many functions in L^2 of Lebesgue measure, being unbounded, cannot be integrated with the classical Riemann integral. So spaces of Riemann integrable functions would not be complete in the L^2 norm, and the orthogonal decomposition would not apply to them. This shows one of the advantages of Lebesgue integration.", Dudley 1989, §5.3, p. 125

[11] For $p \neq 2$, $L^p(\Omega)$ is not a Hilbert space.

[12] A basis of a Hilbert space is not the same thing as a basis in the sense of linear algebra above. For distinction, the latter is then called a Hamel basis.

[13] Although the Fourier series is periodic, the technique can be applied to any L^2 function on an interval by considering the function to be continued periodically outside the interval. See Kreyszig 1988, p. 601

[14] That is to say (BSE-3 2001), the plane passing through the point of contact P such that the distance from a point P_1 on the surface to the plane is infinitesimally small compared to the distance from P_1 to P in the limit as P_1 approaches P along the surface.

[15] That is, there is a homeomorphism from $\pi^{-1}(U)$ to $V \times U$ which restricts to linear isomorphisms between fibers.

[16] A line bundle, such as the tangent bundle of S^1 is trivial if and only if there is a section that vanishes nowhere, see Husemoller 1994, Corollary 8.3. The sections of the tangent bundle are just vector fields.

4.12 Footnotes

[1] Roman 2005, ch. 1, p. 27

[2] van der Waerden 1993, Ch. 19

[3] Bourbaki 1998, §II.1.1. Bourbaki calls the group homomorphisms $f(a)$ *homotheties*.

[4] Bourbaki 1969, ch. "Algèbre linéaire et algèbre multilinéaire", pp. 78–91

[5] Bolzano 1804

[6] Möbius 1827

[7] Hamilton 1853

[8] Grassmann 2000

[9] Peano 1888, ch. IX

[10] Banach 1922

[11] Dorier 1995, Moore 1995

[12] Lang 1987, ch. I.1

[13] Lang 2002, ch. V.1

[14] e.g. Lang 1993, ch. XII.3., p. 335

[15] Lang 1987, ch. IX.1

[16] Lang 1987, ch. VI.3.

[17] Roman 2005, Theorem 1.9, p. 43

[18] Blass 1984

[19] Halpern 1966, pp. 670–673

[20] Artin 1991, Theorem 3.3.13

[21] Braun 1993, Th. 3.4.5, p. 291

[22] Stewart 1975, Proposition 4.3, p. 52

[23] Stewart 1975, Theorem 6.5, p. 74

[24] Roman 2005, ch. 2, p. 45

[25] Lang 1987, ch. IV.4, Corollary, p. 106

[26] Lang 1987, Example IV.2.6

[27] Lang 1987, ch. VI.6

[28] Halmos 1974, p. 28, Ex. 9

[29] Lang 1987, Theorem IV.2.1, p. 95

[30] Roman 2005, Th. 2.5 and 2.6, p. 49

[31] Lang 1987, ch. V.1

[32] Lang 1987, ch. V.3., Corollary, p. 106

[33] Lang 1987, Theorem VII.9.8, p. 198

[34] Roman 2005, ch. 8, p. 135–156

[35] Lang 1987, ch. IX.4

[36] Roman 2005, ch. 1, p. 29

[37] Roman 2005, ch. 1, p. 35

[38] Roman 2005, ch. 3, p. 64

[39] Lang 1987, ch. IV.3.

[40] Roman 2005, ch. 2, p. 48

[41] Mac Lane 1998

[42] Roman 2005, ch. 1, pp. 31–32

[43] Lang 2002, ch. XVI.1

[44] Roman 2005, Th. 14.3. See also Yoneda lemma.

[45] Schaefer & Wolff 1999, pp. 204–205

[46] Bourbaki 2004, ch. 2, p. 48

[47] Roman 2005, ch. 9

[48] Naber 2003, ch. 1.2

[49] Treves 1967

[50] Bourbaki 1987

[51] Kreyszig 1989, §4.11-5

[52] Kreyszig 1989, §1.5-5

[53] Choquet 1966, Proposition III.7.2

[54] Treves 1967, p. 34–36

[55] Lang 1983, Cor. 4.1.2, p. 69

[56] Treves 1967, ch. 11

[57] Treves 1967, Theorem 11.2, p. 102

[58] Evans 1998, ch. 5

[59] Treves 1967, ch. 12

[60] Dennery 1996, p.190

[61] Lang 1993, Th. XIII.6, p. 349

[62] Lang 1993, Th. III.1.1

[63] Choquet 1966, Lemma III.16.11

[64] Kreyszig 1999, Chapter 11

[65] Griffiths 1995, Chapter 1

[66] Lang 1993, ch. XVII.3

[67] Lang 2002, ch. III.1, p. 121

[68] Eisenbud 1995, ch. 1.6

[69] Varadarajan 1974

[70] Lang 2002, ch. XVI.7

[71] Lang 2002, ch. XVI.8

[72] Luenberger 1997, §7.13

[73] See representation theory and group representation.

[74] Lang 1993, Ch. XI.1

[75] Evans 1998, Th. 6.2.1

[76] Folland 1992, p. 349 *ff*

[77] Gasquet & Witomski 1999, p. 150

[78] Gasquet & Witomski 1999, §4.5

[79] Gasquet & Witomski 1999, p. 57

[80] Loomis 1953, Ch. VII

[81] Ashcroft & Mermin 1976, Ch. 5

[82] Kreyszig 1988, p. 667

[83] Fourier 1822

[84] Gasquet & Witomski 1999, p. 67

[85] Ifeachor & Jervis 2002, pp. 3–4, 11

[86] Wallace Feb 1992

[87] Ifeachor & Jervis 2002, p. 132

[88] Gasquet & Witomski 1999, §10.2

[89] Ifeachor & Jervis 2002, pp. 307–310

[90] Gasquet & Witomski 1999, §10.3

[91] Schönhage & Strassen 1971

[92] Spivak 1999, ch. 3

[93] Jost 2005. See also Lorentzian manifold.

[94] Misner, Thorne & Wheeler 1973, ch. 1.8.7, p. 222 and ch. 2.13.5, p. 325

[95] Jost 2005, ch. 3.1

[96] Varadarajan 1974, ch. 4.3, Theorem 4.3.27

[97] Kreyszig 1991, §34, p. 108

[98] Eisenberg & Guy 1979

[99] Atiyah 1989

[100] Artin 1991, ch. 12

[101] Grillet, Pierre Antoine. Abstract algebra. Vol. 242. Springer Science & Business Media, 2007.

[102] Meyer 2000, Example 5.13.5, p. 436

[103] Meyer 2000, Exercise 5.13.15–17, p. 442

[104] Coxeter 1987

4.13   References

4.13.1   Algebra

- Artin, Michael (1991), *Algebra*, Prentice Hall, ISBN 978-0-89871-510-1

- Blass, Andreas (1984), "Existence of bases implies the axiom of choice", *Axiomatic set theory (Boulder, Colorado, 1983)*, Contemporary Mathematics **31**, Providence, R.I.: American Mathematical Society, pp. 31–33, MR 763890

- Brown, William A. (1991), *Matrices and vector spaces*, New York: M. Dekker, ISBN 978-0-8247-8419-5

- Lang, Serge (1987), *Linear algebra*, Berlin, New York: Springer-Verlag, ISBN 978-0-387-96412-6

- Lang, Serge (2002), *Algebra*, Graduate Texts in Mathematics **211** (Revised third ed.), New York: Springer-Verlag, ISBN 978-0-387-95385-4, MR 1878556

- Mac Lane, Saunders (1999), *Algebra* (3rd ed.), pp. 193–222, ISBN 0-8218-1646-2

- Meyer, Carl D. (2000), *Matrix Analysis and Applied Linear Algebra*, SIAM, ISBN 978-0-89871-454-8

- Roman, Steven (2005), *Advanced Linear Algebra*, Graduate Texts in Mathematics **135** (2nd ed.), Berlin, New York: Springer-Verlag, ISBN 978-0-387-24766-3

- Spindler, Karlheinz (1993), *Abstract Algebra with Applications: Volume 1: Vector spaces and groups*, CRC, ISBN 978-0-8247-9144-5

- van der Waerden, Bartel Leendert (1993), *Algebra* (in German) (9th ed.), Berlin, New York: Springer-Verlag, ISBN 978-3-540-56799-8

4.13.2   Analysis

- Bourbaki, Nicolas (1987), *Topological vector spaces*, Elements of mathematics, Berlin, New York: Springer-Verlag, ISBN 978-3-540-13627-9

- Bourbaki, Nicolas (2004), *Integration I*, Berlin, New York: Springer-Verlag, ISBN 978-3-540-41129-1

- Braun, Martin (1993), *Differential equations and their applications: an introduction to applied mathematics*, Berlin, New York: Springer-Verlag, ISBN 978-0-387-97894-9

- BSE-3 (2001), "Tangent plane", in Hazewinkel, Michiel, *Encyclopedia of Mathematics*, Springer, ISBN 978-1-55608-010-4

- Choquet, Gustave (1966), *Topology*, Boston, MA: Academic Press

- Dennery, Philippe; Krzywicki, Andre (1996), *Mathematics for Physicists*, Courier Dover Publications, ISBN 978-0-486-69193-0

- Dudley, Richard M. (1989), *Real analysis and probability*, The Wadsworth & Brooks/Cole Mathematics Series, Pacific Grove, CA: Wadsworth & Brooks/Cole Advanced Books & Software, ISBN 978-0-534-10050-6

- Dunham, William (2005), *The Calculus Gallery*, Princeton University Press, ISBN 978-0-691-09565-3

- Evans, Lawrence C. (1998), *Partial differential equations*, Providence, R.I.: American Mathematical Society, ISBN 978-0-8218-0772-9

- Folland, Gerald B. (1992), *Fourier Analysis and Its Applications*, Brooks-Cole, ISBN 978-0-534-17094-3

- Gasquet, Claude; Witomski, Patrick (1999), *Fourier Analysis and Applications: Filtering, Numerical Computation, Wavelets*, Texts in Applied Mathematics, New York: Springer-Verlag, ISBN 0-387-98485-2

- Ifeachor, Emmanuel C.; Jervis, Barrie W. (2001), *Digital Signal Processing: A Practical Approach* (2nd ed.), Harlow, Essex, England: Prentice-Hall (published 2002), ISBN 0-201-59619-9

- Krantz, Steven G. (1999), *A Panorama of Harmonic Analysis*, Carus Mathematical Monographs, Washington, DC: Mathematical Association of America, ISBN 0-88385-031-1

- Kreyszig, Erwin (1988), *Advanced Engineering Mathematics* (6th ed.), New York: John Wiley & Sons, ISBN 0-471-85824-2

- Kreyszig, Erwin (1989), *Introductory functional analysis with applications*, Wiley Classics Library, New York: John Wiley & Sons, ISBN 978-0-471-50459-7, MR 992618

- Lang, Serge (1983), *Real analysis*, Addison-Wesley, ISBN 978-0-201-14179-5

- Lang, Serge (1993), *Real and functional analysis*, Berlin, New York: Springer-Verlag, ISBN 978-0-387-94001-4

- Loomis, Lynn H. (1953), *An introduction to abstract harmonic analysis*, Toronto-New York–London: D. Van Nostrand Company, Inc., pp. x+190

- Schaefer, Helmut H.; Wolff, M.P. (1999), *Topological vector spaces* (2nd ed.), Berlin, New York: Springer-Verlag, ISBN 978-0-387-98726-2

- Treves, François (1967), *Topological vector spaces, distributions and kernels*, Boston, MA: Academic Press

4.13.3 Historical references

- Banach, Stefan (1922), "Sur les opérations dans les ensembles abstraits et leur application aux équations intégrales (On operations in abstract sets and their application to integral equations)" (PDF), *Fundamenta Mathematicae* (in French) **3**, ISSN 0016-2736

- Bolzano, Bernard (1804), *Betrachtungen über einige Gegenstände der Elementargeometrie (Considerations of some aspects of elementary geometry)* (in German)

- Bourbaki, Nicolas (1969), *Éléments d'histoire des mathématiques (Elements of history of mathematics)* (in French), Paris: Hermann

- Dorier, Jean-Luc (1995), "A general outline of the genesis of vector space theory", *Historia Mathematica* **22** (3): 227–261, doi:10.1006/hmat.1995.1024, MR 1347828

- Fourier, Jean Baptiste Joseph (1822), *Théorie analytique de la chaleur* (in French), Chez Firmin Didot, père et fils

- Grassmann, Hermann (1844), *Die Lineale Ausdehnungslehre - Ein neuer Zweig der Mathematik* (in German), O. Wigand, reprint: Hermann Grassmann. Translated by Lloyd C. Kannenberg. (2000), Kannenberg, L.C., ed., *Extension Theory*, Providence, R.I.: American Mathematical Society, ISBN 978-0-8218-2031-5

- Hamilton, William Rowan (1853), *Lectures on Quaternions*, Royal Irish Academy

- Möbius, August Ferdinand (1827), *Der Barycentrische Calcul : ein neues Hülfsmittel zur analytischen Behandlung der Geometrie (Barycentric calculus: a new utility for an analytic treatment of geometry)* (in German)

- Moore, Gregory H. (1995), "The axiomatization of linear algebra: 1875–1940", *Historia Mathematica* **22** (3): 262–303, doi:10.1006/hmat.1995.1025

- Peano, Giuseppe (1888), *Calcolo Geometrico secondo l'Ausdehnungslehre di H. Grassmann preceduto dalle Operazioni della Logica Deduttiva* (in Italian), Turin

4.13.4   Further references

- Ashcroft, Neil; Mermin, N. David (1976), *Solid State Physics*, Toronto: Thomson Learning, ISBN 978-0-03-083993-1

- Atiyah, Michael Francis (1989), *K-theory*, Advanced Book Classics (2nd ed.), Addison-Wesley, ISBN 978-0-201-09394-0, MR 1043170

- Bourbaki, Nicolas (1998), *Elements of Mathematics : Algebra I Chapters 1-3*, Berlin, New York: Springer-Verlag, ISBN 978-3-540-64243-5

- Bourbaki, Nicolas (1989), *General Topology. Chapters 1-4*, Berlin, New York: Springer-Verlag, ISBN 978-3-540-64241-1

- Coxeter, Harold Scott MacDonald (1987), *Projective Geometry* (2nd ed.), Berlin, New York: Springer-Verlag, ISBN 978-0-387-96532-1

- Eisenberg, Murray; Guy, Robert (1979), "A proof of the hairy ball theorem", *The American Mathematical Monthly* (Mathematical Association of America) **86** (7): 572–574, doi:10.2307/2320587, JSTOR 2320587

- Eisenbud, David (1995), *Commutative algebra*, Graduate Texts in Mathematics **150**, Berlin, New York: Springer-Verlag, ISBN 978-0-387-94269-8, MR 1322960

- Goldrei, Derek (1996), *Classic Set Theory: A guided independent study* (1st ed.), London: Chapman and Hall, ISBN 0-412-60610-0

- Griffiths, David J. (1995), *Introduction to Quantum Mechanics*, Upper Saddle River, NJ: Prentice Hall, ISBN 0-13-124405-1

- Halmos, Paul R. (1974), *Finite-dimensional vector spaces*, Berlin, New York: Springer-Verlag, ISBN 978-0-387-90093-3

- Halpern, James D. (Jun 1966), "Bases in Vector Spaces and the Axiom of Choice", *Proceedings of the American Mathematical Society* (American Mathematical Society) **17** (3): 670–673, doi:10.2307/2035388, JSTOR 2035388

- Husemoller, Dale (1994), *Fibre Bundles* (3rd ed.), Berlin, New York: Springer-Verlag, ISBN 978-0-387-94087-8

- Jost, Jürgen (2005), *Riemannian Geometry and Geometric Analysis* (4th ed.), Berlin, New York: Springer-Verlag, ISBN 978-3-540-25907-7

- Kreyszig, Erwin (1991), *Differential geometry*, New York: Dover Publications, pp. xiv+352, ISBN 978-0-486-66721-8

- Kreyszig, Erwin (1999), *Advanced Engineering Mathematics* (8th ed.), New York: John Wiley & Sons, ISBN 0-471-15496-2

- Luenberger, David (1997), *Optimization by vector space methods*, New York: John Wiley & Sons, ISBN 978-0-471-18117-0

- Mac Lane, Saunders (1998), *Categories for the Working Mathematician* (2nd ed.), Berlin, New York: Springer-Verlag, ISBN 978-0-387-98403-2

- Misner, Charles W.; Thorne, Kip; Wheeler, John Archibald (1973), *Gravitation*, W. H. Freeman, ISBN 978-0-7167-0344-0

- Naber, Gregory L. (2003), *The geometry of Minkowski spacetime*, New York: Dover Publications, ISBN 978-0-486-43235-9, MR 2044239

- Schönhage, A.; Strassen, Volker (1971), "Schnelle Multiplikation großer Zahlen (Fast multiplication of big numbers)" (PDF), *Computing* (in German) **7**: 281–292, doi:10.1007/bf02242355, ISSN 0010-485X

- Spivak, Michael (1999), *A Comprehensive Introduction to Differential Geometry (Volume Two)*, Houston, TX: Publish or Perish

- Stewart, Ian (1975), *Galois Theory*, Chapman and Hall Mathematics Series, London: Chapman and Hall, ISBN 0-412-10800-3

- Varadarajan, V. S. (1974), *Lie groups, Lie algebras, and their representations*, Prentice Hall, ISBN 978-0-13-535732-3

- Wallace, G.K. (Feb 1992), "The JPEG still picture compression standard", *IEEE Transactions on Consumer Electronics* **38** (1): xviii–xxxiv, doi:10.1109/30.125072, ISSN 0098-3063

- Weibel, Charles A. (1994), *An introduction to homological algebra*, Cambridge Studies in Advanced Mathematics **38**, Cambridge University Press, ISBN 978-0-521-55987-4, OCLC 36131259, MR 1269324

4.14 External links

- Hazewinkel, Michiel, ed. (2001), "Vector space", *Encyclopedia of Mathematics*, Springer, ISBN 978-1-55608-010-4

- A lecture about fundamental concepts related to vector spaces (given at MIT)

- A graphical simulator for the concepts of span, linear dependency, base and dimension

Chapter 5

Minkowski space

For spacetime graphics, see Minkowski diagram. For Minkowski space associated to a number field, see Minkowski space (number field). For geometry of the Minkowski plane, see Minkowski plane .

In mathematical physics, **Minkowski space** or **Minkowski spacetime** is a combination of Euclidean space and time into a four-dimensional manifold where the spacetime interval between any two events is independent of the inertial frame of reference in which they are recorded. Although initially developed by mathematician Hermann Minkowski for Maxwell's equations of electromagnetism, the mathematical structure of Minkowski spacetime was shown to be an immediate consequence of the postulates of special relativity.[1]

Minkowski space is closely associated with Einstein's theory of special relativity, and is the most common mathematical structure on which special relativity is formulated. While the individual components in Euclidean space and time will often differ due to length contraction and time dilation, in Minkowski spacetime, all frames of reference will agree on the total distance in spacetime between events.[nb 1] Because it treats time differently than the three spacial dimensions, Minkowski space differs from four-dimensional Euclidean space.[nb 2]

The isometry group, preserving Euclidean distances of a Euclidean space equipped with the regular inner product is the Euclidean group. The analogous isometry group for Minkowski apace, preserving intervals of spacetime equipped with the associated non-positive definite bilinear form (here called the **Minkowski inner product**,[nb 3]) is the Poincaré group. The Minkowski inner product is defined as to yield the spacetime interval between two events when given their coordinate difference vector as argument.

5.1 History

5.1.1 Four-dimensional Euclidean spacetime

See also: Four-dimensional space

In 1905, with the publication in 1906, Henri Poincaré showed that by taking time to be an imaginary fourth spacetime coordinate ($\sqrt{-1}\,c\,t$), a Lorentz transformation can be regarded as a rotation of coordinates in a four-dimensional Euclidean space with three real coordinates representing space, and one imaginary coordinate, representing time, as the fourth dimension. Since the space is then a pseudo-Euclidean space, the rotation is a representation of a hyperbolic rotation, although Poincaré did not give this interpretation, his purpose being only to explain the Lorentz transformation in terms of the familiar Euclidean rotation.[2]

This idea was elaborated by Hermann Minkowski,[3] who used it to restate the Maxwell equations in four dimensions, showing directly their invariance under the Lorentz transformation. He further reformulated in four dimensions the then-recent theory of special relativity of Einstein. From this he concluded that time and space should be treated equally, and

Hermann Minkowski (1864 – 1909) was a German mathematician. He found that the theory of special relativity, introduced by his former student Albert Einstein, could best be understood in a four-dimensional space, since known as the Minkowski spacetime.

so arose his concept of events taking place in a unified four-dimensional spacetime continuum.

5.1.2   Minkowski space

In a further development,[4] he gave an alternative formulation of this idea that used a real time coordinate instead of an imaginary one, representing the four variables (x, y, z, t) of space and time in coordinate form in a four dimensional affine space. Points in this space correspond to events in spacetime. In this space, there is a defined light-cone associated with each point (see diagram above), and events not on the light-cone are classified by their relation to the apex as *spacelike* or *timelike*. It is principally this view of spacetime that is current nowadays, although the older view involving imaginary time has also influenced special relativity. Minkowski, aware of the fundamental restatement of the theory which he had made, said

> The views of space and time which I wish to lay before you have sprung from the soil of experimental physics, and therein lies their strength. They are radical. Henceforth space by itself, and time by itself, are doomed to fade away into mere shadows, and only a kind of union of the two will preserve an independent reality.
> — Hermann Minkowski, 1907[4]

For further historical information see references Galison (1979), Corry (1997) and Walter (1999).

5.2   Mathematical structure

For an overview, Minkowski space is a 4-dimensional real vector space equipped with a nondegenerate, symmetric bilinear form on the tangent space at each point in spacetime, here simply called the Minkowski inner product, with signature either (−,+,+,+) or (+,−,−,−). In practice, one need not be concerned with the tangent spaces. The vector space nature of Minkowski space allows for the canonical identification of vectors in tangent spaces at points (events) with vectors (points, events) in Minkowski space itself.[5] For some purposes it is desirable to identify tangent vectors at a point p with *displacement vectors* at p, which is, of course, admissible by essentially the same canonical identification.[6]

The signature refers to which sign the Minkowski inner product yields when given space and time basis vectors as arguments. In general, mathematicians and general relativists prefer the former while particle physicists tend to use the latter. Arguments for the former (pure space vectors yield positive "norm-squared") include "continuity" from the Euclidean case corresponding to the non-relativistic limit $c \to \infty$. Arguments for the latter (pure space vectors yield negative "norm-squared") include that otherwise ubiquitous minus signs in particle physics go away.

Mathematically associated to this bilinear form is a tensor of type (0,2) at each point in spacetime, called the Minkowski metric. The Minkowski metric, the bilinear form, and the Minkowski inner product are actually all the very same object. In coordinates, this is the 4×4 matrix representing the bilinear form. Keeping this in mind may facilitate reading what follows.

For comparison, in general relativity, a Lorentzian manifold L is likewise equipped with a metric tensor g, which is a nondegenerate symmetric bilinear form on the tangent space T_pL at each point p of L. In coordinates, it may be represented by a 4×4 matrix *depending on spacetime position*. Minkowski space is thus a comparatively simple special case of a Lorentzian manifold. Its metric tensor, called the Minkowski metric, is in coordinates the same symmetric matrix at every point of M, and its arguments can, per above, be taken as vectors in spacetime itself.

Introducing more terminology (but not more structure), Minkowski space is thus a pseudo-Euclidean space with total dimension $n = 4$ and signature (3, 1) or (1, 3). Elements of Minkowski space are called events. Minkowski space is often denoted $\mathbf{R}^{3,1}$ or $\mathbf{R}^{1,3}$ to emphasize the chosen signature, or just M. It is perhaps the simplest example of a pseudo-Riemannian manifold.

5.2.1 Pseudo-Euclidean metric generalities

Main article: Pseudo-Euclidean space

The Minkowski metric[nb 4] η is the metric tensor of Minkowski space. It is a Pseudo-Euclidean metric. As such it is a nondegenerate symmetric bilinear form, a type (0,2) tensor. It accepts two arguments *up*, *vp*, vectors in TpM, $p \in M$, the tangent space at p in M. Due to the above mentioned canonical identification of TpM with M itself, it accepts arguments u, v with both u and v in M.

As a notational convention, vectors v in M, called 4-vectors, are denoted in sans-serif italics, and not, as is common in the Eucliedean setting, with boldface **v**. The latter is generally reserved for the 3-vector part (to be introduced below) of a 4-vector.

The definition

$$u \cdot v = \eta(u, v)$$

yields an inner product-like structure on M, previously and also henceforth, called the Minkowski inner product, similar to the Euclidean inner product, but it describes a different geometry. It has the following properties.

- $\eta(au + v, w) = a\eta(u, w) + \eta(v, w). \quad \forall u, v \in M, \forall a \in \mathbb{R} \qquad$ slot) first in (linearity

- $\eta(u, v) = \eta(v, u) \qquad$ (symmetry)

- $\eta(u, v) = 0 \quad \forall v \in M \Rightarrow u = 0 \qquad$ (non-degeneracy)

The first two conditions imply bilinearity. The defining *difference* between a pseudo-inner product and an inner product proper is that the former is *not* required to be positive definite, that is, $\eta(u, u) < 0$ is allowed.

Two vectors *v* and *w* are said to be orthogonal if $\eta(v, w) = 0$.

A vector *e* is called a unit vector if $\eta(e, e) = \pm 1$. A basis for M consisting of mutually orthogonal unit vectors is called an orthonormal basis.

For a given inertial frame, an orthonormal basis in space, combined by the unit time vector, forms an orthonormal basis in Minkowski space. The number of positive and negative unit vectors in any such basis is a fixed pair of numbers, equal to the signature of the bilinear form associated with the inner product. This is Sylvester's law of inertia.

More terminology (but not more structure): The Minkowski metric is a pseudo-Riemannian metric, more specifically, a Lorentzian metric, even more specifically, *the* Lorentz metric, reserved for 4-dimensional flat spacetime with the remaining ambiguity only being the signature convention.

5.2.2 Minkowski metric

From the two postulates of special relativity follows that the spacetime interval between two events 1, 2,

$$\pm \left[c^2(t_1 - t_2)^2 - (x_1 - x_2)^2 - (y_1 - y_2)^2 - (z_1 - z_2)^2 \right],$$

is independent of the inertial frame chosen. The factor $\pm$ simply means that the choice of signature is left open. The numerical values of η, viewed as a matrix representing the Minkowski inner product, follow from the theory of bilinear forms.

Just as the signature of the metric is differently defined in the literature, this quantity is not consistently named. The interval (as defined here) is sometimes referred to as the interval squared.[7] Even the square root of the present interval occurs.[8] When signature and interval are fixed, ambiguity still remains as which coordinate is the time coordinate. It may be the fourth, or it may be the zeroth. This is not an exhaustive list of notational inconsistencies. It is a fact of life that one has to check out the definitions first thing when one consults the relativity literature.

The invariance of the interval under coordinate transformations between inertial frames follows from the invariance of

$$\pm \left[c^2 t^2 - x^2 - y^2 - z^2 \right]$$

(with either sign $\pm$ preserved), provided the transformations are linear. This quadratic form can be used to define a bilinear form

$$u \cdot v = \pm \left[c^2 t_1 t_2 - x_1 x_2 - y_1 y_2 - z_1 z_2 \right].$$

via the polarization identity. This bilinear form can in turn be written as

$$u \cdot v = u^{\mathrm{T}} [\eta] v,$$

where $[\eta]$ is a 4×4 matrix associated with η. Possibly confusingly, denote $[\eta]$ with just η as is common practice. The matrix is read off from the explicit bilinear form as

$$\eta = \pm \begin{pmatrix} -1 & 0 & 0 & 0 \\ 0 & 1 & 0 & 0 \\ 0 & 0 & 1 & 0 \\ 0 & 0 & 0 & 1 \end{pmatrix},$$

and the bilinear form

$$u \cdot v = \eta(u, v),$$

with which this section started by assuming its existence, is now identified.

For definiteness and shorter presentation, the signature $(-,+,+,+)$ is adopted below. The choice has no (known) physical implications. The symmetry group preserving the bilinear form with one choice of signature is isomorphic (under the map given here) with the symmetry group preserving the other choice of signature. This means that both choices are in accord with the two postulates of relativity.

5.2.3 Standard basis

A standard basis for Minkowski space is a set of four mutually orthogonal vectors $\{\, e_0, e_1, e_2, e_3 \,\}$ such that

$$-\eta(e_0, e_0) = \eta(e_1, e_1) = \eta(e_2, e_2) = \eta(e_3, e_3) = 1.$$

These conditions can be written compactly in the form

$$\eta(e_\mu, e_\nu) = \eta_{\mu\nu}.$$

Relative to a standard basis, the components of a vector v are written (v^0, v^1, v^2, v^3) where the Einstein notation is used to write $v = v^\mu e_\mu$. The component v^0 is called the **timelike component** of v while the other three components are called the **spatial components**. The spatial components of a 4-vector v may be identified with a 3-vector $\mathbf{v} = (v_1, v_2, v_3)$.

In terms of components, the Minkowski inner product between two vectors v and w is given by

$$\eta(v, w) = \eta_{\mu\nu} v^\mu w^\nu = v^0 w_0 + v^1 w_1 + v^2 w_2 + v^3 w_3 = v^\mu w_\mu = v_\mu w^\mu,$$

and

$$\eta(v, v) = \eta_{\mu\nu} v^\mu v^\nu = v^0 v_0 + v^1 v_1 + v^2 v_2 + v^3 v_3 = v^\mu v_\mu.$$

Here **lowering of an index** with the metric was used. Technically, a non-degenerate bilinear form provides a map between a vector space and its dual, in this context, the map is between the tangent spaces of M and the cotangent spaces of M. At a point in M, the tangent and cotangent spaces are dual. Just as an authentic inner product on a vector space with one argument fixed, by Riesz representation theorem, may be expressed as the action of a linear functional on the vector space, the same holds for the Minkowski inner product of Minkowski space.

Thus if v^μ are the components of a vector in a tangent space, then $\eta_{\mu\nu} v^\mu = v\nu$ are the components of a vector in the cotangent space (a linear functional). Due to the identification of vectors in tangent spaces with vectors in M itself, this is mostly ignored, and vectors with lower indices are referred to as **covariant vectors**. In this latter interpretation, the covariant vectors are (almost always implicitly) identified with vectors (linear functionals) in the dual of Minkowski space. The ones with upper indices are **contravariant vectors**. In the same fashion, the inverse of the map from tangent to cotangent spaces, explicitly given by the inverse of η in matrix representation, can be used to define **raising of an index**. The components of this inverse are denoted $\eta^{\mu\nu}$. It happens that $\eta^{\mu\nu} = \eta\mu\nu$. These maps between a vector space and its dual can be denoted $\eta^\flat$ (eta-flat) and $\eta^\sharp$ (eta-sharp) by the musical analogy.[9]

The time-proven robustness of the formalism itself, sometimes referred to as index gymnastics, ensures that moving vectors around and changing from contravariant to covariant vectors and vice versa is mathematically sound. Incorrect expressions tend to reveal themselves quickly.

5.2.4 Geometry

5.3 Lorentz transformations and symmetry

The Poincaré group is the group of all transformations preserving the interval. The interval is quite easily seen to be preserved by the translation group in 4 dimensions. The other transformations are those that preserve the interval and leave the origin fixed. Given the bilinear form associated with the Minkowski metric, the appropriate group follows directly from the theory (in particular the definition) of classical groups. In the linked article, one should identify η (in its a matrix representation) with the matrix Φ.

The appropriate group is O(3,1), in this context called the Lorentz group. Its elements are called (homogeneous) Lorentz transformations. For other methods of derivation, with a more physical twist, see derivations of the Lorentz transformations.

Among the simplest Lorentz transformations is a Lorentz boost. For reference, a boost in the x-direction is given by

$$\begin{bmatrix} U_0' \\ U_1' \\ U_2' \\ U_3' \end{bmatrix} = \begin{bmatrix} \gamma & -\beta\gamma & 0 & 0 \\ -\beta\gamma & \gamma & 0 & 0 \\ 0 & 0 & 1 & 0 \\ 0 & 0 & 0 & 1 \end{bmatrix} \begin{bmatrix} U_0 \\ U_1 \\ U_2 \\ U_3 \end{bmatrix},$$

where

$$\gamma = \frac{1}{\sqrt{1 - \frac{v^2}{c^2}}}$$

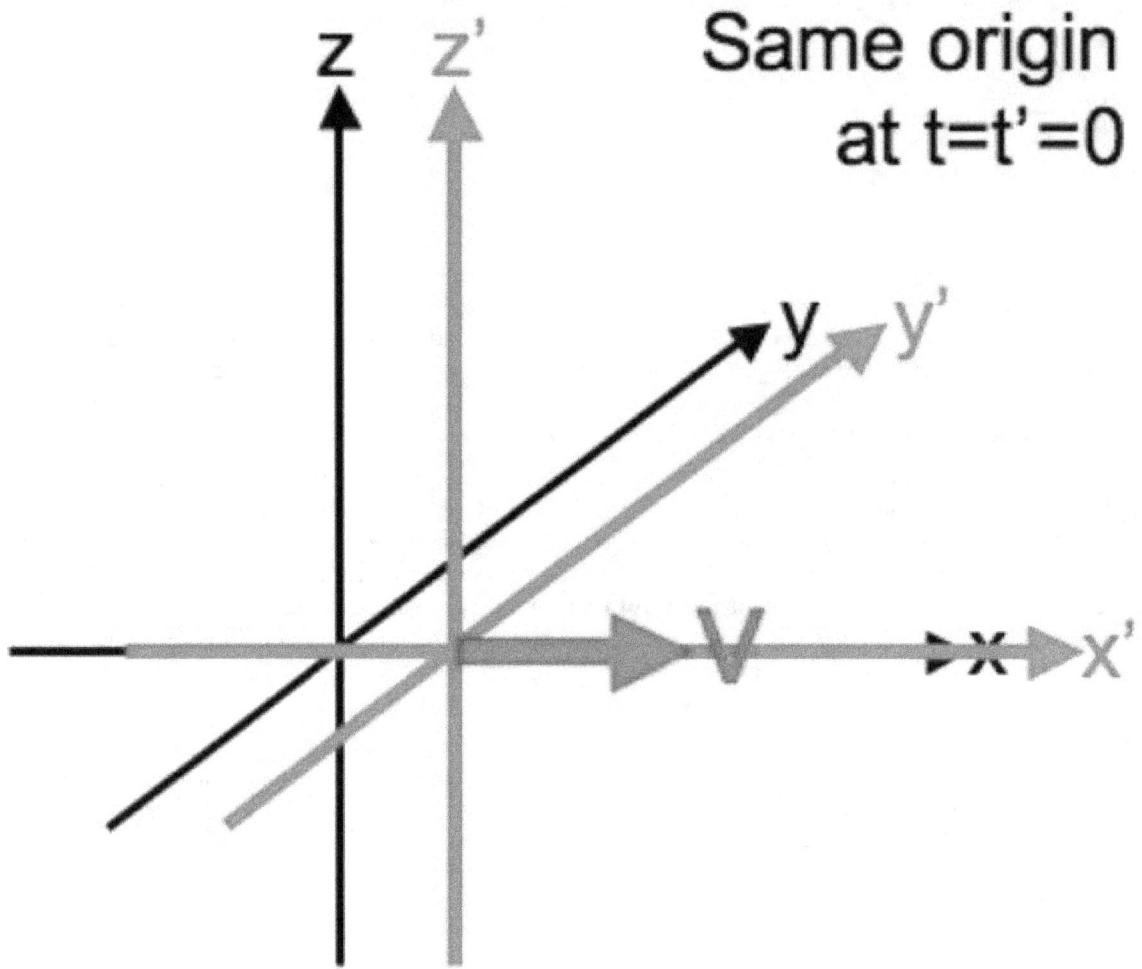

Standard configuration of coordinate systems for Lorentz transformations.

is the Lorentz factor, and

$$\beta = \frac{v}{c}.$$

Other Lorentz transformations are pure rotations, and hence elements of the SO(3) subgroup of O(3,1). A general homogeneous Lorentz transformation is a product of a pure boost and a pure rotation. An *inhomogeneous* Lorentz transformation is a homogeneous transformation followed by a translation in space and time. Special transformations are those that invert the space coordinates (P) and time coordinate (T) respectively, or both (PT).

All four-vectors in Minkowski space transform, by definition, according to the same formula under Lorentz transformations. Minkowski diagrams illustrate Lorentz transformations.

5.4   Causal structure

Main article: Causal structure

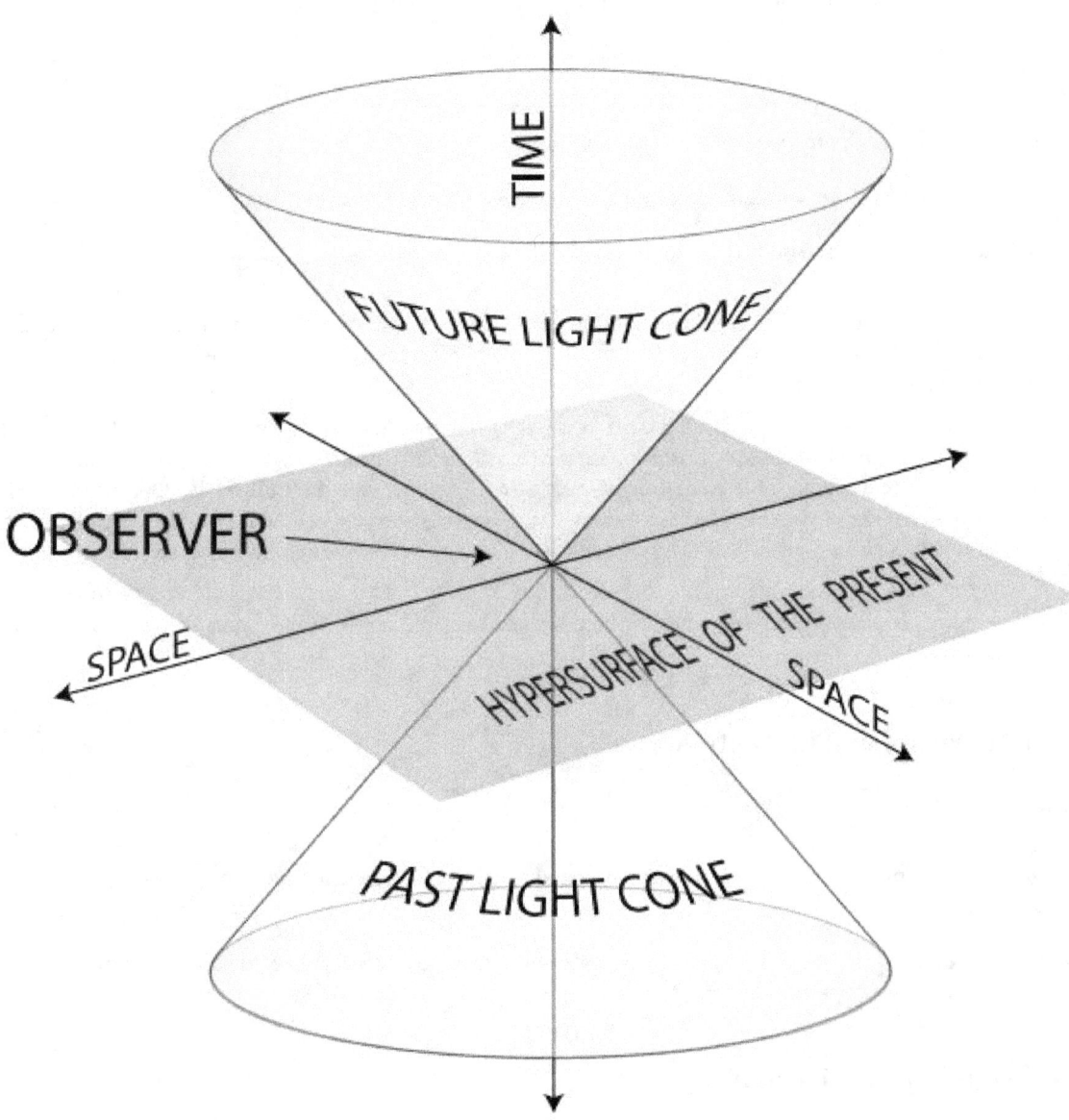

*Subdivision of Minkowski spacetime with respect to an event in four disjoint sets. The light cone, the **absolute future**, the **absolute past**,*
*and **elsewhere**. The terminology is from Sard (1970).*

Vectors $v = (ct, x, y, z) = (ct, \mathbf{r})$ are classified according to the sign of $c^2 t^2 - r^2$. A vector is **timelike** if $c^2 t^2 > r^2$, **spacelike**
if $c^2 t^2 < r^2$, and **null** or **lightlike** if $c^2 t^2 = r^2$. This can be expressed in terms of the sign of $\eta(v,v)$ as well, but depends
on the signature. The classification of any vector will be the same in all frames of reference, because of the invariance of
the interval.

The set of all null vectors at an event[nb 5] of Minkowski space constitutes the light cone of that event. Given a timelike
vector v, there is a worldline of constant velocity associated with it, represented by a straight line in a Minkowski diagram.

Once a direction of time is chosen,[nb 6] timelike and null vectors can be further decomposed into various classes. For
timelike vectors one has

1. future-directed timelike vectors whose first component is positive, (tip of vector located in absolute future in figure)
   and

2. past-directed timelike vectors whose first component is negative (absolute past).

Null vectors fall into three classes:

1. the zero vector, whose components in any basis are (0,0,0,0) (origin),

2. future-directed null vectors whose first component is positive (upper light cone), and

3. past-directed null vectors whose first component is negative (lower light cone).

Spacelike vectors are in elsewhere. The terminology stems from the fact that spacelike separated events are connected by vectors requiring faster-than-light travel, and so cannot possibly influence each other. Together with spacelike and lightlike vectors there are 7 classes in all.

An orthonormal basis for Minkowski space necessarily consists of one timelike and three spacelike unit vectors. If one wishes to work with non-orthonormal bases it is possible to have other combinations of vectors. For example, one can easily construct a (non-orthonormal) basis consisting entirely of null vectors, called a **null basis**. Over the reals, if two null vectors are orthogonal (zero Minkowski tensor value), then they must be proportional. However, allowing complex numbers, one can obtain a null tetrad, which is a basis consisting of null vectors, some of which are orthogonal to each other.

Vector fields are called timelike, spacelike or null if the associated vectors are timelike, spacelike or null at each point where the field is defined.

5.4.1   Chronological and causality relations

Let $x, y \in M$. We say that

1. x *chronologically precedes* y if $y - x$ is future-directed timelike. This relation has the transitive property and so can be written x < y.

2. x *causally precedes* y if $y - x$ is future-directed null or future-directed timelike. It gives a partial ordering of space-time and so can be written x ≤ y.

5.4.2   Reversed triangle inequality

If v and w are both future-directed timelike four-vectors, then in the (+ - - -) sign convention for norm,

$$\|v + w\| \geq \|v\| + \|w\| .$$

5.5   Relationships to other formulations

5.5.1   Different number of dimensions

Strictly speaking, Minkowski space refers to a mathematical formulation in four dimensions. However, the mathematics can easily be extended or simplified to create an analogous "Minkowski space" in any number of dimensions. If $n \geq 2$, n-dimensional Minkowski space is a vector space of real dimension n on which there is a constant Lorentz metric of signature $(n - 1, 1)$ or $(1, n - 1)$. These generalizations are used in theories where spacetime is assumed to have more or less than 4 dimensions. String theory and M-theory are two examples where $n > 4$. In string theory, there appears conformal field theories with $1 + 1$ spacetime dimensions.

5.5.2   Flat versus curved space

As a *flat spacetime*, the three spatial components of Minkowski spacetime always obey the Pythagorean Theorem. Minkowski space is a suitable basis for special relativity, a good description of physical systems over finite distances in systems without significant gravitation. However, in order to take gravity into account, physicists use the theory of general relativity, which is formulated in the mathematics of a non-Euclidean geometry. When this geometry is used as a model of physical space, it is known as curved space.

Even in curved space, Minkowski space is still a good description in an infinitesimal region surrounding any point (barring gravitational singularities).[nb 7] More abstractly, we say that in the presence of gravity spacetime is described by a curved 4-dimensional manifold for which the tangent space to any point is a 4-dimensional Minkowski space. Thus, the structure of Minkowski space is still essential in the description of general relativity.

5.6   See also

- Causal structure
- Euclidean space
- Four vector
- Hyperboloid model
- Introduction to mathematics of general relativity
- Lorentzian manifold
- Metric tensor
- Minkowski diagram
- Minkowski plane
- Speed of light
- Super Minkowski space
- World line

5.7   Remarks

[1] This makes spacetime distance an invariant.

[2] Minkowski space can be formulated as an equivalent 4-D Euclidean space if you assume time is always an imaginary number. This is how the spacetime was first formulated, but since Minkowski reworked the structure, time is almost always required to be a real number.

[3] Consistent use of the term "Minkowski inner product" is intended for the bilinear form here, since it is in widespread use. It is by no means "standard" in the literature, but no such standard seems to exist.

[4] The Minkowski inner product is not an inner product, since it is not positive-definite, i.e. the quadratic form $\eta(v, v)$ need not be positive for nonzero v. The positive-definite condition has been replaced by the weaker condition of non-degeneracy. The bilinear form is said to be *indefinite*.

[5] Translate the coordinate system so that the event is the new origin.

[6] This corresponds to the time coordinate either increasing or decreasing when proper time for any particle increases. An application of T flips this direction.

[7] This similarity between flat and curved space at infinitesimally small distance scales is foundational to the definition of a manifold in general.

5.8   Notes

[1] Landau & Lifshitz 2002, p. 5

[2] Poincaré 1905–1906, pp. 129–176 Wikisource translation: On the Dynamics of the Electron

[3] Minkowski 1907–1908, pp. 53–111 *Wikisource translation: The Fundamental Equations for Electromagnetic Processes in Moving Bodies.

[4] Minkowski 1907–1909, pp. 75–88 Various English translations on Wikisource: Space and Time.

[5] Lee 2003, Proposition 3.8. The identification is routinely done in mathematics.

[6] Lee 2003, See Lee's discussion on geometric tangent vectors early in chapter 3.

[7] Sard 1970, p. 71

[8] Landau & Lifshitz 2002, p. 4

[9] Lee 2003, The tangent-cotangent isomorphism p. 282.

5.9   References

- Corry, L. (1997). "Hermann Minkowski and the postulate of relativity". *Arch. Hist. Exact Sci.* (Springer-Verlag) **51** (4): 273–314. doi:10.1007/BF00518231. ISSN 0003-9519. (subscription required (help)).

- Catoni, F.; et al. (2008). *Mathematics of Minkowski Space*. Frontiers in Mathematics. Basel: Birkhäuser Verlag. doi:10.1007/978-3-7643-8614-6. ISBN 978-3-7643-8613-9. ISSN 1660-8046.

- Galison, P. L. (1979). R McCormach; et al., eds. *Minkowski's Space-Time: from visual thinking to the absolute world*. Historical Studies in the Physical Sciences **10**. Johns Hopkins University Press. pp. 85–121. doi:10.2307/27757388. (subscription required (help)).

- Landau, L.D.; Lifshitz, E.M. (2002) [1939]. *The Classical Theory of Fields*. Course of Theoretical Physics **2** (4th ed.). Butterworth–Heinemann. ISBN 0 7506 2768 9.

- Lee, J. M. (2003). *Introduction to Smooth manifolds*. Springer Graduate Texts in Mathematics **218**. ISBN 0-387-95448-1.

- Minkowski, Hermann (1907–1908), "Die Grundgleichungen für die elektromagnetischen Vorgänge in bewegten Körpern" [The Fundamental Equations for Electromagnetic Processes in Moving Bodies], *Nachrichten von der Gesellschaft der Wissenschaften zu Göttingen, Mathematisch-Physikalische Klasse*: 53–111 *Wikisource translation: The Fundamental Equations for Electromagnetic Processes in Moving Bodies

- Minkowski, Hermann (1907–1909), "Raum und Zeit" [Space and Time], *Physikalische Zeitschrift* **10**: 75–88 Various English translations on Wikisource: Space and Time

- Naber, G. L. (1992). *The Geometry of Minkowski Spacetime*. New York: Springer-Verlag. ISBN 0-387-97848-8.

- Penrose, Roger (2005). "18 Minkowskian geometry". *Road to Reality : A Complete Guide to the Laws of the Universe*. Alfred A. Knopf. ISBN 9780679454434.

- Poincaré, Henri (1905–1906), "Sur la dynamique de l'électron" [On the Dynamics of the Electron], *Rendiconti del Circolo matematico di Palermo* **21**: 129–176, doi:10.1007/BF03013466 Wikisource translation: On the Dynamics of the Electron

- Sard, R. D. (1970). *Relativistic Mechanics - Special Relativity and Classical Particle Dynamics*. New York: W. A. Benjamin. ISBN 978-0805384918.

- Shaw, R. (1982). "§ 6.6 Minkowski space, § 6.7.8 Canonical forms pp 221–242". *Linear Algebra and Group Representations*. Academic Press. ISBN 0-12-639201-3.

- Walter, Scott (1999). "Minkowski, Mathematicians, and the Mathematical Theory of Relativity". In Goenner, Hubert *et al.* (ed.). *The Expanding Worlds of General Relativity*. Boston: Birkhäuser. pp. 45–86. ISBN 0-8176-4060-6.

5.10 External links

Media related to Minkowski diagrams at Wikimedia Commons

- Animation clip on YouTube visualizing Minkowski space in the context of special relativity.
- The Geometry of Special Relativity: The Minkowski Space - Time Light Cone

Chapter 6

String theory

For the study of strings of characters, see Concatenation theory.
For a more accessible and less technical introduction to this topic, see Introduction to M-theory.

In physics, **string theory** is a theoretical framework in which the point-like particles of particle physics are replaced by one-dimensional objects called strings. String theory describes how these strings propagate through space and interact with each other. On distance scales larger than the string scale, a string looks just like an ordinary particle, with its mass, charge, and other properties determined by the vibrational state of the string. In string theory, one of the many vibrational states of the string corresponds to the graviton, a quantum mechanical particle that carries gravitational force. Thus string theory is a theory of quantum gravity.

String theory is a broad and varied subject that attempts to address a number of deep questions of fundamental physics. String theory has been applied to a variety of problems in black hole physics, early universe cosmology, nuclear physics, and condensed matter physics, and it has stimulated a number of major developments in pure mathematics. Because string theory potentially provides a unified description of gravity and particle physics, it is a candidate for a theory of everything, a self-contained mathematical model that describes all fundamental forces and forms of matter. Despite much work on these problems, it is not known to what extent string theory describes the real world or how much freedom the theory allows to choose the details.

String theory was first studied in the late 1960s as a theory of the strong nuclear force, before being abandoned in favor of quantum chromodynamics. Subsequently, it was realized that the very properties that made string theory unsuitable as a theory of nuclear physics made it a promising candidate for a quantum theory of gravity. The earliest version of string theory, bosonic string theory, incorporated only the class of particles known as bosons. It later developed into superstring theory, which posits a connection called supersymmetry between bosons and the class of particles called fermions. Five consistent versions of superstring theory were developed before it was conjectured in the mid-1990s that they were all different limiting cases of a single theory in eleven dimensions known as M-theory. In late 1997, theorists discovered an important relationship called the AdS/CFT correspondence, which relates string theory to another type of physical theory called a quantum field theory.

One of the challenges of string theory is that the full theory does not yet have a satisfactory definition in all circumstances. Another issue is that the theory is thought to describe an enormous landscape of possible universes, and this has complicated efforts to develop theories of particle physics based on string theory. These issues have led some in the community to criticize these approaches to physics and question the value of continued research on string theory unification.

6.1 Fundamentals

In the twentieth century, two theoretical frameworks emerged for formulating the laws of physics. One of these frameworks was Albert Einstein's general theory of relativity, a theory that explains the force of gravity and the structure of space and time. The other was quantum mechanics, a radically different formalism for describing physical phenomena

The fundamental objects of string theory are open and closed strings.

using probability. By the late 1970s, these two frameworks had proven to be sufficient to explain most of the observed features of the universe, from elementary particles to atoms to the evolution of stars and the universe as a whole.[1]

In spite of these successes, there are still many problems that remain to be solved. One of the deepest problems in modern physics is the problem of quantum gravity.[2] The general theory of relativity is formulated within the framework of classical physics, whereas the other fundamental forces are described within the framework of quantum mechanics. A quantum theory of gravity is needed in order to reconcile general relativity with the principles of quantum mechanics, but difficulties arise when one attempts to apply the usual prescriptions of quantum theory to the force of gravity.[3] In addition to the problem of developing a consistent theory of quantum gravity, there are many other fundamental problems in the physics of atomic nuclei, black holes, and the early universe.[lower-alpha 1]

String theory is a theoretical framework that attempts to address these questions and many others. The starting point for string theory is the idea that the point-like particles of particle physics can also be modeled as one-dimensional objects called strings. String theory describes how strings propagate through space and interact with each other. In a given version of string theory, there is only one kind of string, which may look like a small loop or segment of ordinary string, and it can vibrate in different ways. On distance scales larger than the string scale, a string will look just like an ordinary particle, with its mass, charge, and other properties determined by the vibrational state of the string. In this way, all of the different elementary particles may be viewed as vibrating strings. In string theory, one of the vibrational states of the string gives rise to the graviton, a quantum mechanical particle that carries gravitational force. Thus string theory is a theory of quantum gravity.[4]

One of the main developments of the past several decades in string theory was the discovery of certain "dualities", mathematical transformations that identify one physical theory with another. Physicists studying string theory have discovered a number of these dualities between different versions of string theory, and this has led to the conjecture that all consistent versions of string theory are subsumed in a single framework known as M-theory.[5]

Studies of string theory have also yielded a number of results on the nature of black holes and the gravitational interaction. There are certain paradoxes that arise when one attempts to understand the quantum aspects of black holes, and work on string theory has attempted to clarify these issues. In late 1997 this line of work culminated in the discovery of the anti-de Sitter/conformal field theory correspondence or AdS/CFT.[6] This is a theoretical result which relates string theory to other physical theories which are better understood theoretically. The AdS/CFT correspondence has implications for the study of black holes and quantum gravity, and it has been applied to other subjects, including nuclear[7] and condensed matter physics.[8][9]

Since string theory incorporates all of the fundamental interactions, including gravity, many physicists hope that it fully describes our universe, making it a theory of everything. One of the goals of current research in string theory is to find a solution of the theory that reproduces the observed spectrum of elementary particles, with a small cosmological constant, containing dark matter and a plausible mechanism for cosmic inflation. While there has been progress toward these goals, it is not known to what extent string theory describes the real world or how much freedom the theory allows to choose the details.[10]

One of the challenges of string theory is that the full theory does not yet have a satisfactory definition in all circumstances. The scattering of strings is most straightforwardly defined using the techniques of perturbation theory, but it is not known in general how to define string theory nonperturbatively.[11] It is also not clear whether there is any principle by which string theory selects its vacuum state, the physical state that determines the properties of our universe.[12] These problems have led some in the community to criticize these approaches to the unification of physics and question the value of continued research on these problems.[13]

6.1.1 Strings

Main article: String (physics)

The application of quantum mechanics to physical objects such as the electromagnetic field, which are extended in space

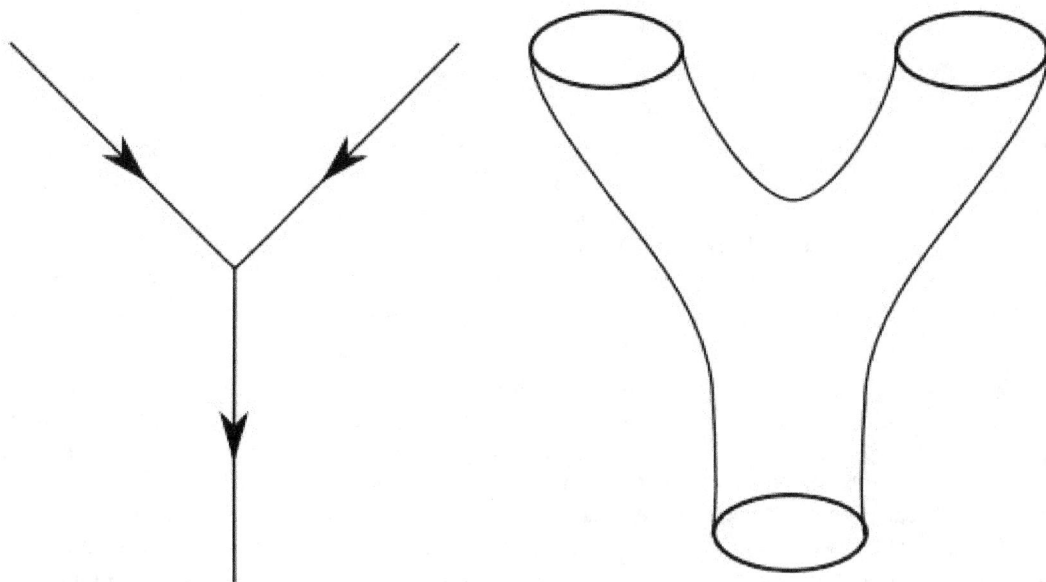

Interaction in the quantum world: worldlines of point-like particles or a worldsheet swept up by closed strings in string theory.

and time, is known as quantum field theory. In particle physics, quantum field theories form the basis for our understanding of elementary particles, which are modeled as excitations in the fundamental fields.[14]

In quantum field theory, one typically computes the probabilities of various physical events using the techniques of perturbation theory. Developed by Richard Feynman and others in the first half of the twentieth century, perturbative quantum field theory uses special diagrams called Feynman diagrams to organize computations. One imagines that these diagrams depict the paths of point-like particles and their interactions.[15]

The starting point for string theory is the idea that the point-like particles of quantum field theory can also be modeled as one-dimensional objects called strings.[16] The interaction of strings is most straightforwardly defined by generalizing the perturbation theory used in ordinary quantum field theory. At the level of Feynman diagrams, this means replacing the one-dimensional diagram representing the path of a point particle by a two-dimensional surface representing the motion of a string.[17] Unlike in quantum field theory, string theory does not yet have a full non-perturbative definition, so many of the theoretical questions that physicists would like to answer remain out of reach.[18]

In theories of particle physics based on string theory, the characteristic length scale of strings is assumed to be on the order of the Planck length, or 10^{-35} meters, the scale at which the effects of quantum gravity are believed to become significant.[19] On much larger length scales, such as the scales visible in physics laboratories, such objects would be indistinguishable from zero-dimensional point particles, and the vibrational state of the string would determine the type of particle. One of the vibrational states of a string corresponds to the graviton, a quantum mechanical particle that carries the gravitational force.[20]

The original version of string theory was bosonic string theory, but this version described only bosons, a class of particles which transmit forces between the matter particles, or fermions. Bosonic string theory was eventually superseded by theories called superstring theories. These theories describe both bosons and fermions, and they incorporate a theoretical idea called supersymmetry. This is a mathematical relation that exists in certain physical theories between the bosons and fermions. In theories with supersymmetry, each boson has a counterpart which is a fermion, and vice versa.[21]

There are several versions of superstring theory: type I, type IIA, type IIB, and two flavors of heterotic string theory ($SO(32)$ and $E_8 \times E_8$). The different theories allow different types of strings, and the particles that arise at low energies exhibit different symmetries. For example, the type I theory includes both open strings (which are segments with endpoints) and closed strings (which form closed loops), while types IIA and IIB include only closed strings.[22]

6.1.2 Extra dimensions

In everyday life, there are three familiar dimensions of space: height, width and length. Einstein's general theory of relativity treats time as a dimension on par with the three spatial dimensions; in general relativity, space and time are not modeled as separate entities but are instead unified to a four-dimensional spacetime. In this framework, the phenomenon of gravity is viewed as a consequence of the geometry of spacetime.[23]

In spite of the fact that the universe is well described by four-dimensional spacetime, there are several reasons why physicists consider theories in other dimensions. In some cases, by modeling spacetime in a different number of dimensions, a theory becomes more mathematically tractable, and one can perform calculations and gain general insights more easily.[lower-alpha 2] There are also situations where theories in two or three spacetime dimensions are useful for describing phenomena in condensed matter physics.[24] Finally, there exist scenarios in which there could actually be more than four dimensions of spacetime which have nonetheless managed to escape detection.[25]

One notable feature of string theories is that these theories require extra dimensions of spacetime for their mathematical consistency. In bosonic string theory, spacetime is 26-dimensional, while in superstring theory it is ten-dimensional. In order to describe real physical phenomena using string theory, one must therefore imagine scenarios in which these extra dimensions would not be observed in experiments.[26]

Compactification is one way of modifying the number of dimensions in a physical theory. In compactification, some of the extra dimensions are assumed to "close up" on themselves to form circles.[27] In the limit where these curled up dimensions become very small, one obtains a theory in which spacetime has effectively a lower number of dimensions. A standard analogy for this is to consider a multidimensional object such as a garden hose. If the hose is viewed from a sufficient distance, it appears to have only one dimension, its length. However, as one approaches the hose, one discovers that it contains a second dimension, its circumference. Thus, an ant crawling on the surface of the hose would move in two dimensions.[28]

Compactification can be used to construct models in which spacetime is effectively four-dimensional. However, not every way of compactifying the extra dimensions produces a model with the right properties to describe nature. In a viable model of particle physics, the compact extra dimensions must be shaped like a Calabi–Yau manifold.[29] A Calabi–Yau manifold is a special space which is typically taken to be six-dimensional in applications to string theory. It is named after

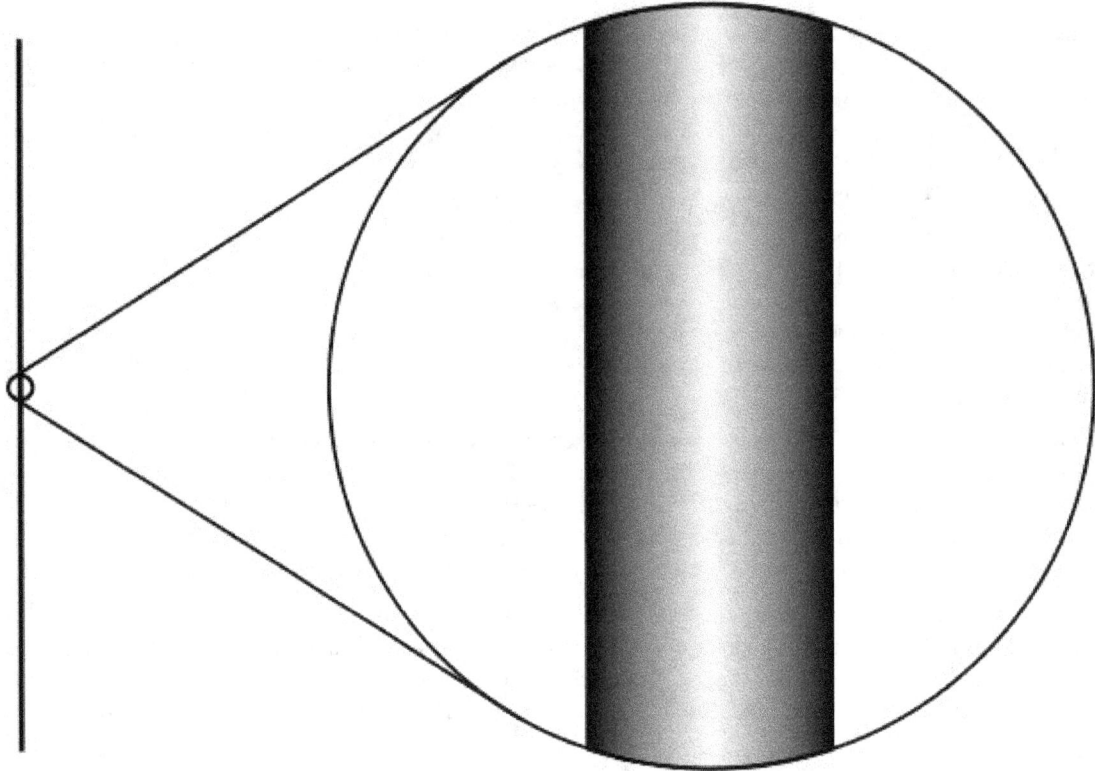

An example of compactification: At large distances, a two dimensional surface with one circular dimension looks one-dimensional.

mathematicians Eugenio Calabi and Shing-Tung Yau.[30]

Another approach to reducing the number of dimensions is the so called brane-world scenario. In this approach, physicists assume that the observable universe is a four-dimensional subspace of a higher dimensional space. In such models, the force-carrying bosons of particle physics arise from open strings with endpoints attached to the four-dimensional subspace, while gravity arises from closed strings propagating through the larger ambient space. This idea plays an important role in attempts to develop models of real world physics based on string theory, and it provides a natural explanation for the weakness of gravity compared to the other fundamental forces.[31]

6.1.3   Dualities

Main articles: S-duality and T-duality

One notable fact about string theory is that the different versions of the theory all turn out to be related in highly nontrivial ways. One of the relationships that can exist between different string theories is called S-duality. This is a relationship which says that a collection of strongly interacting particles in one theory can, in some cases, be viewed as a collection of weakly interacting particles in a completely different theory. Roughly speaking, a collection of particles is said to be strongly interacting if they combine and decay often and weakly interacting if they do so infrequently. Type I string theory turns out to be equivalent by S-duality to the $SO(32)$ heterotic string theory. Similarly, type IIB string theory is related to itself in a nontrivial way by S-duality.[32]

Another relationship between different string theories is T-duality. Here one considers strings propagating around a circular extra dimension. T-duality states that a string propagating around a circle of radius R is equivalent to a string propagating around a circle of radius $1/R$ in the sense that all observable quantities in one description are identified with

A cross section of a quintic Calabi–Yau manifold

quantities in the dual description. For example, a string has momentum as it propagates around a circle, and it can also wind around the circle one or more times. The number of times the string winds around a circle is called the winding number. If a string has momentum p and winding number n in one description, it will have momentum n and winding number p in the dual description. For example, type IIA string theory is equivalent to type IIB string theory via T-duality, and the two versions of heterotic string theory are also related by T-duality.[33]

In general, the term *duality* refers to a situation where two seemingly different physical systems turn out to be equivalent in a nontrivial way. Two theories related by a duality need not be string theories. For example, Montonen–Olive duality is example of an S-duality relationship between quantum field theories. The AdS/CFT correspondence is example of a duality which relates string theory to a quantum field theory. If two theories are related by a duality, it means that one theory can be transformed in some way so that it ends up looking just like the other theory. The two theories are then said to be *dual* to one another under the transformation. Put differently, the two theories are mathematically different descriptions of the same phenomena.[34]

6.1.4 Branes

Main article: Brane

In string theory and related theories, a brane is a physical object that generalizes the notion of a point particle to higher dimensions. For example, a point particle can be viewed as a brane of dimension zero, while a string can be viewed as a brane of dimension one. It is also possible to consider higher-dimensional branes. In dimension p, these are called p-branes. The word brane comes from the word "membrane" which refers to a two-dimensional brane.[35]

Branes are dynamical objects which can propagate through spacetime according to the rules of quantum mechanics. They have mass and can have other attributes such as charge. A p-brane sweeps out a $(p+1)$-dimensional volume in spacetime called its *worldvolume*. Physicists often study fields analogous to the electromagnetic field which live on the worldvolume of a brane.[36]

In string theory, D-branes are an important class of branes that arise when one considers open strings. As an open string propagates through spacetime, its endpoints are required to lie on a D-brane. The letter "D" in D-brane refers to a certain mathematical condition on the system known as the Dirichlet boundary condition. The study of D-branes in string theory has led to important results such as the AdS/CFT correspondence, which has shed light on many problems in quantum field theory.[37]

Branes are also frequently studied from a purely mathematical point of view. Mathematically, branes can be described as objects of certain categories, such as the derived category of coherent sheaves on a complex algebraic variety, or the Fukaya category of a symplectic manifold.[38] The connection between the physical notion of a brane and the mathematical notion of a category has led to important mathematical insights in the fields of algebraic and symplectic geometry[39] and representation theory.[40]

6.2   M-theory

Main article: M-theory

Prior to 1995, theorists believed that there were five consistent versions of superstring theory (type I, type IIA, type IIB, and two versions of heterotic string theory). This understanding changed in 1995 when Edward Witten suggested that the five theories were just special limiting cases of an eleven-dimensional theory called M-theory. Witten's conjecture was based on the work of a number of other physicists, including Ashoke Sen, Chris Hull, Paul Townsend, and Michael Duff. His announcement led to a flurry of research activity now known as the second superstring revolution.[41]

6.2.1   Unification of superstring theories

In the 1970s, many physicists became interested in supergravity theories, which combine general relativity with supersymmetry. Whereas general relativity makes sense in any number of dimensions, supergravity places an upper limit on the number of dimensions.[42] In 1978, work by Werner Nahm showed that the maximum spacetime dimension in which one can formulate a consistent supersymmetric theory is eleven.[43] In the same year, Eugene Cremmer, Bernard Julia, and Joel Scherk of the École Normale Supérieure showed that supergravity not only permits up to eleven dimensions but is in fact most elegant in this maximal number of dimensions.[44][45]

Initially, many physicists hoped that by compactifying eleven-dimensional supergravity, it might be possible to construct realistic models of our four-dimensional world. The hope was that such models would provide a unified description of the four fundamental forces of nature: electromagnetism, the strong and weak nuclear forces, and gravity. Interest in eleven-dimensional supergravity soon waned as various flaws in this scheme were discovered. One of the problems was that the laws of physics appear to distinguish between clockwise and counterclockwise, a phenomenon known as chirality. Edward Witten and others observed this chirality property cannot be readily derived by compactifying from eleven dimensions.[46]

In the first superstring revolution in 1984, many physicists turned to string theory as a unified theory of particle physics and quantum gravity. Unlike supergravity theory, string theory was able to accommodate the chirality of the standard model, and it provided a theory of gravity consistent with quantum effects.[47] Another feature of string theory that many physicists were drawn to in the 1980s and 1990s was its high degree of uniqueness. In ordinary particle theories, one can consider any collection of elementary particles whose classical behavior is described by an arbitrary Lagrangian. In string theory, the possibilities are much more constrained: by the 1990s, physicists had argued that there were only five consistent supersymmetric versions of the theory.[48]

Although there were only a handful of consistent superstring theories, it remained a mystery why there was not just one consistent formulation.[49] However, as physicists began to examine string theory more closely, they realized that these theories are related in intricate and nontrivial ways. They found that a system of strongly interacting strings can, in some cases, be viewed as a system of weakly interacting strings. This phenomenon is known as S-duality. It was studied by Ashoke Sen in the context of heterotic strings in four dimensions[50][51] and by Chris Hull and Paul Townsend in the context of the type IIB theory.[52] Theorists also found that different string theories may be related by T-duality. This duality implies that strings propagating on completely different spacetime geometries may be physically equivalent.[53]

At around the same time, as many physicists were studying the properties of strings, a small group of physicists was examining the possible applications of higher dimensional objects. In 1987, Eric Bergshoeff, Ergin Sezgin, and Paul Townsend showed that eleven-dimensional supergravity includes two-dimensional branes.[54] Intuitively, these objects look like sheets or membranes propagating through the eleven-dimensional spacetime. Shortly after this discovery, Michael Duff, Paul Howe, Takeo Inami, and Kellogg Stelle considered a particular compactification of eleven-dimensional supergravity with one of the dimensions curled up into a circle.[55] In this setting, one can imagine the membrane wrapping around the circular dimension. If the radius of the circle is sufficiently small, then this membrane looks just like a string in ten-dimensional spacetime. In fact, Duff and his collaborators showed that this construction reproduces exactly the strings appearing in type IIA superstring theory.[56]

Speaking at a string theory conference in 1995, Edward Witten made the surprising suggestion that all five superstring theories were in fact just different limiting cases of a single theory in eleven spacetime dimensions. Witten's announcement drew together all of the previous results on S- and T-duality and the appearance of higher dimensional branes in string theory.[57] In the months following Witten's announcement, hundreds of new papers appeared on the Internet confirming different parts of his proposal.[58] Today this flurry of work is known as the second superstring revolution.[59]

Initially, some physicists suggested that the new theory was a fundamental theory of membranes, but Witten was skeptical of the role of membranes in the theory. In a paper from 1996, Hořava and Witten wrote "As it has been proposed that the eleven-dimensional theory is a supermembrane theory but there are some reasons to doubt that interpretation, we will non-committally call it the M-theory, leaving to the future the relation of M to membranes."[60] In the absence of an understanding of the true meaning and structure of M-theory, Witten has suggested that the M should stand for "magic", "mystery", or "membrane" according to taste, and the true meaning of the title should be decided when a more fundamental formulation of the theory is known.[61]

6.2.2 Matrix theory

Main article: Matrix theory (physics)

In mathematics, a matrix is a rectangular array of numbers or other data. In physics, a matrix model is a particular kind of physical theory whose mathematical formulation involves the notion of a matrix in an important way. A matrix model describes the behavior of a set of matrices within the framework of quantum mechanics.[62]

One important example of a matrix model is the BFSS matrix model proposed by Tom Banks, Willy Fischler, Stephen Shenker, and Leonard Susskind in 1997. This theory describes the behavior of a set of nine large matrices. In their original paper, these authors showed, among other things, that the low energy limit of this matrix model is described by eleven-dimensional supergravity. These calculations led them to propose that the BFSS matrix model is exactly equivalent to M-theory. The BFSS matrix model can therefore be used as a prototype for a correct formulation of M-theory and a tool for investigating the properties of M-theory in a relatively simple setting.[63]

The development of the matrix model formulation of M-theory has led physicists to consider various connections between string theory and a branch of mathematics called noncommutative geometry. This subject is a generalization of ordinary geometry in which mathematicians define new geometric notions using tools from noncommutative algebra.[64] In a paper from 1998, Alain Connes, Michael R. Douglas, and Albert Schwarz showed that some aspects of matrix models and M-theory are described by a noncommutative quantum field theory, a special kind of physical theory in which spacetime is described mathematically using noncommutative geometry.[65] This established a link between matrix models and M-theory on the one hand, and noncommutative geometry on the other hand. It quickly led to the discovery of other important links between noncommutative geometry and various physical theories.[66][67]

6.3   Black holes

In general relativity, a black hole is defined as a region of spacetime in which the gravitational field is so strong that no particle or radiation can escape. In the currently accepted models of stellar evolution, black holes are thought to arise when massive stars undergo gravitational collapse, and many galaxies are thought to contain supermassive black holes at their centers. Black holes are also important for theoretical reasons, as they present profound challenges for theorists attempting to understand the quantum aspects of gravity. String theory has proved to be an important tool for investigating the theoretical properties of black holes because it provides a framework in which theorists can study their thermodynamics.[68]

6.3.1   Bekenstein–Hawking formula

In the branch of physics called statistical mechanics, entropy is a measure of the randomness or disorder of a physical system. This concept was studied in the 1870s by the Austrian physicist Ludwig Boltzmann, who showed that the thermodynamic properties of a gas could be derived from the combined properties of its many constituent molecules. Boltzmann argued that by averaging the behaviors of all the different molecules in a gas, one can understand macroscopic properties such as volume, temperature, and pressure. In addition, this perspective led him to give a precise definition of entropy as the natural logarithm of the number of different states of the molecules (also called *microstates*) that give rise to the same macroscopic features.[69]

In the twentieth century, physicists began to apply the same concepts to black holes. In most systems such as gases, the entropy scales with the volume. In the 1970s, the physicist Jacob Bekenstein suggested that the entropy of a black hole is instead proportional to the *surface area* of its event horizon, the boundary beyond which matter and radiation is lost to its gravitational attraction.[70] When combined with ideas of the physicist Stephen Hawking,[71] Bekenstein's work yielded a precise formula for the entropy of a black hole. The formula expresses the entropy S as

$$S = \frac{c^3 k A}{4 \hbar G}$$

where c is the speed of light, k is Boltzmann's constant, $\hbar$ is the reduced Planck constant, G is Newton's constant, and A is the surface area of the event horizon.[72]

Like any physical system, a black hole has an entropy defined in terms of the number of different microstates that lead to the same macroscopic features. The Bekenstein–Hawking entropy formula gives the expected value of the entropy of a black hole, but by the 1990s, physicists still lacked a derivation of this formula by counting microstates in a theory of quantum gravity. Finding such a derivation of this formula was considered an important test of the viability of any theory of quantum gravity such as string theory.[73]

6.3.2   Derivation within string theory

In a paper from 1996, Andrew Strominger and Cumrun Vafa showed how to derive the Beckenstein–Hawking formula for certain black holes in string theory.[74] Their calculation was based on the observation that D-branes—which look like fluctuating membranes when they are weakly interacting—become dense, massive objects with event horizons when the interactions are strong. In other words, a system of strongly interacting D-branes in string theory is indistinguishable from a black hole. Strominger and Vafa analyzed such D-brane systems and calculated the number of different ways of placing D-branes in spacetime so that their combined mass and charge is equal to a given mass and charge for the resulting black hole. Their calculation reproduced the Bekenstein–Hawking formula exactly, including the factor of 1/4.[75] Subsequent work by Strominger, Vafa, and others refined the original calculations and gave the precise values of the "quantum corrections" needed to describe very small black holes.[76][77]

The black holes that Strominger and Vafa considered in their original work were quite different from real astrophysical black holes. One difference was that Strominger and Vafa considered only extremal black holes in order to make the calculation tractable. These are defined as black holes with the lowest possible mass compatible with a given charge.[78] Strominger and Vafa also restricted attention to black holes in five-dimensional spacetime with unphysical supersymmetry.[79]

Although it was originally developed in this very particular and physically unrealistic context in string theory, the entropy calculation of Strominger and Vafa has led to a qualitative understanding of how black hole entropy can be accounted for in any theory of quantum gravity. Indeed, in 1998, Strominger argued that the original result could be generalized to an arbitrary consistent theory of quantum gravity without relying on strings or supersymmetry.[80] In collaboration with several other authors in 2010, he showed that some results on black hole entropy could be extended to non-extremal astrophysical black holes.[81][82]

6.4 AdS/CFT correspondence

Main article: AdS/CFT correspondence

One approach to formulating string theory and studying its properties is provided by the anti-de Sitter/conformal field theory (AdS/CFT) correspondence. This is a theoretical result which implies that string theory is in some cases equivalent to a quantum field theory. In addition to providing insights into the mathematical structure of string theory, the AdS/CFT correspondence has shed light on many aspects of quantum field theory in regimes where traditional calculational techniques are ineffective.[83] The AdS/CFT correspondence was first proposed by Juan Maldacena in late 1997.[84] Important aspects of the correspondence were elaborated in articles by Steven Gubser, Igor Klebanov, and Alexander Markovich Polyakov,[85] and by Edward Witten.[86] By 2010, Maldacena's article had over 7000 citations, becoming the most highly cited article in the field of high energy physics.[lower-alpha 3]

6.4.1 Overview of the correspondence

In the AdS/CFT correspondence, the geometry of spacetime is described in terms of a certain vacuum solution of Einstein's equation called anti-de Sitter space.[87] In very elementary terms, anti-de Sitter space is a mathematical model of spacetime in which the notion of distance between points (the metric) is different from the notion of distance in ordinary Euclidean geometry. It is closely related to hyperbolic space, which can be viewed as a disk as illustrated on the left.[88] This image shows a tessellation of a disk by triangles and squares. One can define the distance between points of this disk in such a way that all the triangles and squares are the same size and the circular outer boundary is infinitely far from any point in the interior.[89]

One can imagine a stack of hyperbolic disks where each disk represents the state of the universe at a given time. The resulting geometric object is three-dimensional anti-de Sitter space.[90] It looks like a solid cylinder in which any cross section is a copy of the hyperbolic disk. Time runs along the vertical direction in this picture. The surface of this cylinder plays an important role in the AdS/CFT correspondence. As with the hyperbolic plane, anti-de Sitter space is curved in such a way that any point in the interior is actually infinitely far from this boundary surface.[91]

This construction describes a hypothetical universe with only two space dimensions and one time dimension, but it can be generalized to any number of dimensions. Indeed, hyperbolic space can have more than two dimensions and one can "stack up" copies of hyperbolic space to get higher-dimensional models of anti-de Sitter space.[92]

An important feature of anti-de Sitter space is its boundary (which looks like a cylinder in the case of three-dimensional anti-de Sitter space). One property of this boundary is that, within a small region on the surface around any given point, it looks just like Minkowski space, the model of spacetime used in nongravitational physics.[93] One can therefore consider an auxiliary theory in which "spacetime" is given by the boundary of anti-de Sitter space. This observation is the starting point for AdS/CFT correspondence, which states that the boundary of anti-de Sitter space can be regarded as the "spacetime" for a quantum field theory. The claim is that this quantum field theory is equivalent to a gravitational theory, such as string theory, in the bulk anti-de Sitter space in the sense that there is a "dictionary" for translating entities and calculations in one theory into their counterparts in the other theory. For example, a single particle in the gravitational theory might correspond to some collection of particles in the boundary theory. In addition, the predictions in the two theories are quantitatively identical so that if two particles have a 40 percent chance of colliding in the gravitational theory, then the corresponding collections in the boundary theory would also have a 40 percent chance of colliding.[94]

6.4.2   Applications to quantum gravity

The discovery of the AdS/CFT correspondence was a major advance in physicists' understanding of string theory and quantum gravity. One reason for this is that the correspondence provides a formulation of string theory in terms of quantum field theory, which is well understood by comparison. Another reason is that it provides a general framework in which physicists can study and attempt to resolve the paradoxes of black holes.[95]

In 1975, Stephen Hawking published a calculation which suggested that black holes are not completely black but emit a dim radiation due to quantum effects near the event horizon.[96] At first, Hawking's result posed a problem for theorists because it suggested that black holes destroy information. More precisely, Hawking's calculation seemed to conflict with one of the basic postulates of quantum mechanics, which states that physical systems evolve in time according to the Schrödinger equation. This property is usually referred to as unitarity of time evolution. The apparent contradiction between Hawking's calculation and the unitarity postulate of quantum mechanics came to be known as the black hole information paradox.[97]

The AdS/CFT correspondence resolves the black hole information paradox, at least to some extent, because it shows how a black hole can evolve in a manner consistent with quantum mechanics in some contexts. Indeed, one can consider black holes in the context of the AdS/CFT correspondence, and any such black hole corresponds to a configuration of particles on the boundary of anti-de Sitter space.[98] These particles obey the usual rules of quantum mechanics and in particular evolve in a unitary fashion, so the black hole must also evolve in a unitary fashion, respecting the principles of quantum mechanics.[99] In 2005, Hawking announced that the paradox had been settled in favor of information conservation by the AdS/CFT correspondence, and he suggested a concrete mechanism by which black holes might preserve information.[100]

6.4.3   Applications to quantum field theory

Main articles: AdS/QCD correspondence and AdS/CMT correspondence
In addition to its applications to theoretical problems in quantum gravity, the AdS/CFT correspondence has been applied to a variety of problems in quantum field theory. One physical system that has been studied using the AdS/CFT correspondence is the quark–gluon plasma, an exotic state of matter produced in particle accelerators. This state of matter arises for brief instants when heavy ions such as gold or lead nuclei are collided at high energies. Such collisions cause the quarks that make up atomic nuclei to deconfine at temperatures of approximately two trillion kelvins, conditions similar to those present at around 10^{-11} seconds after the Big Bang.[102]

The physics of the quark–gluon plasma is governed by a theory called quantum chromodynamics, but this theory is mathematically intractable in problems involving the quark–gluon plasma.[lower-alpha 4] In an article appearing in 2005, Đàm Thanh Sơn and his collaborators showed that the AdS/CFT correspondence could be used to understand some aspects of the quark–gluon plasma by describing it in the language of string theory.[103] By applying the AdS/CFT correspondence, Sơn and his collaborators were able to describe the quark gluon plasma in terms of black holes in five-dimensional spacetime. The calculation showed that the ratio of two quantities associated with the quark–gluon plasma, the shear viscosity and volume density of entropy, should be approximately equal to a certain universal constant. In 2008, the predicted value of this ratio for the quark–gluon plasma was confirmed at the Relativistic Heavy Ion Collider at Brookhaven National Laboratory.[104][105]

The AdS/CFT correspondence has also been used to study aspects of condensed matter physics. Over the decades, experimental condensed matter physicists have discovered a number of exotic states of matter, including superconductors and superfluids. These states are described using the formalism of quantum field theory, but some phenomena are difficult to explain using standard field theoretic techniques. Some condensed matter theorists including Subir Sachdev hope that the AdS/CFT correspondence will make it possible to describe these systems in the language of string theory and learn more about their behavior.[106]

So far some success has been achieved in using string theory methods to describe the transition of a superfluid to an insulator. A superfluid is a system of electrically neutral atoms that flows without any friction. Such systems are often produced in the laboratory using liquid helium, but recently experimentalists have developed new ways of producing artificial superfluids by pouring trillions of cold atoms into a lattice of criss-crossing lasers. These atoms initially behave as a superfluid, but as experimentalists increase the intensity of the lasers, they become less mobile and then suddenly transition to an insulating state. During the transition, the atoms behave in an unusual way. For example, the atoms slow to a halt at a rate that depends on the temperature and on Planck's constant, the fundamental parameter of quantum mechanics, which does not enter into the description of the other phases. This behavior has recently been understood by considering a dual description where properties of the fluid are described in terms of a higher dimensional black hole.[107]

6.5   Phenomenology

Main article: String phenomenology

In addition to being an idea of considerable theoretical interest, string theory provides a framework for constructing models of real world physics that combine general relativity and particle physics. Phenomenology is the branch of theoretical physics in which physicists construct realistic models of nature from more abstract theoretical ideas. String phenomenology is the part of string theory that attempts to construct realistic models based on string theory.

Partly because of theoretical and mathematical difficulties and partly because of the extremely high energies needed to test these theories experimentally, there is so far no experimental evidence that would unambiguously point to any of these models being a correct fundamental description of nature. This has led some in the community to criticize these approaches to unification and question the value of continued research on these problems.[108]

6.5.1   Particle physics

The currently accepted theory describing elementary particles and their interactions is known as the standard model of particle physics. This theory provides a unified description of three of the fundamental forces of nature: electromagnetism and the strong and weak nuclear forces. Despite its remarkable success in explaining a wide range of physical phenomena, the standard model cannot be a complete description of reality. This is because the standard model fails to incorporate the force of gravity and because of problems such as the hierarchy problem and the inability to explain the structure of fermion masses or dark matter.

String theory has been used to construct a variety of models of particle physics going beyond the standard model. Typically, such models are based on the idea of compactification. Starting with the ten- or eleven-dimensional spacetime of string or M-theory, physicists postulate a shape for the extra dimensions. By choosing this shape appropriately, they can construct models roughly similar to the standard model of particle physics, together with additional undiscovered particles.[109] One popular way of deriving realistic physics from string theory is to start with the heterotic theory in ten dimensions and assume that the six extra dimensions of spacetime are shaped like a six-dimensional Calabi–Yau manifold. Such compactifications offer many ways of extracting realistic physics from string theory. Other similar methods can be used to construct realistic models of our four-dimensional world based on M-theory.[110]

6.5.2   Cosmology

Main article: String cosmology

The Big Bang theory is the prevailing cosmological model for the universe from the earliest known periods through its subsequent large-scale evolution. Despite its success in explaining many observed features of the universe including galactic redshifts, the relative abundance of light elements such as hydrogen and helium, and the existence of a cosmic microwave background, there are several questions that remain unanswered. For example, the standard Big Bang model does not explain why the universe appears to be same in all directions, why it appears flat on very large distance scales, or why certain hypothesized particles such as magnetic monopoles are not observed in experiments.[111]

Currently, the leading candidate for a theory going beyond the Big Bang is the theory of cosmic inflation. Developed by Alan Guth and others in the 1980s, inflation postulates a period of extremely rapid accelerated expansion of the universe prior to the expansion described by the standard Big Bang theory. The theory of cosmic inflation preserves the successes of the Big Bang while providing a natural explanation for some of the mysterious features of the universe.[112] The theory has also received striking support from observations of the cosmic microwave background, the radiation that has filled the sky since around 380,000 years after the Big Bang.[113]

In the theory of inflation, the rapid initial expansion of the universe is caused by a hypothetical particle called the inflaton. The exact properties of this particle are not fixed by the theory but should ultimately be derived from a more fundamental theory such as string theory.[114] Indeed, there have been a number of attempts to identify an inflation within the spectrum of particles described by string theory and to study inflation using string theory. While these approaches might eventually find support in observational data such as measurements of the cosmic microwave background, the application of string theory to cosmology is still in its early stages.[115]

6.6   Connections to mathematics

In addition to influencing research in theoretical physics, string theory has stimulated a number of major developments in pure mathematics. Like many developing ideas in theoretical physics, string theory does not at present have a mathematically rigorous formulation in which all of its concepts can be defined precisely. As a result, physicists who study string theory are often guided by physical intuition to conjecture relationships between the seemingly different mathematical structures that are used to formalize different parts of the theory. These conjectures are later proved by mathematicians, and in this way, string theory serves as a source of new ideas in pure mathematics.[116]

6.6.1   Mirror symmetry

Main article: Mirror symmetry (string theory)
 After Calabi–Yau manifolds had entered physics as a way to compactify extra dimensions in string theory, many physicists began studying these manifolds. In the late 1980s, several physicists noticed that given such a compactification of string theory, it is not possible to reconstruct uniquely a corresponding Calabi–Yau manifold.[117] Instead, two different versions of string theory, type IIA and type IIB, can be compactified on completely different Calabi–Yau manifolds giving rise to the same physics. In this situation, the manifolds are called mirror manifolds, and the relationship between the two physical theories is called mirror symmetry.[118]

Regardless of whether Calabi–Yau compactifications of string theory provide a correct description of nature, the existence of the mirror duality between different string theories has significant mathematical consequences. The Calabi–Yau manifolds used in string theory are of interest in pure mathematics, and mirror symmetry allows mathematicians to solve problems in enumerative geometry, a branch of mathematics concerned with counting the numbers of solutions to geometric questions.[119][120]

Enumerative geometry studies a class of geometric objects called algebraic varieties which are defined by the vanishing of polynomials. For example, the Clebsch cubic illustrated on the right is an algebraic variety defined using a certain polynomial of degree three in four variables. A celebrated result of nineteenth-century mathematicians Arthur Cayley and George Salmon states that there are exactly 27 straight lines that lie entirely on such a surface.[121]

Generalizing this problem, one can ask how many lines can be drawn on a quintic Calabi–Yau manifold, such as the one illustrated above, which is defined by a polynomial of degree five. This problem was solved by the nineteenth-century German mathematician Hermann Schubert, who found that there are exactly 2,875 such lines. In 1986, geometer Sheldon Katz proved that the number of curves, such as circles, that are defined by polynomials of degree two and lie entirely in the quintic is 609,250.[122]

By the year 1991, most of the classical problems of enumerative geometry had been solved and interest in enumerative geometry had begun to diminish.[123] The field was reinvigorated in May 1991 when physicists Philip Candelas, Xenia de la Ossa, Paul Green, and Linda Parks showed that mirror symmetry could be used to translate difficult mathematical questions about one Calabi–Yau manifold into easier questions about its mirror.[124] In particular, they used mirror symmetry to show that a six-dimensional Calabi–Yau manifold can contain exactly 317,206,375 curves of degree three.[125] In addition to counting degree-three curves, Candelas and his collaborators obtained a number of more general results for counting rational curves which went far beyond the results obtained by mathematicians.[126]

Originally, these results of Candelas were justified on physical grounds. However, mathematicians generally prefer rigorous proofs that do not require an appeal to physical intuition. Inspired by physicists' work on mirror symmetry, mathematicians have therefore constructed their own arguments proving the enumerative predictions of mirror symmetry.[lower-alpha 5] Today mirror symmetry is an active area of research in mathematics, and mathematicians are working to develop a more complete mathematical understanding of mirror symmetry based on physicists' intuition.[127] Major approaches to mirror symmetry include the homological mirror symmetry program of Maxim Kontsevich[128] and the SYZ conjecture of Andrew Strominger, Shing-Tung Yau, and Eric Zaslow.[129]

6.6.2 Monstrous moonshine

Main article: Monstrous moonshine

Group theory is the branch of mathematics that studies the concept of symmetry. For example, one can consider a geometric shape such as an equilateral triangle. There are various operations that one can perform on this triangle without changing its shape. One can rotate it through 120°, 240°, or 360°, or one can reflect in any of the lines labeled S_0, S_1, or S_2 in the picture. Each of these operations is called a *symmetry*, and the collection of these symmetries satisfies certain technical properties making it into what mathematicians call a group. In this particular example, the group is known as the dihedral group of order 6 because it has six elements. A general group may describe finitely many or infinitely many symmetries; if there are only finitely many symmetries, it is called a finite group.[130]

Mathematicians often strive for a classification (or list) of all mathematical objects of a given type. It is generally believed that finite groups are too diverse to admit a useful classification. A more modest but still challenging problem is to classify all finite *simple* groups. These are finite groups which may be used as building blocks for constructing arbitrary finite groups in the same way that prime numbers can be used to construct arbitrary whole numbers by taking products.[lower-alpha 6] One of the major achievements of contemporary group theory is the classification of finite simple groups, a mathematical theorem which provides a list of all possible finite simple groups.[131]

This classification theorem identifies several infinite families of groups as well as 26 additional groups which do not fit into any family. The latter groups are called the "sporadic" groups, and each one owes its existence to a remarkable combination of circumstances. The largest sporadic group, the so called monster group, has over 10^{53} elements, more than a thousand times the number of atoms in the Earth.[132]

A seemingly unrelated construction is the *j*-function of number theory. This object belongs to a special class of functions called modular functions, whose graphs form a certain kind of repeating pattern.[133] Although this function appears in a branch of mathematics which seems very different from the theory of finite groups, the two subjects turn out to be intimately related. In the late 1970s, mathematicians John McKay and John Thompson noticed that certain numbers arising in the analysis of the monster group (namely, the dimensions of its irreducible representations) are related to numbers that appear in a formula for the *j*-function (namely, the coefficients of its Fourier series).[134] This relationship was further developed by John Horton Conway and Simon Norton[135] who called it monstrous moonshine because it seemed so far fetched.[136]

In 1992, Richard Borcherds constructed a bridge between the theory of modular functions and finite groups and, in the process, explained the observations of McKay and Thompson.[137][138] Borcherds' work used ideas from string theory in an essential way, extending earlier results of Igor Frenkel, James Lepowsky, and Arne Meurman, who had realized the monster group as the symmetries of a particular version of string theory.[139] In 1998, Borcherds was awarded the Fields medal for his work.[140]

Since the 1990s, the connection between string theory and moonshine has led to further results in mathematics and physics.[141] In 2010, physicists Tohru Eguchi, Hirosi Ooguri, and Yuji Tachikawa discovered connections between a different sporadic group, the Mathieu group M_{24}, and a certain version of string theory.[142] Miranda Cheng, John Duncan, and Jeffrey A. Harvey proposed a generalization of this moonshine phenomenon called umbral moonshine,[143] and their conjecture was proved mathematically by Duncan, Michael Griffin, and Ken Ono.[144] Witten has also speculated that the version of string theory appearing in monstrous moonshine might be related to a certain simplified model of gravity in three spacetime dimensions.[145]

6.7 History

Main article: History of string theory

6.7.1 Early results

Some of the structures reintroduced by string theory arose for the first time much earlier as part of the program of classical unification started by Albert Einstein. The first person to add a fifth dimension to a theory of gravity was Gunnar Nordström in 1914, who noted that gravity in five dimensions describes both gravity and electromagnetism in four. Nordström attempted to unify electromagnetism with his theory of gravitation, which was however superseded by Einstein's general relativity in 1919. Thereafter, German mathematician Theodor Kaluza combined the fifth dimension with general relativity, and only Kaluza is usually credited with the idea. In 1926, the Swedish physicist Oskar Klein gave a physical interpretation of the unobservable extra dimension—it is wrapped into a small circle. Einstein introduced a non-symmetric metric tensor, while much later Brans and Dicke added a scalar component to gravity. These ideas would be revived within string theory, where they are demanded by consistency conditions.

String theory was originally developed during the late 1960s and early 1970s as a never completely successful theory of hadrons, the subatomic particles like the proton and neutron that feel the strong interaction. In the 1960s, Geoffrey Chew and Steven Frautschi discovered that the mesons make families called Regge trajectories with masses related to spins in a way that was later understood by Yoichiro Nambu, Holger Bech Nielsen and Leonard Susskind to be the relationship expected from rotating strings. Chew advocated making a theory for the interactions of these trajectories that did not presume that they were composed of any fundamental particles, but would construct their interactions from self-consistency conditions on the S-matrix. The S-matrix approach was started by Werner Heisenberg in the 1940s as a way of constructing a theory that did not rely on the local notions of space and time, which Heisenberg believed break down at the nuclear scale. While the scale was off by many orders of magnitude, the approach he advocated was ideally suited for a theory of quantum gravity.

Working with experimental data, R. Dolen, D. Horn and C. Schmid developed some sum rules for hadron exchange. When a particle and antiparticle scatter, virtual particles can be exchanged in two qualitatively different ways. In the s-channel, the two particles annihilate to make temporary intermediate states that fall apart into the final state particles. In the t-channel, the particles exchange intermediate states by emission and absorption. In field theory, the two contributions add together, one giving a continuous background contribution, the other giving peaks at certain energies. In the data, it was clear that the peaks were stealing from the background—the authors interpreted this as saying that the t-channel contribution was dual to the s-channel one, meaning both described the whole amplitude and included the other.

The result was widely advertised by Murray Gell-Mann, leading Gabriele Veneziano to construct a scattering amplitude that had the property of Dolen-Horn-Schmid duality, later renamed world-sheet duality. The amplitude needed poles where the particles appear, on straight line trajectories, and there is a special mathematical function whose poles are evenly spaced on half the real line— the Gamma function— which was widely used in Regge theory. By manipulating combinations of Gamma functions, Veneziano was able to find a consistent scattering amplitude with poles on straight lines, with mostly positive residues, which obeyed duality and had the appropriate Regge scaling at high energy. The amplitude could fit near-beam scattering data as well as other Regge type fits, and had a suggestive integral representation that could be used for generalization.

Over the next years, hundreds of physicists worked to complete the bootstrap program for this model, with many surprises. Veneziano himself discovered that for the scattering amplitude to describe the scattering of a particle that appears in the theory, an obvious self-consistency condition, the lightest particle must be a tachyon. Miguel Virasoro and Joel Shapiro found a different amplitude now understood to be that of closed strings, while Ziro Koba and Holger Nielsen generalized Veneziano's integral representation to multiparticle scattering. Veneziano and Sergio Fubini introduced an operator formalism for computing the scattering amplitudes that was a forerunner of world-sheet conformal theory, while Virasoro understood how to remove the poles with wrong-sign residues using a constraint on the states. Claud Lovelace calculated a loop amplitude, and noted that there is an inconsistency unless the dimension of the theory is 26. Charles Thorn, Peter Goddard and Richard Brower went on to prove that there are no wrong-sign propagating states in dimensions less than or equal to 26.

In 1969, Yoichiro Nambu, Holger Bech Nielsen, and Leonard Susskind recognized that the theory could be given a description in space and time in terms of strings. The scattering amplitudes were derived systematically from the action principle by Peter Goddard, Jeffrey Goldstone, Claudio Rebbi, and Charles Thorn, giving a space-time picture to the vertex operators introduced by Veneziano and Fubini and a geometrical interpretation to the Virasoro conditions.

In 1970, Pierre Ramond added fermions to the model, which led him to formulate a two-dimensional supersymmetry to cancel the wrong-sign states. John Schwarz and André Neveu added another sector to the fermi theory a short time later. In the fermion theories, the critical dimension was 10. Stanley Mandelstam formulated a world sheet conformal theory for both the bose and fermi case, giving a two-dimensional field theoretic path-integral to generate the operator formalism. Michio Kaku and Keiji Kikkawa gave a different formulation of the bosonic string, as a string field theory, with infinitely many particle types and with fields taking values not on points, but on loops and curves.

In 1974, Tamiaki Yoneya discovered that all the known string theories included a massless spin-two particle that obeyed the correct Ward identities to be a graviton. John Schwarz and Joel Scherk came to the same conclusion and made the bold leap to suggest that string theory was a theory of gravity, not a theory of hadrons. They reintroduced Kaluza–Klein theory as a way of making sense of the extra dimensions. At the same time, quantum chromodynamics was recognized as the correct theory of hadrons, shifting the attention of physicists and apparently leaving the bootstrap program in the dustbin of history.

String theory eventually made it out of the dustbin, but for the following decade all work on the theory was completely ignored. Still, the theory continued to develop at a steady pace thanks to the work of a handful of devotees. Ferdinando Gliozzi, Joel Scherk, and David Olive realized in 1976 that the original Ramond and Neveu Schwarz-strings were separately inconsistent and needed to be combined. The resulting theory did not have a tachyon, and was proven to have space-time supersymmetry by John Schwarz and Michael Green in 1981. The same year, Alexander Polyakov gave the theory a modern path integral formulation, and went on to develop conformal field theory extensively. In 1979, Daniel Friedan showed that the equations of motions of string theory, which are generalizations of the Einstein equations of General Relativity, emerge from the Renormalization group equations for the two-dimensional field theory. Schwarz and Green discovered T-duality, and constructed two superstring theories—IIA and IIB related by T-duality, and type I theories with open strings. The consistency conditions had been so strong, that the entire theory was nearly uniquely determined, with only a few discrete choices.

6.7.2   First superstring revolution

In the early 1980s, Edward Witten discovered that most theories of quantum gravity could not accommodate chiral fermions like the neutrino. This led him, in collaboration with Luis Alvarez-Gaumé to study violations of the conservation laws in gravity theories with anomalies, concluding that type I string theories were inconsistent. Green and Schwarz discovered a contribution to the anomaly that Witten and Alvarez-Gaumé had missed, which restricted the gauge group of the type I string theory to be SO(32). In coming to understand this calculation, Edward Witten became convinced that string theory was truly a consistent theory of gravity, and he became a high-profile advocate. Following Witten's lead, between 1984 and 1986, hundreds of physicists started to work in this field, and this is sometimes called the first superstring revolution.

During this period, David Gross, Jeffrey Harvey, Emil Martinec, and Ryan Rohm discovered heterotic strings. The gauge group of these closed strings was two copies of E8, and either copy could easily and naturally include the standard model. Philip Candelas, Gary Horowitz, Andrew Strominger and Edward Witten found that the Calabi–Yau manifolds are the compactifications that preserve a realistic amount of supersymmetry, while Lance Dixon and others worked out the physical properties of orbifolds, distinctive geometrical singularities allowed in string theory. Cumrun Vafa generalized T-duality from circles to arbitrary manifolds, creating the mathematical field of mirror symmetry. Daniel Friedan, Emil Martinec and Stephen Shenker further developed the covariant quantization of the superstring using conformal field theory techniques. David Gross and Vipul Periwal discovered that string perturbation theory was divergent. Stephen Shenker showed it diverged much faster than in field theory suggesting that new non-perturbative objects were missing.

In the 1990s, Joseph Polchinski discovered that the theory requires higher-dimensional objects, called D-branes and identified these with the black-hole solutions of supergravity. These were understood to be the new objects suggested by the perturbative divergences, and they opened up a new field with rich mathematical structure. It quickly became clear that D-branes and other p-branes, not just strings, formed the matter content of the string theories, and the physical interpretation of the strings and branes was revealed—they are a type of black hole. Leonard Susskind had incorporated the holographic principle of Gerardus 't Hooft into string theory, identifying the long highly excited string states with ordinary thermal black hole states. As suggested by 't Hooft, the fluctuations of the black hole horizon, the world-sheet or world-volume theory, describes not only the degrees of freedom of the black hole, but all nearby objects too.

6.7.3 Second superstring revolution

In 1995, at the annual conference of string theorists at the University of Southern California (USC), Edward Witten gave a speech on string theory that in essence united the five string theories that existed at the time, and giving birth to a new 11-dimensional theory called M-theory. M-theory was also foreshadowed in the work of Paul Townsend at approximately the same time. The flurry of activity that began at this time is sometimes called the second superstring revolution.[146]

During this period, Tom Banks, Willy Fischler, Stephen Shenker and Leonard Susskind formulated matrix theory, a full holographic description of M-theory using IIA D0 branes.[147] This was the first definition of string theory that was fully non-perturbative and a concrete mathematical realization of the holographic principle. It is an example of a gauge-gravity duality and is now understood to be a special case of the AdS/CFT correspondence. Andrew Strominger and Cumrun Vafa calculated the entropy of certain configurations of D-branes and found agreement with the semi-classical answer for extreme charged black holes.[148] Petr Hořava and Witten found the eleven-dimensional formulation of the heterotic string theories, showing that orbifolds solve the chirality problem. Witten noted that the effective description of the physics of D-branes at low energies is by a supersymmetric gauge theory, and found geometrical interpretations of mathematical structures in gauge theory that he and Nathan Seiberg had earlier discovered in terms of the location of the branes.

In 1997, Juan Maldacena noted that the low energy excitations of a theory near a black hole consist of objects close to the horizon, which for extreme charged black holes looks like an anti-de Sitter space.[149] He noted that in this limit the gauge theory describes the string excitations near the branes. So he hypothesized that string theory on a near-horizon extreme-charged black-hole geometry, an anti-deSitter space times a sphere with flux, is equally well described by the low-energy limiting gauge theory, the N = 4 supersymmetric Yang–Mills theory. This hypothesis, which is called the AdS/CFT correspondence, was further developed by Steven Gubser, Igor Klebanov and Alexander Polyakov,[150] and by Edward Witten,[151] and it is now well-accepted. It is a concrete realization of the holographic principle, which has far-reaching implications for black holes, locality and information in physics, as well as the nature of the gravitational interaction.[152] Through this relationship, string theory has been shown to be related to gauge theories like quantum chromodynamics and this has led to more quantitative understanding of the behavior of hadrons, bringing string theory back to its roots.[153]

6.8 Criticism

6.8.1 Number of solutions

To construct models of particle physics based on string theory, physicists typically begin by specifying a shape for the extra dimensions of spacetime. Each of these different shapes corresponds to a different possible universe, or "vacuum state", with a different collection of particles and forces. String theory as it is currently understood has an enormous number of vacuum states, typically estimated to be around 10^{500}, and these might be sufficiently diverse to accommodate almost any phenomena that might be observed at low energies.[154]

Many critics of string theory have expressed concerns about the large number of possible universes described by string theory. In his book *Not Even Wrong*, Peter Woit, a lecturer in the mathematics department at Columbia University, has argued that the large number of different physical scenarios renders string theory vacuous as a framework for constructing models of particle physics. According to Woit,

The possible existence of, say, 10^{500} consistent different vacuum states for superstring theory probably destroys the hope of using the theory to predict anything. If one picks among this large set just those states whose properties agree with present experimental observations, it is likely there still will be such a large number of these that one can get just about whatever value one wants for the results of any new observation.[155]

Some physicists believe this large number of solutions is actually a virtue because it may allow a natural anthropic explanation of the observed values of physical constants, in particular the small value of the cosmological constant.[156] The anthropic principle is the idea that some of the numbers appearing in the laws of physics are not fixed by any fundamental principle but must be compatible with the evolution of intelligent life. In 1987, Steven Weinberg published an article in which he argued that the cosmological constant could not have been too large, or else galaxies and intelligent life would not have been able to develop.[157] Weinberg suggested that there might be a huge number of possible consistent universes, each with a different value of the cosmological constant, and observations indicate a small value of the cosmological constant only because humans happen to live in a universe that has allowed intelligent life, and hence observers, to exist.[158]

String theorist Leonard Susskind has argued that string theory provides a natural anthropic explanation of the small value of the cosmological constant.[159] According to Susskind, the different vacuum states of string theory might be realized as different universes within a larger multiverse. The fact that the observed universe has a small cosmological constant is just a tautological consequence of the fact that a small value is required for life to exist.[160] Many prominent theorists and critics have disagreed with Susskind's conclusions.[161] According to Woit, "in this case [anthropic reasoning] is nothing more than an excuse for failure. Speculative scientific ideas fail not just when they make incorrect predictions, but also when they turn out to be vacuous and incapable of predicting anything."[162]

6.8.2 Background independence

Main article: Background independence

One of the fundamental principles of Einstein's general theory of relativity is the idea that the laws of physics should be background independent. This means that the geometry of spacetime is not specified from the outset but is instead determined dynamically by the theory. In general relativity, the geometry of spacetime can evolve in time, responding to whatever matter is present.[163]

One of the older criticisms of string theory is that it is not manifestly background independent. In string theory, one must typically specify a fixed reference geometry for spacetime, and all other possible geometries are described as perturbations of this fixed one. In his book *The Trouble With Physics*, physicist Lee Smolin of the Perimeter Institute for Theoretical Physics claims that this is the principal weakness of string theory as a theory of quantum gravity, saying that string theory has failed to incorporate this important insight from general relativity.[164]

Others have disagreed with Smolin's characterization of string theory. In a review of Smolin's book, string theorist Joseph Polchinski writes

> [Smolin] is mistaking an aspect of the mathematical language being used for one of the physics being described. New physical theories are often discovered using a mathematical language that is not the most suitable for them... In string theory it has always been clear that the physics is background-independent even if the language being used is not, and the search for more suitable language continues. Indeed, as Smolin belatedly notes, [AdS/CFT] provides a solution to this problem, one that is unexpected and powerful.[165]

Polchinski notes that an important open problem in quantum gravity is to develop holographic descriptions of gravity which do not require the gravitational field to be asymptotically anti-de Sitter.[166]

6.8.3 Sociological issues

Since the superstring revolutions of the 1980s and 1990s, string theory has become the dominant paradigm of high energy theoretical physics.[167] Some string theorists have expressed the view that there does not exist an equally successful alternative theory addressing the deep questions of fundamental physics. In an interview from 1987, Nobel laureate David Gross made the following controversial comments about the reasons for the popularity of string theory:

> The most important [reason] is that there are no other good ideas around. That's what gets most people into it. When people started to get interested in string theory they didn't know anything about it. In fact, the first reaction of most people is that the theory is extremely ugly and unpleasant, at least that was the case a few years ago when the understanding of string theory was much less developed. It was difficult for people to learn about it and to be turned on. So I think the real reason why people have got attracted by it is because there is no other game in town. All other approaches of constructing grand unified theories, which were more conservative to begin with, and only gradually became more and more radical, have failed, and this game hasn't failed yet.[168]

Several other high profile theorists and commentators have expressed similar views, suggesting that there are no viable alternatives to string theory.[169]

Many critics of string theory have commented on this state of affairs. In his book criticizing string theory, Peter Woit views the status of string theory research as unhealthy and detrimental to the future of fundamental physics. He argues that the extreme popularity of string theory among theoretical physicists is partly a consequence of the financial structure of academia and the fierce competition for scarce resources.[170] In his book *The Road to Reality*, mathematical physicist Roger Penrose expresses similar views, stating "The often frantic competitiveness that this ease of communication engenders leads to 'bandwagon' effects, where researchers fear to be left behind if they do not join in."[171] Penrose also claims that the technical difficulty of modern physics forces young scientists to rely on the preferences of established researchers, rather than forging new paths of their own.[172] Lee Smolin expresses a slightly different position in his critique, claiming that string theory grew out of a tradition of particle physics which discourages speculation about the foundations of physics, while his preferred approach, loop quantum gravity, encourages more radical thinking. According to Smolin,

> String theory is a powerful, well-motivated idea and deserves much of the work that has been devoted to it. If it has so far failed, the principal reason is that its intrinsic flaws are closely tied to its strengths—and, of course, the story is unfinished, since string theory may well turn out to be part of the truth. The real question is not why we have expended so much energy on string theory but why we haven't expended nearly enough on alternative approaches.[173]

Smolin goes on to offer a number of prescriptions for how scientists might encourage a greater diversity of approaches to quantum gravity research.[174]

6.9   References

6.9.1   Notes

[1] For example, physicists are still working to understand the phenomenon of quark confinement, the paradoxes of black holes, and the origin of dark energy.

[2] For example, in the context of the AdS/CFT correspondence, theorists often formulate and study theories of gravity in unphysical numbers of spacetime dimensions.

[3] "Top Cited Articles during 2010 in hep-th". Retrieved 25 July 2013.

[4] More precisely, one cannot apply the methods of perturbative quantum field theory.

[5] Two independent mathematical proofs of mirror symmetry were given by Givental 1996, 1998 and Lian, Liu, Yau 1997, 1999, 2000.

[6] More precisely, a nontrivial group is called *simple* if its only normal subgroups are the trivial group and the group itself. The Jordan–Hölder theorem exhibits finite simple groups as the building blocks for all finite groups.

6.9.2 Citations

[1] Becker, Becker, and Schwarz 2007, p. 1

[2] Becker, Becker, and Schwarz 2007, p. 1

[3] Zwiebach 2009, p. 6

[4] Becker, Becker, and Schwarz 2007, pp. 2–3

[5] Becker, Becker, and Schwarz 2007, pp. 9–12

[6] Becker, Becker, and Schwarz 2007, pp. 14–15

[7] Klebanov and Maldacena 2009

[8] Merali 2011

[9] Sachdev 2013

[10] Becker, Becker, and Schwarz 2007, pp. 3, 15–16

[11] Becker, Becker, and Schwarz 2007, p. 8

[12] Becker, Becker, and Schwarz 13–14

[13] Woit 2006

[14] Zee 2010

[15] Zee 2010

[16] Becker, Becker, and Schwarz 2007, p. 2

[17] Becker, Becker, and Schwarz 2007, p. 6

[18] Zwiebach 2009, p. 12

[19] Becker, Becker, and Schwarz 2007, p. 6

[20] Becker, Becker, and Schwarz 2007, pp. 2–3

[21] Becker, Becker, and Schwarz 2007, p. 4

[22] Zwiebach 2009, p. 324

[23] Wald 1984, p. 4

[24] Zee 2010, Parts V and VI

[25] Zwiebach 2009, p. 9

[26] Zwiebach 2009, p. 8

[27] Yau and Nadis 2010, Ch. 6

[28] Greene 2000, p. 186

[29] Yau and Nadis 2010, Ch. 6

[30] Yau and Nadis 2010, p. ix

[31] Randall and Sundrum 1999

[32] Becker, Becker, and Schwarz 2007

[33] Becker, Becker, and Schwarz 2007

[34] Zwiebach 2009, p. 376

[35] Moore 2005, p. 214

[36] Moore 2005, p. 214

[37] Moore 2005, p. 215

[38] Aspinwall et al. 2009

[39] Kontsevich 1995

[40] Kapustin and Witten 2007

[41] Duff 1998

[42] Duff 1998, p. 64

[43] Nahm 1978

[44] Cremmer, Julia, and Scherk 1978

[45] Duff 1998, p. 65

[46] Duff 1998, p. 65

[47] Duff 1998, p. 65

[48] Duff 1998, p. 65

[49] Duff 1998, p. 65

[50] Sen 1994a

[51] Sen 1994b

[52] Hull and Townsend 1995

[53] Duff 1998, p. 67

[54] Bergshoeff, Sezgin, and Townsend 1987

[55] Duff et al. 1987

[56] Duff 1998, p. 66

[57] Witten 1995

[58] Duff 1998, pp. 67–68

[59] Becker, Becker, and Schwarz 2007, p. 296

[60] Hořava and Witten 1996

[61] Duff 1996, sec. 1

[62] Banks et al. 1997

[63] Banks et al. 1997

[64] Connes 1994

[65] Connes, Douglas, and Schwarz 1998

[66] Nekrasov and Schwarz 1998

[67] Seiberg and Witten 1999

[68] de Haro et al. 2013, p. 2

[69] Yau and Nadis 2010, p. 187–188

[70] Bekenstein 1973

[71] Hawking 1975

[72] Wald 1984, p. 417

[73] Yau and Nadis 2010, p. 189

[74] Strominger and Vafa 1996

[75] Yau and Nadis 2010, pp. 190–192

[76] Maldacena, Strominger, and Witten 1997

[77] Ooguri, Strominger, and Vafa 2004

[78] Yau and Nadis 2010, pp. 192–193

[79] Yau and Nadis 2010, pp. 194–195

[80] Strominger 1998

[81] Guica et al. 2009

[82] Castro, Maloney, and Strominger 2010

[83] Klebanov and Maldacena 2009

[84] Maldacena 1998

[85] Gubser, Klebanov, and Polyakov 1998

[86] Witten 1998

[87] Klebanov and Maldacena 2009, p. 28

[88] Maldacena 2005, p. 60

[89] Maldacena 2005, p. 61

[90] Maldacena 2005, p. 60

[91] Maldacena 2005, p. 61

[92] Maldacena 2005, p. 60

[93] Zwiebach 2009, p. 552

[94] Maldacena 2005, pp. 61–62

[95] de Haro et al. 2013, p. 2

[96] Hawking 1975

[97] Susskind 2008

[98] Zwiebach 2009, p. 554

[99] Maldacena 2005, p. 63

[100] Hawking 2005

[101] Merali 2011

[102] Zwiebach 2009, p. 559

[103] Kovtun, Son, and Starinets 2001

[104] Merali 2011, p. 303

[105] Luzum and Romatschke 2008

[106] Merali 2011, p. 303

[107] Sachdev 2013, p. 51

[108] Woit 2006

[109] Candelas et al. 1985

[110] Yau and Nadis 2010, pp. 147–150

[111] Becker, Becker, and Schwarz 2007, pp. 530–531

[112] Becker, Becker, and Schwarz 2007, p. 531

[113] Becker, Becker, and Schwarz 2007, p. 538

[114] Becker, Becker, and Schwarz 2007, p. 533

[115] Becker, Becker, and Schwarz 2007, pp. 539–543

[116] Deligne et al. 1999, p. 1

[117] Hori et al. 2003, p. xvii

[118] Aspinwall et al. 2009, p. 13

[119] Hori et al. 2003

[120] Aspinwall et al. 2009

[121] Yau and Nadis 2010, p. 167

[122] Yau and Nadis 2010, p. 166

[123] Yau and Nadis 2010, p. 169

[124] Candelas et al. 1991

[125] Yau and Nadis 2010, p. 169

[126] Yau and Nadis 2010, p. 171

[127] Hori et al. 2003, p. xix

[128] Kontsevich 1995

[129] Strominger, Yau, and Zaslow 1996

[130] Dummit and Foote 2004

[131] Dummit and Foote 2004, pp. 102–103

[132] Klarreich 2015

[133] Gannon 2006, p. 2

[134] Gannon 2006, p. 4

[135] Conway and Norton 1979

[136] Gannon 2006, p. 5

[137] Gannon 2006, p. 8

[138] Borcherds 1992

[139] Frenkel, Lepowsky, and Meurman 1988

[140] Gannon 2006, p. 11

[141] Klarreich 2015

[142] Eguchi, Ooguri, and Tachikawa 2010

[143] Cheng, Duncan, and Harvey 2013

[144] Duncan, Griffin, and Ono 2015

[145] Witten 2007

[146] Duff 1998

[147] Banks et al. 1997

[148] Strominger and Vafa 1996

[149] Maldacena 1998

[150] Gubser, Klebanov, and Polyakov 1998

[151] Witten 1998

[152] de Haro et al. 2013, p. 2

[153] Kovtun, Son, and Starinets 2001

[154] Woit 2006, pp. 240–242

[155] Woit 2006, p. 242

[156] Woit 2006, p. 242

[157] Weinberg 1987

[158] Woit 2006, p. 243

[159] Susskind 2005

[160] Woit 2006, pp. 242–243

[161] Woit 2006, p. 240

[162] Woit 2006, p. 249

[163] Smolin 2006, p. 81

[164] Smolin 2006, p. 184

[165] Polchinski 2007

[166] Polchinski 2007

[167] Penrose 2004, p. 1017

[168] Woit 2006, pp. 224–225

[169] Woit 2006, Ch. 16

[170] Woit 2006, p. 239

[171] Penrose 2004, p. 1018

[172] Penrose 2004, pp. 1019–1020

[173] Smolin 2006, p. 349

[174] Smolin 2006, Ch. 20

6.9.3 Bibliography

- Aspinwall, Paul; Bridgeland, Tom; Craw, Alastair; Douglas, Michael; Gross, Mark; Kapustin, Anton; Moore, Gregory; Segal, Graeme; Szendrői, Balázs; Wilson, P.M.H., eds. (2009). *Dirichlet Branes and Mirror Symmetry*. American Mathematical Society. ISBN 978-0-8218-3848-8.

- Banks, Tom; Fischler, Willy; Schenker, Stephen; Susskind, Leonard (1997). "M theory as a matrix model: A conjecture". *Physical Review D* **55**(8): 5112. arXiv:hep-th/9610043. Bibcode:1997PhRvD..55.5112B.doi:10.1103

   - Becker, Katrin; Becker, Melanie; Schwarz, John (2007). *String theory and M-theory: A modern*

   introduction.
   Cambridge University Press. ISBN 978-0-521-86069-7.

- Bekenstein,Jacob(1973). "Black holes and entropy". *Physical Review D* **7**(8): 2333. Bibcode:1973PhRvD...7.2333B. doi:10.1103/PhysRevD.7.2333.

- Bergshoeff, Eric; Sezgin, Ergin; Townsend, Paul (1987). "Supermembranes and eleven-dimensional supergravity". *Physics Letters B* **189** (1): 75–78. Bibcode:1987PhLB..189...75B. doi:10.1016/0370-2693(87)91272-X.

- Borcherds, Richard (1992). "Monstrous moonshine and Lie superalgebras". *Inventiones mathematicae* **109** (1): 405–444. Bibcode:1992InMat.109..405B. doi:10.1007/BF01232032.

- Candelas, Philip; de la Ossa, Xenia; Green, Paul; Parks, Linda (1991). "A pair of Calabi–Yau manifolds as an exactly soluble superconformal field theory". *Nuclear Physics B* **359** (1): 21–74. Bibcode:1991NuPhB.359...21C. doi:10.1016/0550-3213(91)90292-6.

- Candelas, Philip; Horowitz, Gary; Strominger, Andrew; Witten, Edward (1985). "Vacuum configurations for superstrings". *Nuclear Physics B* **258**: 46–74. Bibcode:1985NuPhB.258...46C. doi:10.1016/0550-3213(85)90602-9.

- Castro, Alejandra; Maloney, Alexander; Strominger, Andrew (2010). "Hidden conformal symmetry of the Kerr black hole". *Physical Review D* **82**(2). arXiv:1004.0996. Bibcode:2010PhRvD..82b4008C.doi:10.1103/

   PhysRevD.82.024008.   - Cheng, Miranda; Duncan, John; Harvey, Jeffrey (2013). "Umbral Moonshine". arXiv:

   1204.2779.

- Connes, Alain (1994). *Noncommutative Geometry*. Academic Press. ISBN 978-0-12-185860-5.

- Connes, Alain; Douglas, Michael; Schwarz, Albert (1998). "Noncommutative geometry and matrix theory". *Journal of High Energy Physics*. 19981(2): 003. arXiv:hep-th/9711162. Bibcode:1998JHEP...02..003C.doi:10.1088/1126-6708/1998/02/003.

- Conway, John; Norton, Simon (1979). "Monstrous moonshine". *Bull. London. Math. Soc.* **11** (3): 308–339.

- Cremmer,Eugene;Julia,Bernard;Scherk,Joel(1978). "Supergravity theory in eleven dimensions". *Physics Letters B* **76** (4): 409–412. Bibcode:1978PhLB..76..409C. doi:10.1016/0370-2693(78)90894-8.

- de Haro, Sebastian; Dieks, Dennis; 't Hooft, Gerard; Verlinde, Erik(2013). "Forty Years of String Theory Reflecting on the Foundations". *Foundations of Physics* **43** (1): 1–7. Bibcode:2013FoPh...43....1D. doi:10.1007/s10701-012-9691-3.

- Duff,Michael(1996). "M-theory(the theory formerly known as strings)". *International Journal of Modern Physics*
  A11(32): 6523–41. arXiv:hep-th/9608117. Bibcode:1996IJMPA..11.5623D.doi:10.1142/S0217751X 96002583.

- Duff,Michael(1998). "The theory formerly known as strings". *Scientific American***278**(2): 64–9. doi:10.1038/scientificamerican0298- 64.

- Duff, Michael; Howe, Paul; Inami, Takeo; Stelle, Kellogg (1987). "Superstrings in D=10 from supermembranes in
  D=11". *Nuclear Physics B* **191** (1): 70–74. Bibcode:1987PhLB..191...70D. doi:10.1016/0370-2693(87)91323 -2.

- Dummit, David; Foote, Richard (2004). *Abstract Algebra*. Wiley. ISBN 978-0-471-43334-7.

- Duncan,John;Griffin,Michael;Ono,Ken(2015). "Proof of the Umbral Moonshine Conjecture". arXiv:1503. 01472.

- Eguchi, Tohru; Ooguri, Hirosi; Tachikawa, Yuji (2011). "Notes on the K3 surface and the Mathieu group M_{24}".
  Experimental Mathematics **20** (1): 91–96. doi:10.1080/10586458.2011.544585.

- Frenkel, Igor; Lepowsky, James; Meurman, Arne (1988). *Vertex Operator Algebras and the Monster*. Pure and
  Applied Mathematics **134**. Academic Press. ISBN 0-12-267065-5.

- Gannon, Terry. *Moonshine Beyond the Monster: The Bridge Connecting Algebra, Modular Forms, and Physics.*
  Cambridge University Press.

- Givental, Alexander (1996). "Equivariant Gromov-Witten invariants". *International Mathematics Research Notices*
  1996 (13): 613–663. doi:10.1155/S1073792896000414.

- Givental,Alexander(1998). "A mirror theorem for toric complete intersections". *Topologicalfield theory, primitive*
  forms and related topics: 141–175. doi:10.1007/978-1-4612-0705-4_5. ISBN 978-1-4612-6874-1.

- Gubser, Steven; Klebanov, Igor; Polyakov, Alexander (1998). "Gauge theory correlators from non-critical string
  theory". *Physics Letters B***428**: 105–114. arXiv:hep-th/9802109. Bibcode:1998PhLB..428..105G.doi:10.1016/S0370-
  2693(98)00377-3.

- Guica, Monica; Hartman, Thomas; Song, Wei; Strominger, Andrew (2009). "The Kerr/CFT Correspondence".
  *Physical Review D***80**(12). arXiv:0809.4266. Bibcode:2009PhRvD..80l4008G.doi:10.1103/PhysRevD.80. 124008.

- Hawking, Stephen (1975). "Particle creation by black holes". *Communications in mathematical physics* **43** (3):
  199–220. Bibcode:1975CMaPh..43..199H. doi:10.1007/BF02345020.

- Kapustin, Anton; Witten, Edward (2007). "Electric-magnetic duality and the geometric Langlands program". *Communications in Number Theory and Physics* **1** (1): 1–236. arXiv:hep-th/0604151. Bibcode:2007CNTP....1....1K. doi:10.4310/cntp.2007.v1.n1.a1.

- Klarreich, Erica. "Mathematicians chase moonshine's shadow". *Quanta Magazine*. Retrieved March 2015.

- Klebanov, Igor; Maldacena, Juan (2009). "Solving Quantum Field Theories via Curved Spacetimes" (PDF). *Physics Today* **62**: 28. Bibcode:2009PhT....62a..28K. doi:10.1063/1.3074260. Retrieved May 2013.

- Kontsevich, Maxim (1995). "Homological algebra of mirror symmetry". *Proceedings of the International Congress of Mathematicians*: 120–139. arXiv:alg-geom/9411018. Bibcode:1994alg.geom.11018K.

- Kovtun, P. K.; Son, Dam T.; Starinets, A. O. (2001). "Viscosity in strongly interacting quantum field theories from black hole physics". *Physical review letters* **94** (11): 111601. arXiv:hep-th/0405231. Bibcode:2005PhRvL..94k1601K. doi:10.1103/PhysRevLett.94.111601. PMID 15903845.

- Lian, Bong; Liu, Kefeng; Yau, Shing-Tung (1997). "Mirror principle, I". *Asian Journal of Math* **1**: 729–763. arXiv:alg-geom/9712011. Bibcode:1997alg.geom.12011L.

- Lian, Bong; Liu, Kefeng; Yau, Shing-Tung (1999a). "Mirror principle, II". *Asian Journal of Math* **3**: 109–146. arXiv:math/9905006. Bibcode:1999math......5006L.

- Lian, Bong; Liu, Kefeng; Yau, Shing-Tung (1999b). "Mirror principle, III". *Asian Journal of Math* **3**: 771–800. arXiv:math/9912038. Bibcode:1999math.....12038L.

- Lian, Bong; Liu, Kefeng; Yau, Shing-Tung (2000). "Mirror principle, IV". *Surveys in Differential Geometry*: 475–496. arXiv:math/0007104. Bibcode:2000math......7104L.

- Luzum, Matthew; Romatschke, Paul (2008). "Conformal relativistic viscous hydrodynamics: Applications to RHIC results at $\sqrt{s}NN$=200 GeV". *Physical Review C* **78** (3). arXiv:0804.4015. doi:10.1103/PhysRevC.78.034915.

- Maldacena, Juan (1998). "The Large N limit of superconformal field theories and supergravity". *Advances in Theoretical and Mathematical Physics* **2**: 231–252. arXiv:hep-th/9711200. Bibcode:1998AdTMP...2..231M. doi:10.1063/1.59653.

- Maldacena, Juan (2005). "The Illusion of Gravity" (PDF). *Scientific American* **293** (5): 56–63. Bibcode:2005SciAm.293e..56M. doi:10.1038/scientificamerican1105-56. PMID 16318027. Retrieved July 2013.

- Maldacena, Juan; Strominger, Andrew; Witten, Edward (1997). "Black hole entropy in M-theory". *Journal of High Energy Physics* **1997** (12).

- Merali, Zeeya (2011). "Collaborative physics: string theory finds a bench mate". *Nature* **478** (7369): 302–304. Bibcode:2011Natur.478..302M. doi:10.1038/478302a. PMID 22012369.

- Moore, Gregory (2005). "What is ... a Brane?" (PDF). *Notices of the AMS* **52**: 214. Retrieved June 2013.

- Nahm, Walter (1978). "Supersymmetries and their representations". *Nuclear Physics B* **135** (1): 149–166. Bibcode:1978NuPhB.135..149N. doi:10.1016/0550-3213(78)90218-3.

- Nekrasov, Nikita; Schwarz, Albert (1998). "Instantons on noncommutative $\mathbf{R}^4$ and (2,0) superconformal six dimensional theory". *Communications in Mathematical Physics* **198** (3): 689–703. arXiv:hep-th/9802068. Bibcode:1998CMaPh.198..689N. doi:10.1007/s002200050490.

- Ooguri, Hirosi; Strominger, Andrew; Vafa, Cumrun (2004). "Black hole attractors and the topological string ". *Physical Review D* **70** (10).

- Polchinski, Joseph (2007). "All Strung Out?". *American Scientist*. Retrieved April 2015.

- Penrose, Roger (2005). *The Road to Reality: A Complete Guide to the Laws of the Universe*. Knopf. ISBN 0-679-45443-8.

- Randall, Lisa; Sundrum, Raman (1999). "An alternative to compactification". *Physical Review Letters* **83** (23): 4690. arXiv:hep-th/9906064. Bibcode:1999PhRvL..83.4690R. doi:10.1103/PhysRevLett.83.4690.

- Sachdev, Subir (2013). "Strange and stringy". *Scientific American* **308** (44): 44. Bibcode:2012SciAm.308a..44S. doi:10.1038/scientificamerican0113-44.

- Seiberg, Nathan; Witten, Edward (1999). "String Theory and Noncommutative Geometry". *Journal of High Energy Physics* **1999**(9): 032. arXiv:hep-th/9908142. Bibcode:1999JHEP...09..032S.doi:10.1088/1126-6708/1999/09/032.

- Sen, Ashoke (1994a). "Strong-weak coupling duality in four-dimensional string theory". *International Journal of Modern Physics A***9**(21): 3707–3750. arXiv:hep-th/9402002. Bibcode:1994IJMPA...9.3707S.doi:10.1142/S0217751X94001497.

- Sen, Ashoke (1994b). "Dyon-monopole bound states, self-dual harmonic forms on the multi-monopole moduli space, and $SL(2,\mathbf{Z})$ invariance in string theory". *Physics Letters B* **329** (2): 217–221. arXiv:hep-th/9402032 Bibcode:1994PhLB..329..217S. doi:10.1016/0370-2693(94)90763-3.

- Smolin, Lee (2006). *The Trouble with Physics: The Rise of String Theory, the Fall of a Science, and What Comes Next*. New York: Houghton Mifflin Co. ISBN 0-618-55105-0.

- Strominger, Andrew (1998). "Black hole entropy from near-horizon microstates". *Journal of High Energy Physics* **1998** (2). arXiv:hep-th/9712251. Bibcode:1998JHEP...02..009S. doi:10.1088/1126-6708/1998/02/009.

- Strominger, Andrew; Vafa, Cumrun (1996). "Microscopic origin of the Bekenstein–Hawking entropy". *Physics Letters B***379**(1): 99–104. arXiv:hep-th/9601029. Bibcode:1996PhLB..379...99S.doi:10.1016/0370-2693(96)00345-0.

- Strominger, Andrew; Yau, Shing-Tung; Zaslow, Eric (1996). "Mirror symmetry is T-duality". *Nuclear Physics B* **479** (1): 243–259. arXiv:hep-th/9606040. Bibcode:1996NuPhB.479..243S. doi:10.1016/0550-3213(96)00434-

- Witten, Edward (1998). "Anti-de Sitter space and holography". *Advances in Theoretical and Mathematical Physics* **2**: 253–291. arXiv:hep-th/9802150. Bibcode:1998AdTMP...2..253W.

- Witten, Edward (2007). "Three-dimensional gravity revisited". arXiv:0706.3359 [hep-th].

- Woit, Peter (2006). *Not Even Wrong: The Failure of String Theory and the Search for Unity in Physical Law*. Basic Books. p. 105. ISBN 0-465-09275-6.

- Yau, Shing-Tung; Nadis, Steve (2010). *The Shape of Inner Space: String Theory and the Geometry of the Universe's Hidden Dimensions*. Basic Books. ISBN 978-0-465-02023-2.

- Zee, Anthony (2010). *Quantum Field Theory in a Nutshell* (2nd ed.). Princeton University Press. ISBN 978-0-691-14034-6.

- Zwiebach, Barton (2009). *A First Course in String Theory*. Cambridge University Press. ISBN 978-0-521-88032-9.

6.10   Further reading

6.10.1   Popularizations

General

- Greene, Brian (2003). *The Elegant Universe: Superstrings, Hidden Dimensions, and the Quest for the Ultimate Theory*. New York: W.W. Norton & Company. ISBN 0-393-05858-1.

- Greene, Brian (2004). *The Fabric of the Cosmos: Space, Time, and the Texture of Reality*. New York: Alfred A. Knopf. ISBN 0-375-41288-3.

Critical

- Penrose, Roger (2005). *The Road to Reality: A Complete Guide to the Laws of the Universe*. Knopf. ISBN 0-679-45443-8.

- Smolin, Lee (2006). *The Trouble with Physics: The Rise of String Theory, the Fall of a Science, and What Comes Next*. New York: Houghton Mifflin Co. ISBN 0-618-55105-0.

- Woit, Peter (2006). *Not Even Wrong: The Failure of String Theory And the Search for Unity in Physical Law*. London: Jonathan Cape &: New York: Basic Books. ISBN 978-0-465-09275-8.

6.10.2   Textbooks

For physicists

- Becker, Katrin; Becker, Melanie; Schwarz, John (2007). *String Theory and M-theory: A Modern Introduction*. Cambridge University Press. ISBN 978-0-521-86069-7.

- Green, Michael; Schwarz, John; Witten, Edward (2012). *Superstring theory. Vol. 1: Introduction*. Cambridge University Press. ISBN 978-1107029118.

- Green, Michael; Schwarz, John; Witten, Edward (2012). *Superstring theory. Vol. 2: Loop amplitudes, anomalies and phenomenology.* Cambridge University Press. ISBN 978-1107029132.

- Polchinski, Joseph (1998). *String Theory Vol. 1: An Introduction to the Bosonic String.* Cambridge University Press. ISBN 0-521-63303-6.

- Polchinski, Joseph (1998). *String Theory Vol. 2: Superstring Theory and Beyond.* Cambridge University Press. ISBN 0-521-63304-4.

- Zwiebach, Barton (2009). *A First Course in String Theory.* Cambridge University Press. ISBN 978-0-521-88032-9.

For mathematicians

- Deligne, Pierre; Etingof, Pavel; Freed, Daniel; Jeffery, Lisa; Kazhdan, David; Morgan, John; Morrison, David; Witten, Edward, eds. (1999). *Quantum Fields and Strings: A Course for Mathematicians, Vol. 2.* American Mathematical Society. ISBN 978-0821819883.

6.11 External links

- *The Elegant Universe*—A three-hour miniseries with Brian Greene by *NOVA* (original PBS Broadcast Dates: October 28, 8–10 p.m. and November 4, 8–9 p.m., 2003). Various images, texts, videos and animations explaining string theory.

- Not Even Wrong—A blog critical of string theory

- The Official String Theory Web Site

- Why String Theory—An introduction to string theory.

A diagram of string theory dualities. Yellow arrows indicate S-duality. Blue arrows indicate T-duality.

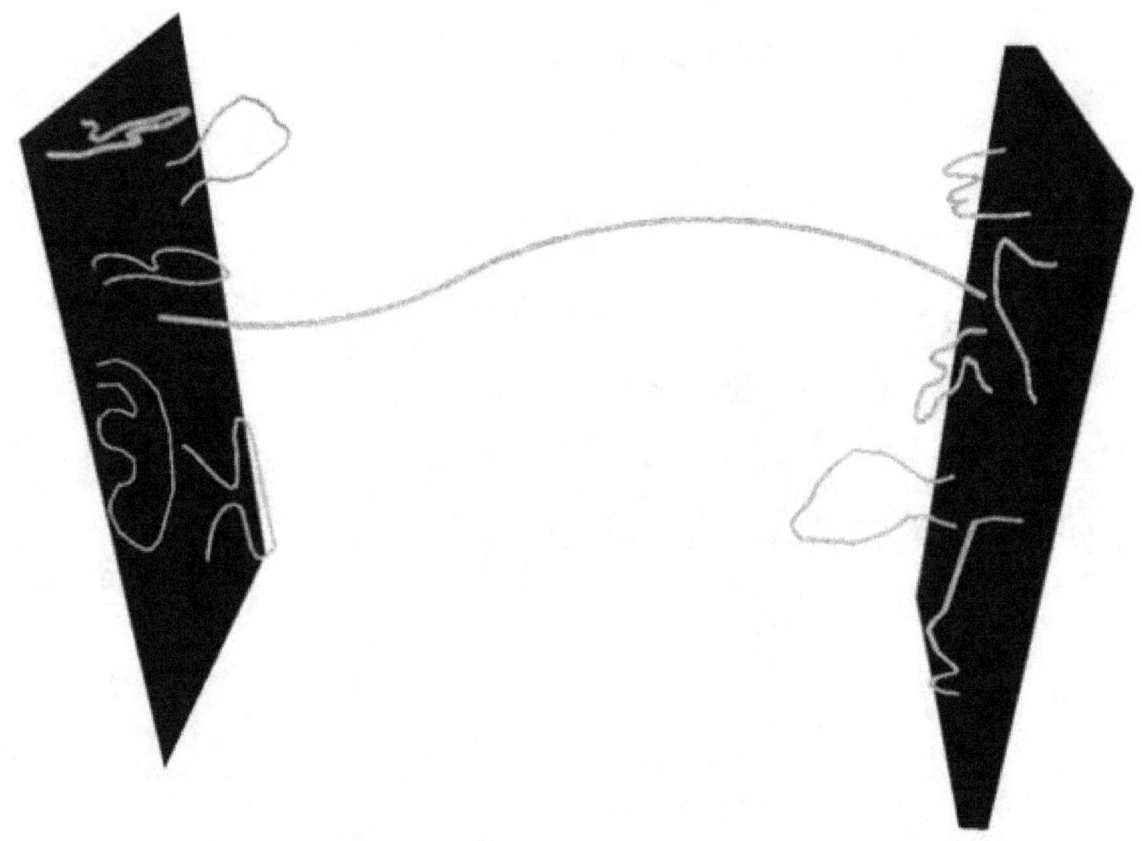

Open strings attached to a pair of D-branes

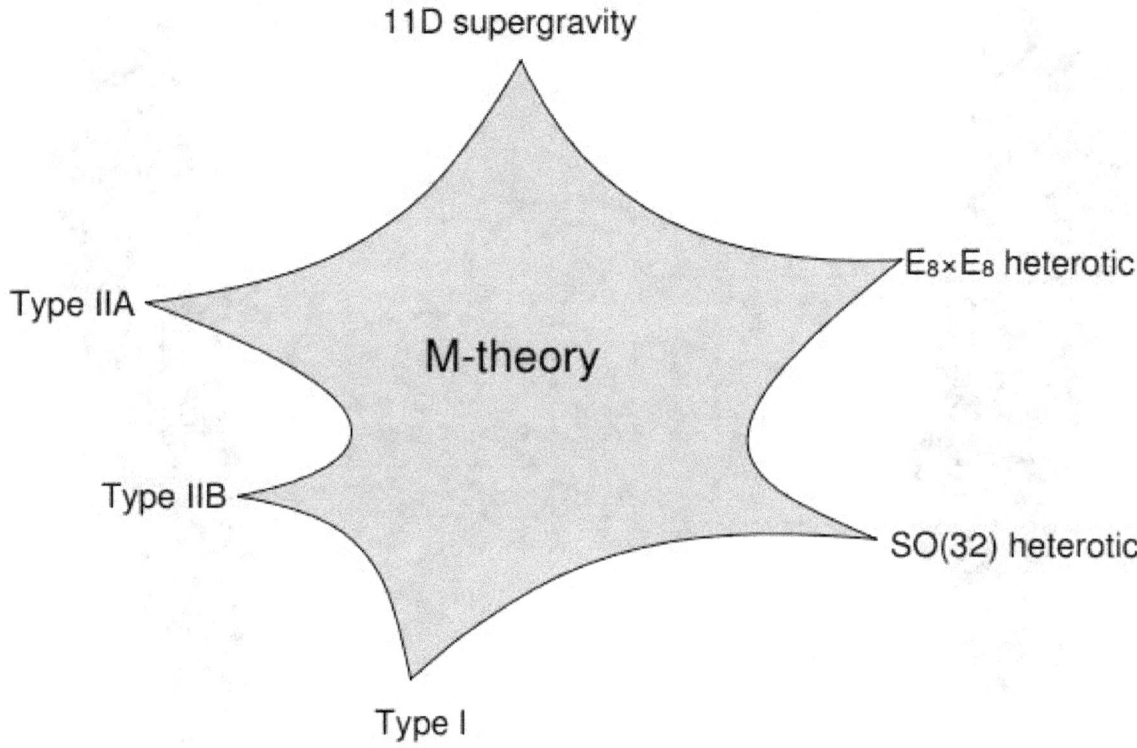

A schematic illustration of the relationship between M-theory, the five superstring theories, and eleven-dimensional supergravity. The shaded region represents a family of different physical scenarios that are possible in M-theory. In certain limiting cases corresponding to the cusps, it is natural to describe the physics using one of the six theories labeled there.

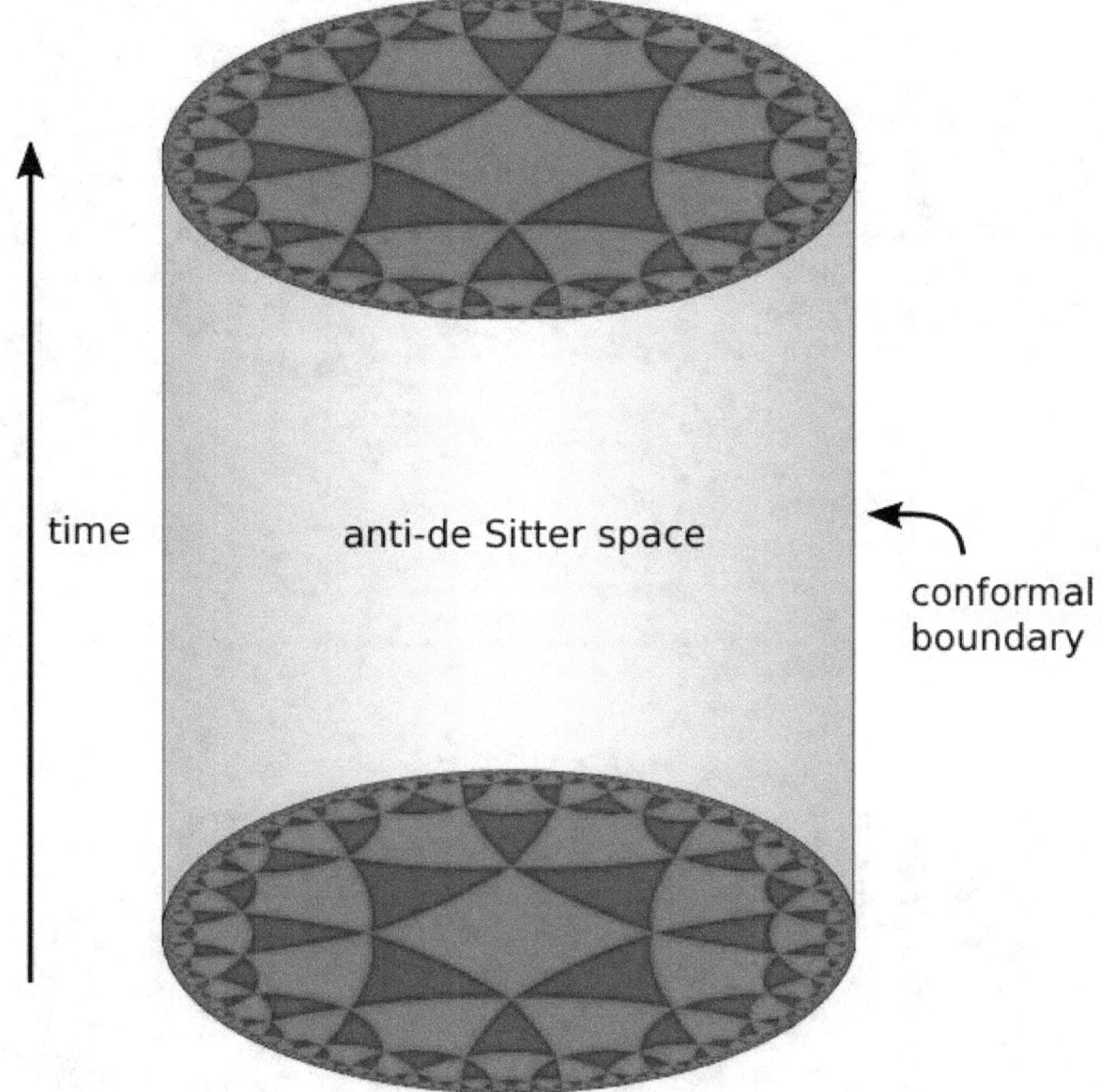

Three-dimensional anti-de Sitter space is like a stack of hyperbolic disks, each one representing the state of the universe at a given time. The resulting spacetime looks like a solid cylinder.

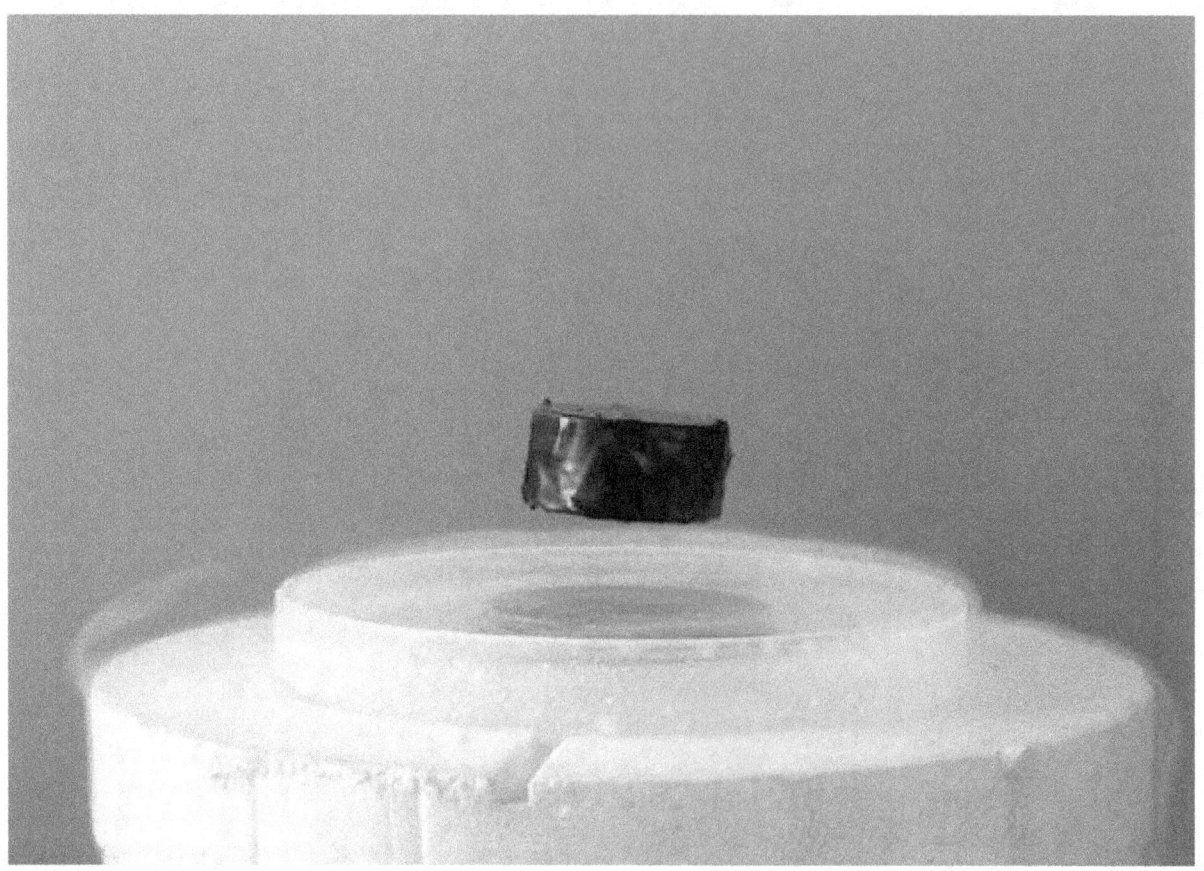

A magnet levitating above a high-temperature superconductor. Today some physicists are working to understand high-temperature super-conductivity using the AdS/CFT correspondence.[101]

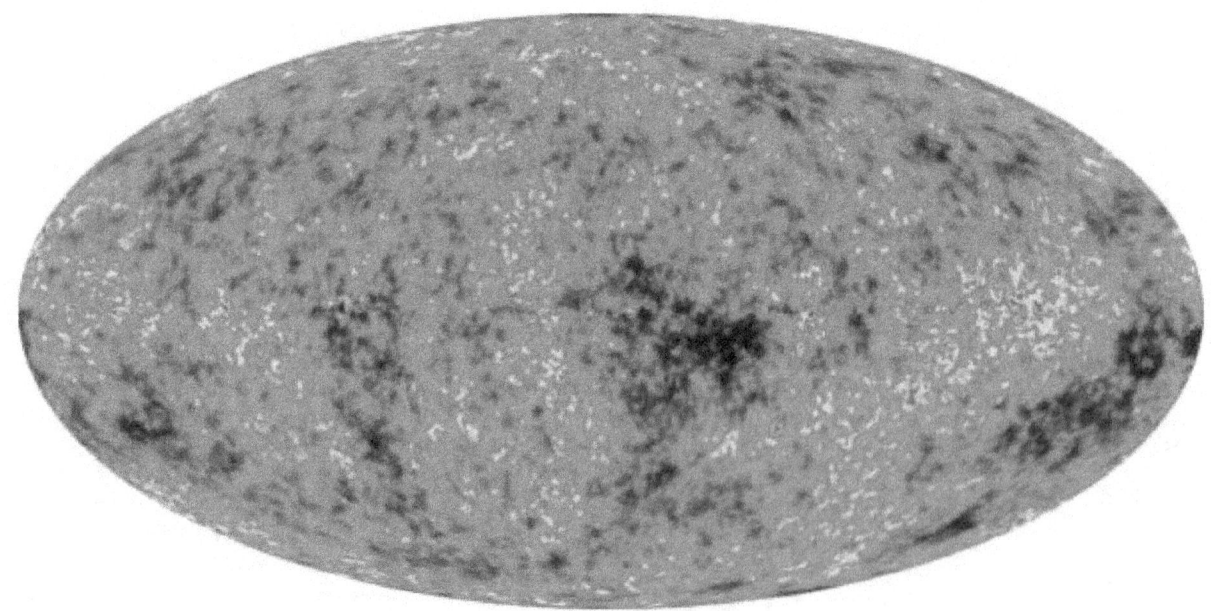

A map of the cosmic microwave background produced by the Wilkinson Microwave Anisotropy Probe

The Clebsch cubic is an example of a kind of geometric object called an algebraic variety. A classical result of enumerative geometry states that there are exactly 27 straight lines that lie entirely on this surface.

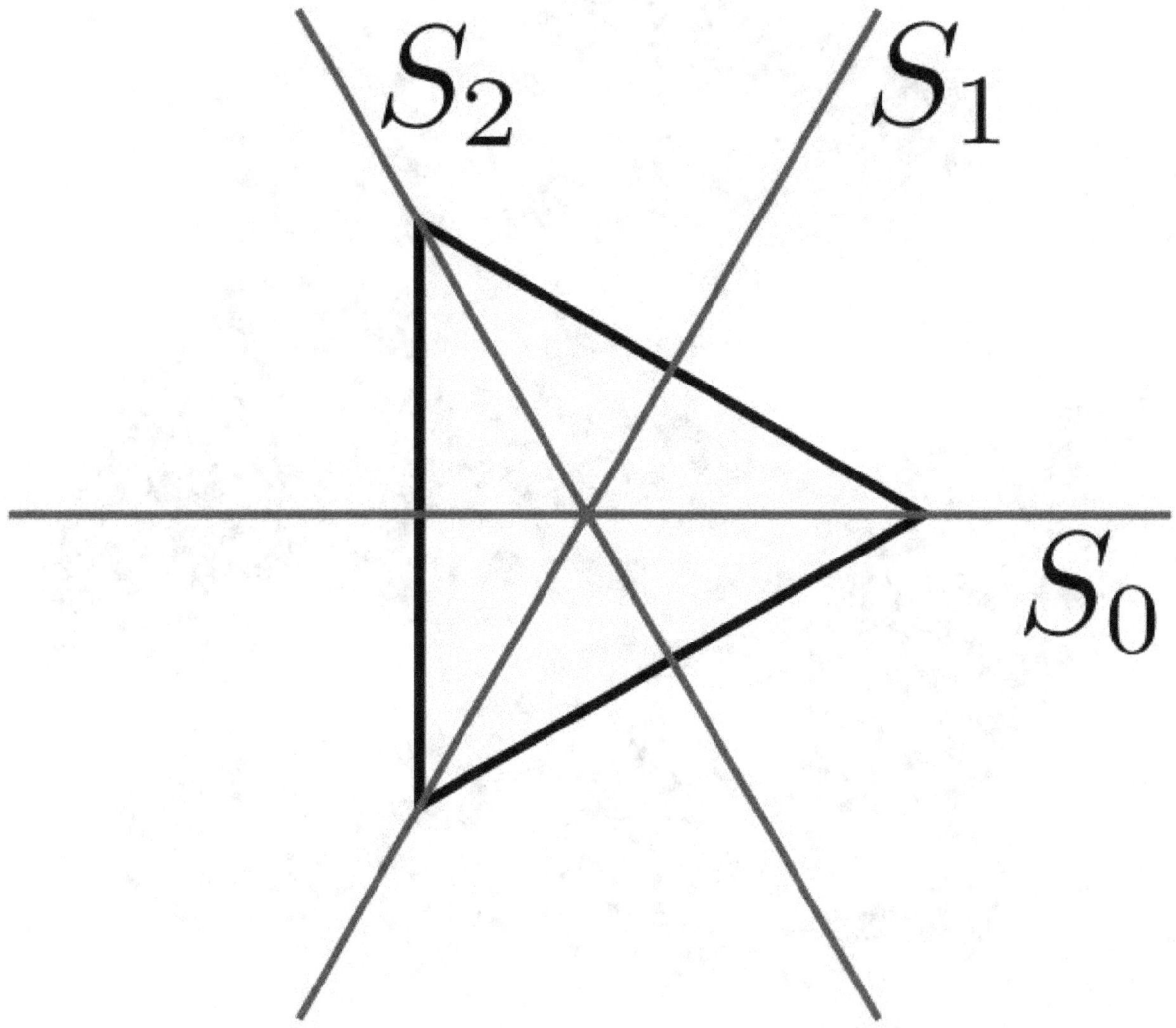

An equilateral triangle can be rotated through 120°, 240°, or 360°, or reflected in any of the three lines pictured without changing its shape.

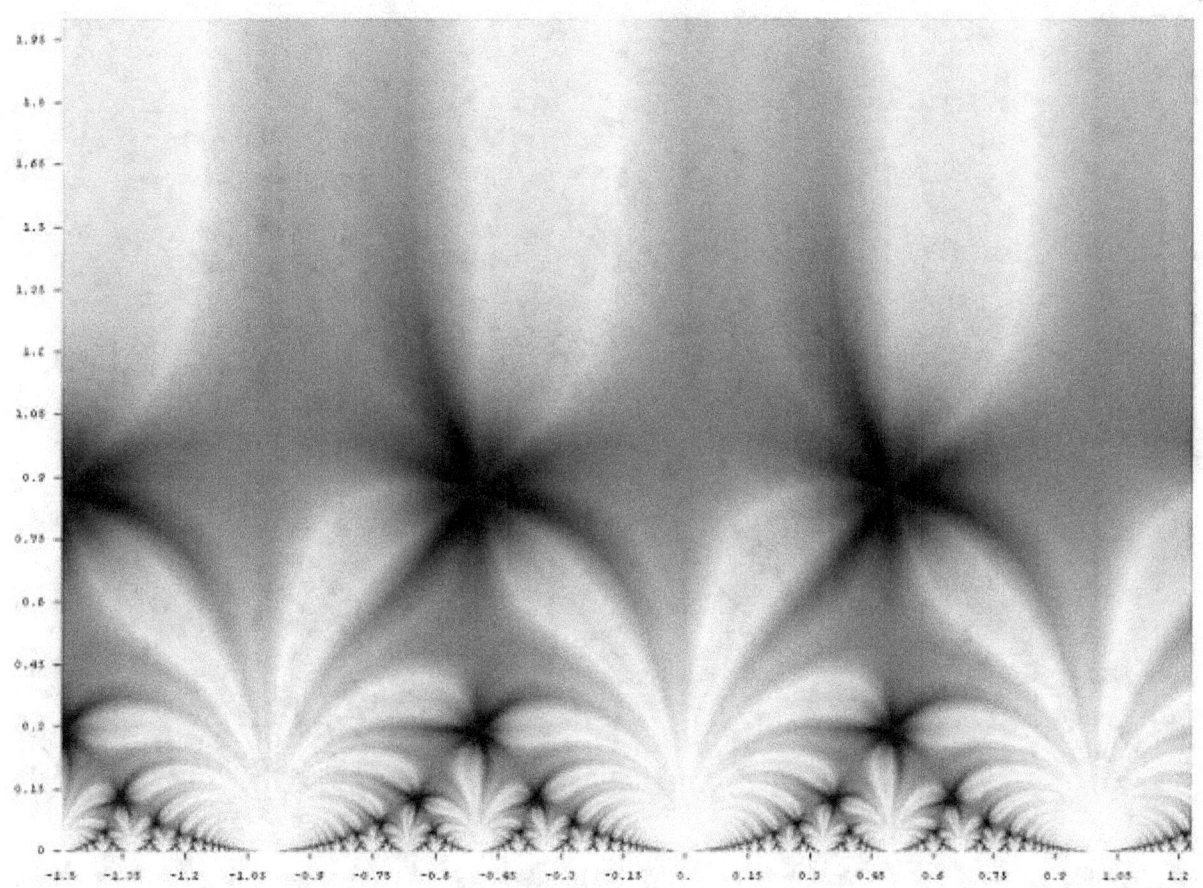

A graph of the j-function in the complex plane

Leonard Susskind

Gabriele Veneziano

John Schwarz

Edward Witten

Joseph Polchinski

Juan Maldacena

Chapter 7

Topological string theory

In theoretical physics, **topological string theory** is a simplified version of string theory. The operators in topological string theory represent the algebra of operators in the full string theory that preserve a certain amount of supersymmetry. Topological string theory is obtained by a topological twist of the worldsheet description of ordinary string theory: the operators are given different spins. The operation is fully analogous to the construction of topological field theory which is a related concept. Consequently, there are no local degrees of freedom in topological string theory.

There are two main versions of topological string theory: the topological A-model and the topological B-model. The results of the calculations in topological string theory generically encode all holomorphic quantities within the full string theory whose values are protected by spacetime supersymmetry. Various calculations in topological string theory are closely related to Chern–Simons theory, Gromov–Witten invariants, mirror symmetry, geometric Langlands Program, and many other topics.

Topological string theory was established and is studied by physicists such as Edward Witten and Cumrun Vafa.

7.1 Admissible spacetimes

The fundamental strings of string theory are two-dimensional surfaces. A quantum field theory known as the $N = (1,1)$ sigma model is defined on each surface. This theory consist of maps from the surface to a supermanifold. Physically the supermanifold is interpreted as spacetime and each map is interpreted as the embedding of the string in spacetime.

Only special spacetimes admit topological strings. Classically one must choose a spacetime such that the theory respects an additional pair of supersymmetries, and so is in fact an $N = (2,2)$ sigma model. This will be the case for example if the spacetime is a Kähler manifold and the H-flux is identically equal to zero, although there are more general cases in which the target is a generalized Kähler manifold and the H-flux is nontrivial.

So far we have described ordinary strings on special backgrounds. These strings are never topological. To make these strings topological, one needs to modify the sigma model via a procedure called a topological twist which was invented by Edward Witten in 1988. The central observation is that these theories have two U(1) symmetries known as R-symmetries, and one may modify the Lorentz symmetry by mixing rotations and R-symmetries. One may use either of the two R-symmetries, leading to two different theories, called the A model and the B model. After this twist the action of the theory is BRST exact, and as a result the theory has no dynamics, instead all observables depend on the topology of a configuration. Such theories are known as topological theories.

While classically this procedure is always possible, quantum mechanically the U(1) symmetries may be anomalous. In this case the twisting is not possible. For example, in the Kähler case with $H = 0$ the twist leading to the A-model is always possible but that leading to the B-model is only possible when the first Chern class of the spacetime vanishes, implying that the spacetime is Calabi-Yau. More generally (2,2) theories have two complex structures and the B model exists when the first Chern classes of associated bundles sum to zero whereas the A model exists when the difference of the Chern classes is zero. In the Kähler case the two complex structures are the same and so the difference is always zero, which is why the A model always exists.

There is no restriction on the number of dimensions of spacetime, other than that it must be even because spacetime is generalized Kähler. However all correlation functions with worldsheets that are not spheres vanish unless the complex dimension of the spacetime is three, and so spacetimes with complex dimension three are the most interesting. This is fortunate for phenomenology, as phenomenological models often use a physical string theory compactified on a 3 complex-dimensional space. The topological string theory is not equivalent to the physical string theory, even on the same space, but certain supersymmetric quantities agree in the two theories.

7.2 Objects

7.2.1 A-model

The topological A-model comes with a target space which is a 6 real-dimensional generalized Kähler spacetime. In the case in which the spacetime is Kähler, the theory describes two objects. There are fundamental strings, which wrap two real-dimensional holomorphic curves. Amplitudes for the scattering of these strings depend only on the Kähler form of the spacetime, and not on the complex structure. Classically these correlation functions are determined by the cohomology ring. There are quantum mechanical instanton effects which correct these and yield Gromov–Witten invariants, which measure the cup product in a deformed cohomology ring called the quantum cohomology. The string field theory of the A-model closed strings is known as Kähler gravity, and was introduced by Michael Bershadsky and Vladimir Sadov in Theory of Kähler Gravity.

In addition, there are D2-branes which wrap Lagrangian submanifolds of spacetime. These are submanifolds whose dimensions are one half that of space time, and such that the pullback of the Kähler form to the submanifold vanishes. The worldvolume theory on a stack of N D2-branes is the string field theory of the open strings of the A-model, which is a U(N) Chern–Simons theory.

The fundamental topological strings may end on the D2-branes. While the embedding of a string depends only on the Kähler form, the embeddings of the branes depends entirely on the complex structure. In particular, when a string ends on a brane the intersection will always be orthogonal, as the wedge product of the Kähler form and the holomorphic 3-form is zero. In the physical string this is necessary for the stability of the configuration, but here it is a property of Lagrangian and holomorphic cycles on a Kahler manifold.

There may also be coisotropic branes in various dimensions other than half dimensions of Lagrangian submanifolds. These were first introduced by Anton Kapustin and Dmitri Orlov in Remarks on A-Branes, Mirror Symmetry, and the Fukaya Category

7.2.2 B-model

The B-model also contains fundamental strings, but their scattering amplitudes depend entirely upon the complex structure and are independent of the Kähler structure. In particular, they are insensitive to worldsheet instanton effects and so can often be calculated exactly. Mirror symmetry then relates them to A model amplitudes, allowing one to compute Gromov–Witten invariants. The string field theory of the closed strings of the B-model is known as the Kodaira–Spencer theory of gravity and was developed by Michael Bershadsky, Sergio Cecotti, Hirosi Ooguri and Cumrun Vafa in Kodaira–Spencer Theory of Gravity and Exact Results for Quantum String Amplitudes.

The B-model also comes with D(−1), D1, D3 and D5-branes, which wrap holomorphic 0, 2, 4 and 6-submanifolds respectively. The 6-submanifold is a connected component of the spacetime. The theory on a D5-brane is known as holomorphic Chern–Simons theory. The Lagrangian density is the wedge product of that of ordinary Chern–Simons theory with the holomorphic (3,0)-form, which exists in the Calabi-Yau case. The Lagrangian densities of the theories on the lower-dimensional branes may be obtained from holomorphic Chern–Simons theory by dimensional reductions.

7.2.3 Topological M-theory

Topological M-theory, which enjoys a seven-dimensional spacetime, is not a topological string theory, as it contains no topological strings. However topological M-theory on a circle bundle over a 6-manifold has been conjectured to be equivalent to the topological A-model on that 6-manifold.

In particular, the D2-branes of the A-model lift to points at which the circle bundle degenerates, or more precisely Kaluza–Klein monopoles. The fundamental strings of the A-model lift to membranes named M2-branes in topological M-theory.

One special case that has attracted much interest is topological M-theory on a space with G_2 holonomy and the A-model on a Calabi-Yau. In this case, the M2-branes wrap associative 3-cycles. Strictly speaking, the topological M-theory conjecture has only been made in this context, as in this case functions introduced by Nigel Hitchin in The Geometry of Three-Forms in Six and Seven Dimensions and Stable Forms and Special Metrics provide a candidate low energy effective action.

These functions are called "Hitchin functional" and Topological string is closely related to Hitchin's ideas on generalized complex structure, Hitchin system, and ADHM construction etc..

7.3 Observables

7.3.1 The topological twist

The 2-dimensional worldsheet theory is an $N = (2,2)$ supersymmetric sigma model, the $(2,2)$ supersymmetry means that the fermionic generators of the supersymmetry algebra, called supercharges, may be assembled into a single Dirac spinor, which consists of two Majorana–Weyl spinors of each chirality. This sigma model is topologically twisted, which means that the Lorentz symmetry generators that appear in the supersymmetry algebra simultaneously rotate the physical spacetime and also rotate the fermionic directions via the action of one of the R-symmetries. The R-symmetry group of a 2-dimensional $N = (2,2)$ field theory is $U(1) \times U(1)$, twists by the two different factors lead to the A and B models respectively. The topological twisted construction of topological string theories was introduced by Edward Witten in his 1988 paper Topological Sigma Models.

7.3.2 What do the correlators depend on?

The topological twist leads to a topological theory because the stress–energy tensor may be written as an anticommutator of a supercharge and another field. As the stress–energy tensor measures the dependence of the action on the metric tensor, this implies that all correlation functions of Q-invariant operators are independent of the metric. In this sense, the theory is topological.

More generally, any D-term in the action, which is any term which may be expressed as an integral over all of superspace, is an anticommutator of a supercharge and so does not affect the topological observables. Yet more generally, in the B model any term which may be written as an integral over the fermionic $\overline{\theta}^{\pm}$ coordinates does not contribute, whereas in the A-model any term which is an integral over θ^- or over $\overline{\theta}^+$ does not contribute. This implies that A model observables are independent of the superpotential (as it may be written as an integral over just $\overline{\theta}^{\pm}$) but depend holomorphically on the twisted superpotential, and vice versa for the B model.

7.4 Dualities

7.4.1 Dualities between TSTs

A number of dualities relate the above theories. The A-model and B-model on two mirror manifolds are related by mirror symmetry, which has been described as a T-duality on a three-torus. The A-model and B-model on the same manifold are conjectured to be related by S-duality, which implies the existence of several new branes, called NS branes by analogy with the NS5-brane, which wrap the same cycles as the original branes but in the opposite theory. Also a combination of the A-model and a sum of the B-model and its conjugate are related to topological M-theory by a kind of dimensional reduction. Here the degrees of freedom of the A-model and the B-models appear to not be simultaneously observable, but rather to have a relation similar to that between position and momentum in quantum mechanics.

The holomorphic anomaly

The sum of the B-model and its conjugate appears in the above duality because it is the theory whose low energy effective action is expected to be described by Hitchin's formalism. This is because the B-model suffers from a holomorphic anomaly, which states that the dependence on complex quantities, while classically holomorphic, receives nonholomorphic quantum corrections. In Quantum Background Independence in String Theory, Edward Witten argued that this structure is analogous to a structure that one finds geometrically quantizing the space of complex structures. Once this space has been quantized, only half of the dimensions simultaneously commute and so the number of degrees of freedom has been halved. This halving depends on an arbitrary choice, called a polarization. The conjugate model contains the missing degrees of freedom, and so by tensoring the B-model and its conjugate one reobtains all of the missing degrees of freedom and also eliminates the dependence on the arbitrary choice of polarization.

7.4.2 Geometric transitions

There are also a number of dualities that relate configurations with D-branes, which are described by open strings, to those with branes the branes replaced by flux and with the geometry described by the near-horizon geometry of the lost branes. The latter are described by closed strings.

Perhaps the first such duality is the Gopakumar-Vafa duality, which was introduced by Rajesh Gopakumar and Cumrun Vafa in On the Gauge Theory/Geometry Correspondence.This relates a stack of N D2-branes on a 3-sphere in the A-model on the deformed conifold to the closed string theory of the A-model on a resolved conifold with a B field equal to N times the string coupling constant. The open strings in the A model are described by a U(N) Chern–Simons theory, while the closed string theory on the A-model is described by the Kähler gravity.

Although the conifold is said to be resolved, the area of the blown up two-sphere is zero, it is only the B-field, which is often considered to be the complex part of the area, which is nonvanishing. In fact, as the Chern–Simons theory is topological, one may shrink the volume of the deformed three-sphere to zero and so arrive at the same geometry as in the dual theory.

The mirror dual of this duality is another duality, which relates open strings in the B model on a brane wrapping the 2-cycle in the resolved conifold to closed strings in the B model on the deformed conifold. Open strings in the B-model are described by dimensional reductions of homolomorphic Chern–Simons theory on the branes on which they end, while closed strings in the B model are described by Kodaira–Spencer gravity.

7.4.3 Dualities with other theories

Crystal melting, quantum foam and U(1) gauge theory

In the paper Quantum Calabi-Yau and Classical Crystals, Andrei Okounkov, Nicolai Reshetikhin and Cumrun Vafa conjectured that the quantum A-model is dual to a classical melting crystal at a temperature equal to the inverse of the string coupling constant. This conjecture was interpreted in Quantum Foam and Topological Strings, by Amer Iqbal, Nikita Nekrasov, Andrei Okounkov and Cumrun Vafa. They claim that the statistical sum over melting crystal configurations is equivalent to a path integral over changes in spacetime topology supported in small regions with area of order the product of the string coupling constant and α'.

Such configurations, with spacetime full of many small bubbles, dates back to John Archibald Wheeler in 1964, but has rarely appeared in string theory as it is notoriously difficult to make precise. However in this duality the authors are able to cast the dynamics of the quantum foam in the familiar language of a topologically twisted U(1) gauge theory, whose field strength is linearly related to the Kähler form of the A-model. In particular this suggests that the A-model Kähler form should be quantized.

7.5   Applications

A-model topological string theory amplitudes are used to compute prepotentials in N=2 supersymmetric gauge theories in four and five dimensions. The amplitudes of the topological B-model, with fluxes and or branes, are used to compute superpotentials in N=1 supersymmetric gauge theories in four dimensions. Perturbative A model calculations also count BPS states of spinning black holes in five dimensions.

7.6   See also

- Quantum topology
- Topological defect
- Topological entropy in physics
- Topological order
- Topological quantum field theory
- Topological quantum number
- Introduction to M-theory

7.7   References

- Topological Strings and their Physical Applications by Andrew Neitzke and Cumrun Vafa.

- Topological M-theory as Unification of Form Theories of Gravity by Robbert Dijkgraaf, Sergei Gukov, Andrew Neitzke and Cumrun Vafa.

- Topological string theory on arxiv.org

- Naqvi, Asad (2006). "Topological Strings" (PDF-MICROSOFT POWERPOINT). *Asad Naqvi - University of Wales, Swansea, United Kingdom.* National Center for Physics. Retrieved 2010.

Chapter 8

Quantum field theory

"Relativistic quantum field theory" redirects here. For other uses, see Relativity.

In theoretical physics, **quantum field theory** (QFT) is a theoretical framework for constructing quantum mechanical models of subatomic particles in particle physics and quasiparticles in condensed matter physics. A QFT treats particles as excited states of an underlying physical field, so these are called field quanta.

In quantum field theory, quantum mechanical interactions between particles are described by interaction terms between the corresponding underlying fields.

8.1 Definition

Quantum electrodynamics (QED) has one electron field and one photon field; quantum chromodynamics (QCD) has one field for each type of quark; and, in condensed matter, there is an atomic displacement field that gives rise to phonon particles. Edward Witten describes QFT as "by far" the most difficult theory in modern physics.[1]

8.1.1 Dynamics

See also: Relativistic dynamics

Ordinary quantum mechanical systems have a fixed number of particles, with each particle having a finite number of degrees of freedom. In contrast, the excited states of a QFT can represent any number of particles. This makes quantum field theories especially useful for describing systems where the particle count/number may change over time, a crucial feature of relativistic dynamics.

8.1.2 States

QFT interaction terms are similar in spirit to those between charges with electric and magnetic fields in Maxwell's equations. However, unlike the classical fields of Maxwell's theory, fields in QFT generally exist in quantum superpositions of states and are subject to the laws of quantum mechanics.

Because the fields are continuous quantities over space, there exist excited states with arbitrarily large numbers of particles in them, providing QFT systems with an effectively infinite number of degrees of freedom. Infinite degrees of freedom can easily lead to divergences of calculated quantities (i.e., the quantities become infinite). Techniques such as renormalization of QFT parameters or discretization of spacetime, as in lattice QCD, are often used to avoid such infinities so as to yield physically meaningful results.

8.1.3   Fields and radiation

The gravitational field and the electromagnetic field are the only two fundamental fields in nature that have infinite range and a corresponding classical low-energy limit, which greatly diminishes and hides their "particle-like" excitations. Albert Einstein in 1905, attributed "particle-like" and discrete exchanges of momenta and energy, characteristic of "field quanta", to the electromagnetic field. Originally, his principal motivation was to explain the thermodynamics of radiation. Although the photoelectric effect and Compton scattering strongly suggest the existence of the photon, it might alternately be explained by a mere quantization of emission; more definitive evidence of the quantum nature of radiation is now taken up into modern quantum optics as in the antibunching effect.[2]

8.2   Theories

There is currently no complete quantum theory of the remaining fundamental force, gravity. Many of the proposed theories to describe gravity as a QFT postulate the existence of a graviton particle that mediates the gravitational force. Presumably, the as yet unknown correct quantum field-theoretic treatment of the gravitational field will behave like Einstein's general theory of relativity in the low-energy limit. Quantum field theory of the fundamental forces itself has been postulated to be the low-energy effective field theory limit of a more fundamental theory such as superstring theory.

Most theories in standard particle physics are formulated as **relativistic quantum field theories**, such as QED, QCD, and the Standard Model. QED, the quantum field-theoretic description of the electromagnetic field, approximately reproduces Maxwell's theory of electrodynamics in the low-energy limit, with small non-linear corrections to the Maxwell equations required due to virtual electron–positron pairs.

In the perturbative approach to quantum field theory, the full field interaction terms are approximated as a perturbative expansion in the number of particles involved. Each term in the expansion can be thought of as forces between particles being mediated by other particles. In QED, the electromagnetic force between two electrons is caused by an exchange of photons. Similarly, intermediate vector bosons mediate the weak force and gluons mediate the strong force in QCD. The notion of a force-mediating particle comes from perturbation theory, and does not make sense in the context of non-perturbative approaches to QFT, such as with bound states.

8.3   History

Main article: History of quantum field theory

8.3.1   Foundations

The early development of the field involved Dirac, Fock, Pauli, Heisenberg and Bogolyubov. This phase of development culminated with the construction of the theory of quantum electrodynamics in the 1950s.

8.3.2   Gauge theory

Gauge theory was formulated and quantized, leading to the **unification of forces** embodied in the standard model of particle physics. This effort started in the 1950s with the work of Yang and Mills, was carried on by Martinus Veltman and a host of others during the 1960s and completed by the 1970s through the work of Gerard 't Hooft, Frank Wilczek, David Gross and David Politzer.

8.3.3 Grand synthesis

Parallel developments in the understanding of phase transitions in condensed matter physics led to the study of the renormalization group. This in turn led to the **grand synthesis** of theoretical physics, which unified theories of particle and condensed matter physics through quantum field theory. This involved the work of Michael Fisher and Leo Kadanoff in the 1970s, which led to the seminal reformulation of quantum field theory by Kenneth G. Wilson in 1975.

8.4 Principles

8.4.1 Classical and quantum fields

Main article: Classical field theory

A classical field is a function defined over some region of space and time.[3] Two physical phenomena which are described by classical fields are Newtonian gravitation, described by Newtonian gravitational field $\mathbf{g}(\mathbf{x}, t)$, and classical electromagnetism, described by the electric and magnetic fields $\mathbf{E}(\mathbf{x}, t)$ and $\mathbf{B}(\mathbf{x}, t)$. Because such fields can in principle take on distinct values at each point in space, they are said to have infinite degrees of freedom.[3]

Classical field theory does not, however, account for the quantum-mechanical aspects of such physical phenomena. For instance, it is known from quantum mechanics that certain aspects of electromagnetism involve discrete particles—photons—rather than continuous fields. The business of *quantum* field theory is to write down a field that is, like a classical field, a function defined over space and time, but which also accommodates the observations of quantum mechanics. This is a *quantum field*.

It is not immediately clear *how* to write down such a quantum field, since quantum mechanics has a structure very unlike a field theory. In its most general formulation, quantum mechanics is a theory of abstract operators (observables) acting on an abstract state space (Hilbert space), where the observables represent physically observable quantities and the state space represents the possible states of the system under study.[4] For instance, the fundamental observables associated with the motion of a single quantum mechanical particle are the position and momentum operators $\hat{x}$ and $\hat{p}$. Field theory, in contrast, treats x as a way to index the field rather than as an operator.[5]

There are two common ways of developing a quantum field: the path integral formalism and canonical quantization.[6] The latter of these is pursued in this article.

Lagrangian formalism

Quantum field theory frequently makes use of the Lagrangian formalism from classical field theory. This formalism is analogous to the Lagrangian formalism used in classical mechanics to solve for the motion of a particle under the influence of a field. In classical field theory, one writes down a Lagrangian density, $\mathcal{L}$, involving a field, $\varphi(\mathbf{x}, t)$, and possibly its first derivatives ($\partial\varphi/\partial t$ and $\nabla\varphi$), and then applies a field-theoretic form of the Euler–Lagrange equation. Writing coordinates $(t, \mathbf{x}) = (x^0, x^1, x^2, x^3) = x^\mu$, this form of the Euler–Lagrange equation is[3]

$$\frac{\partial}{\partial x^\mu}\left[\frac{\partial \mathcal{L}}{\partial(\partial\phi/\partial x^\mu)}\right] - \frac{\partial \mathcal{L}}{\partial\phi} = 0.$$

where a sum over μ is performed according to the rules of Einstein notation.

By solving this equation, one arrives at the "equations of motion" of the field.[3] For example, if one begins with the Lagrangian density

$$\mathcal{L}(\phi, \nabla\phi) = -\rho(t, \mathbf{x})\,\phi(t, \mathbf{x}) - \frac{1}{8\pi G}|\nabla\phi|^2,$$

and then applies the Euler–Lagrange equation, one obtains the equation of motion

$$4\pi G\rho(t, \mathbf{x}) = \nabla^2 \phi.$$

This equation is Newton's law of universal gravitation, expressed in differential form in terms of the gravitational potential $\varphi(t, \mathbf{x})$ and the mass density $\rho(t, \mathbf{x})$. Despite the nomenclature, the "field" under study is the gravitational potential, φ, rather than the gravitational field, $\mathbf{g}$. Similarly, when classical field theory is used to study electromagnetism, the "field" of interest is the electromagnetic four-potential $(V/c, \mathbf{A})$, rather than the electric and magnetic fields $\mathbf{E}$ and $\mathbf{B}$.

Quantum field theory uses this same Lagrangian procedure to determine the equations of motion for quantum fields. These equations of motion are then supplemented by commutation relations derived from the canonical quantization procedure described below, thereby incorporating quantum mechanical effects into the behavior of the field.

8.4.2   Single- and many-particle quantum mechanics

Main articles: Quantum mechanics and First quantization

In quantum mechanics, a particle (such as an electron or proton) is described by a complex wavefunction, $\psi(x, t)$, whose time-evolution is governed by the Schrödinger equation:

$$-\frac{\hbar^2}{2m}\frac{\partial^2}{\partial x^2}\psi(x, t) + V(x)\psi(x, t) = i\hbar\frac{\partial}{\partial t}\psi(x, t).$$

Here m is the particle's mass and $V(x)$ is the applied potential. Physical information about the behavior of the particle is extracted from the wavefunction by constructing expected values for various quantities; for example, the expected value of the particle's position is given by integrating $\psi^*(x)\, x\, \psi(x)$ over all space, and the expected value of the particle's momentum is found by integrating $-i\hbar\psi^*(x)\mathrm{d}\psi/\mathrm{d}x$. The quantity $\psi^*(x)\psi(x)$ is itself in the Copenhagen interpretation of quantum mechanics interpreted as a probability density function. This treatment of quantum mechanics, where a particle's wavefunction evolves against a classical background potential $V(x)$, is sometimes called *first quantization*.

This description of quantum mechanics can be extended to describe the behavior of multiple particles, so long as the number and the type of particles remain fixed. The particles are described by a wavefunction $\psi(x_1, x_2, \ldots, xN, t)$, which is governed by an extended version of the Schrödinger equation.

Often one is interested in the case where N particles are all of the same type (for example, the 18 electrons orbiting a neutral argon nucleus). As described in the article on identical particles, this implies that the state of the entire system must be either symmetric (bosons) or antisymmetric (fermions) when the coordinates of its constituent particles are exchanged. This is achieved by using a Slater determinant as the wavefunction of a fermionic system (and a Slater permanent for a bosonic system), which is equivalent to an element of the symmetric or antisymmetric subspace of a tensor product.

For example, the general quantum state of a system of N bosons is written as

$$|\phi_1 \cdots \phi_N\rangle = \sqrt{\frac{\prod_j N_j!}{N!}} \sum_{p \in S_N} |\phi_{p(1)}\rangle \otimes \cdots \otimes |\phi_{p(N)}\rangle,$$

where $|\phi_i\rangle$ are the single-particle states, Nj is the number of particles occupying state j, and the sum is taken over all possible permutations p acting on N elements. In general, this is a sum of $N!$ (N factorial) distinct terms. $\sqrt{\frac{\prod_j N_j!}{N!}}$ is a normalizing factor.

There are several shortcomings to the above description of quantum mechanics, which are addressed by quantum field theory. First, it is unclear how to extend quantum mechanics to include the effects of special relativity.[7] Attempted replacements for the Schrödinger equation, such as the Klein–Gordon equation or the Dirac equation, have many unsatisfactory qualities; for instance, they possess energy eigenvalues that extend to $-\infty$, so that there seems to be no easy definition of a ground state. It turns out that such inconsistencies arise from relativistic wavefunctions not having a well-defined probabilistic interpretation in position space, as probability conservation is not a relativistically covariant concept. The second shortcoming, related to the first, is that in quantum mechanics there is no mechanism to describe particle creation and annihilation;[8] this is crucial for describing phenomena such as pair production, which result from the conversion between mass and energy according to the relativistic relation $E = mc^2$.

8.4.3 Second quantization

Main article: Second quantization

In this section, we will describe a method for constructing a quantum field theory called **second quantization**. This basically involves choosing a way to index the quantum mechanical degrees of freedom in the space of multiple identical-particle states. It is based on the Hamiltonian formulation of quantum mechanics.

Several other approaches exist, such as the Feynman path integral,[9] which uses a Lagrangian formulation. For an overview of some of these approaches, see the article on quantization.

Bosons

For simplicity, we will first discuss second quantization for bosons, which form perfectly symmetric quantum states. Let us denote the mutually orthogonal single-particle states which are possible in the system by $|\phi_1\rangle, |\phi_2\rangle, |\phi_3\rangle$, and so on. For example, the 3-particle state with one particle in state $|\phi_1\rangle$ and two in state $|\phi_2\rangle$ is

$$\frac{1}{\sqrt{3}} \left[|\phi_1\rangle|\phi_2\rangle|\phi_2\rangle + |\phi_2\rangle|\phi_1\rangle|\phi_2\rangle + |\phi_2\rangle|\phi_2\rangle|\phi_1\rangle \right].$$

The first step in second quantization is to express such quantum states in terms of **occupation numbers**, by listing the number of particles occupying each of the single-particle states $|\phi_1\rangle, |\phi_2\rangle$, etc. This is simply another way of labelling the states. For instance, the above 3-particle state is denoted as

$$|1, 2, 0, 0, 0, \ldots\rangle.$$

An N-particle state belongs to a space of states describing systems of N particles. The next step is to combine the individual N-particle state spaces into an extended state space, known as Fock space, which can describe systems of any number of particles. This is composed of the state space of a system with no particles (the so-called vacuum state, written as $|0\rangle$), plus the state space of a 1-particle system, plus the state space of a 2-particle system, and so forth. States describing a definite number of particles are known as Fock states: a general element of Fock space will be a linear combination of Fock states. There is a one-to-one correspondence between the occupation number representation and valid boson states in the Fock space.

At this point, the quantum mechanical system has become a quantum field in the sense we described above. The field's elementary degrees of freedom are the occupation numbers, and each occupation number is indexed by a number j indicating which of the single-particle states $|\phi_1\rangle, |\phi_2\rangle, \ldots, |\phi_j\rangle, \ldots$ it refers to:

$$|N_1, N_2, N_3, \ldots, N_j, \ldots\rangle.$$

The properties of this quantum field can be explored by defining creation and annihilation operators, which add and subtract particles. They are analogous to ladder operators in the quantum harmonic oscillator problem, which added and subtracted energy quanta. However, these operators literally create and annihilate particles of a given quantum state. The bosonic annihilation operator a_2 and creation operator $a_2^\dagger$ are easily defined in the occupation number representation as having the following effects:

$$a_2 | N_1, N_2, N_3, \ldots \rangle = \sqrt{N_2} \; | \, N_1, (N_2 - 1), N_3, \ldots \rangle,$$

$$a_2^\dagger | N_1, N_2, N_3, \ldots \rangle = \sqrt{N_2 + 1} \; | \, N_1, (N_2 + 1), N_3, \ldots \rangle.$$

It can be shown that these are operators in the usual quantum mechanical sense, i.e. linear operators acting on the Fock space. Furthermore, they are indeed Hermitian conjugates, which justifies the way we have written them. They can be shown to obey the commutation relation

$$[a_i, a_j] = 0 \quad , \quad \left[a_i^\dagger, a_j^\dagger \right] = 0 \quad , \quad \left[a_i, a_j^\dagger \right] = \delta_{ij},$$

where δ stands for the Kronecker delta. These are precisely the relations obeyed by the ladder operators for an infinite set of independent quantum harmonic oscillators, one for each single-particle state. Adding or removing bosons from each state is therefore analogous to exciting or de-exciting a quantum of energy in a harmonic oscillator.

Applying an annihilation operator a_k followed by its corresponding creation operator $a_k^\dagger$ returns the number N_k of particles in the kth single-particle eigenstate:

$$a_k^\dagger a_k | \ldots, N_k, \ldots \rangle = N_k | \ldots, N_k, \ldots \rangle.$$

The combination of operators $a_k^\dagger a_k$ is known as the number operator for the kth eigenstate.

The Hamiltonian operator of the quantum field (which, through the Schrödinger equation, determines its dynamics) can be written in terms of creation and annihilation operators. For instance, for a field of free (non-interacting) bosons, the total energy of the field is found by summing the energies of the bosons in each energy eigenstate. If the kth single-particle energy eigenstate has energy E_k and there are N_k bosons in this state, then the total energy of these bosons is $E_k N_k$. The energy in the *entire* field is then a sum over k :

$$E_{\text{tot}} = \sum_k E_k N_k$$

This can be turned into the Hamiltonian operator of the field by replacing N_k with the corresponding number operator, $a_k^\dagger a_k$. This yields

$$H = \sum_k E_k \, a_k^\dagger a_k.$$

Fermions

It turns out that a different definition of creation and annihilation must be used for describing fermions. According to the Pauli exclusion principle, fermions cannot share quantum states, so their occupation numbers N_i can only take on the value 0 or 1. The fermionic annihilation operators c and creation operators $c^\dagger$ are defined by their actions on a Fock state thus

$$c_j | N_1, N_2, \ldots, N_j = 0, \ldots \rangle = 0$$

$$c_j | N_1, N_2, \ldots, N_j = 1, \ldots \rangle = (-1)^{(N_1 + \cdots + N_{j-1})} | N_1, N_2, \ldots, N_j = 0, \ldots \rangle$$

$$c_j^\dagger | N_1, N_2, \ldots, N_j = 0, \ldots \rangle = (-1)^{(N_1 + \cdots + N_{j-1})} | N_1, N_2, \ldots, N_j = 1, \ldots \rangle$$

$$c_j^\dagger | N_1, N_2, \ldots, N_j = 1, \ldots \rangle = 0.$$

These obey an anticommutation relation:

$$\{c_i, c_j\} = 0 \quad , \quad \left\{ c_i^\dagger, c_j^\dagger \right\} = 0 \quad , \quad \left\{ c_i, c_j^\dagger \right\} = \delta_{ij}.$$

One may notice from this that applying a fermionic creation operator twice gives zero, so it is impossible for the particles to share single-particle states, in accordance with the exclusion principle.

Field operators

We have previously mentioned that there can be more than one way of indexing the degrees of freedom in a quantum field. Second quantization indexes the field by enumerating the single-particle quantum states. However, as we have discussed, it is more natural to think about a "field", such as the electromagnetic field, as a set of degrees of freedom indexed by position.

To this end, we can define *field operators* that create or destroy a particle at a particular point in space. In particle physics, these operators turn out to be more convenient to work with, because they make it easier to formulate theories that satisfy the demands of relativity.

Single-particle states are usually enumerated in terms of their momenta (as in the particle in a box problem.) We can construct field operators by applying the Fourier transform to the creation and annihilation operators for these states. For example, the bosonic field annihilation operator $\phi(\mathbf{r})$ is

$$\phi(\mathbf{r}) \overset{\text{def}}{=} \sum_j e^{i\mathbf{k}_j \cdot \mathbf{r}} a_j.$$

The bosonic field operators obey the commutation relation

$$[\phi(\mathbf{r}), \phi(\mathbf{r}')] = 0 \quad , \quad [\phi^\dagger(\mathbf{r}), \phi^\dagger(\mathbf{r}')] = 0 \quad , \quad [\phi(\mathbf{r}), \phi^\dagger(\mathbf{r}')] = \delta^3(\mathbf{r} - \mathbf{r}')$$

where $\delta(x)$ stands for the Dirac delta function. As before, the fermionic relations are the same, with the commutators replaced by anticommutators.

The field operator is not the same thing as a single-particle wavefunction. The former is an operator acting on the Fock space, and the latter is a quantum-mechanical amplitude for finding a particle in some position. However, they are closely related, and are indeed commonly denoted with the same symbol. If we have a Hamiltonian with a space representation, say

$$H = -\frac{\hbar^2}{2m} \sum_i \nabla_i^2 + \sum_{i<j} U(|\mathbf{r}_i - \mathbf{r}_j|)$$

where the indices i and j run over all particles, then the field theory Hamiltonian (in the non-relativistic limit and for negligible self-interactions) is

$$H = -\frac{\hbar^2}{2m} \int d^3r \, \phi^\dagger(\mathbf{r}) \nabla^2 \phi(\mathbf{r}) + \frac{1}{2} \int d^3r \int d^3r' \, \phi^\dagger(\mathbf{r}) \phi^\dagger(\mathbf{r}') U(|\mathbf{r} - \mathbf{r}'|) \phi(\mathbf{r}') \phi(\mathbf{r}).$$

This looks remarkably like an expression for the expectation value of the energy, with ϕ playing the role of the wavefunction. This relationship between the field operators and wavefunctions makes it very easy to formulate field theories starting from space-projected Hamiltonians.

8.4.4   Dynamics

Once the Hamiltonian operator is obtained as part of the canonical quantization process, the time dependence of the state is described with the Schrödinger equation, just as with other quantum theories. Alternatively, the Heisenberg picture can be used where the time dependence is in the operators rather than in the states.

8.4.5   Implications

Unification of fields and particles

The "second quantization" procedure that we have outlined in the previous section takes a set of single-particle quantum states as a starting point. Sometimes, it is impossible to define such single-particle states, and one must proceed directly to quantum field theory. For example, a quantum theory of the electromagnetic field *must* be a quantum field theory, because it is impossible (for various reasons) to define a wavefunction for a single photon.[10] In such situations, the quantum field theory can be constructed by examining the mechanical properties of the classical field and guessing the corresponding quantum theory. For free (non-interacting) quantum fields, the quantum field theories obtained in this way have the same properties as those obtained using second quantization, such as well-defined creation and annihilation operators obeying commutation or anticommutation relations.

Quantum field theory thus provides a unified framework for describing "field-like" objects (such as the electromagnetic field, whose excitations are photons) and "particle-like" objects (such as electrons, which are treated as excitations of an underlying electron field), so long as one can treat interactions as "perturbations" of free fields. There are still unsolved problems relating to the more general case of interacting fields that may or may not be adequately described by perturbation theory. For more on this topic, see Haag's theorem.

Physical meaning of particle indistinguishability

The second quantization procedure relies crucially on the particles being identical. We would not have been able to construct a quantum field theory from a distinguishable many-particle system, because there would have been no way of separating and indexing the degrees of freedom.

Many physicists prefer to take the converse interpretation, which is that *quantum field theory explains what identical particles are*. In ordinary quantum mechanics, there is not much theoretical motivation for using symmetric (bosonic) or antisymmetric (fermionic) states, and the need for such states is simply regarded as an empirical fact. From the point of view of quantum field theory, particles are identical if and only if they are excitations of the same underlying quantum field. Thus, the question "why are all electrons identical?" arises from mistakenly regarding individual electrons as fundamental objects, when in fact it is only the electron field that is fundamental.

Particle conservation and non-conservation

During second quantization, we started with a Hamiltonian and state space describing a fixed number of particles (N), and ended with a Hamiltonian and state space for an arbitrary number of particles. Of course, in many common situations N is an important and perfectly well-defined quantity, e.g. if we are describing a gas of atoms sealed in a box. From the point of view of quantum field theory, such situations are described by quantum states that are eigenstates of the number operator $\hat{N}$, which measures the total number of particles present. As with any quantum mechanical observable, $\hat{N}$ is conserved if it commutes with the Hamiltonian. In that case, the quantum state is trapped in the N-particle subspace of the total Fock space, and the situation could equally well be described by ordinary N-particle quantum mechanics. (Strictly speaking, this is only true in the noninteracting case or in the low energy density limit of renormalized quantum field theories)

For example, we can see that the free-boson Hamiltonian described above conserves particle number. Whenever the Hamiltonian operates on a state, each particle destroyed by an annihilation operator a_k is immediately put back by the creation operator $a_k^\dagger$.

On the other hand, it is possible, and indeed common, to encounter quantum states that are *not* eigenstates of $\hat{N}$, which do not have well-defined particle numbers. Such states are difficult or impossible to handle using ordinary quantum mechanics, but they can be easily described in quantum field theory as quantum superpositions of states having different values of N. For example, suppose we have a bosonic field whose particles can be created or destroyed by interactions with a fermionic field. The Hamiltonian of the combined system would be given by the Hamiltonians of the free boson and free fermion fields, plus a "potential energy" term such as

$$H_I = \sum_{k,q} V_q(a_q + a_{-q}^\dagger)c_{k+q}^\dagger c_k.$$

where $a_k^\dagger$ and a_k denotes the bosonic creation and annihilation operators, $c_k^\dagger$ and c_k denotes the fermionic creation and annihilation operators, and V_q is a parameter that describes the strength of the interaction. This "interaction term" describes processes in which a fermion in state k either absorbs or emits a boson, thereby being kicked into a different eigenstate $k + q$. (In fact, this type of Hamiltonian is used to describe interaction between conduction electrons and phonons in metals. The interaction between electrons and photons is treated in a similar way, but is a little more complicated because the role of spin must be taken into account.) One thing to notice here is that even if we start out with a fixed number of bosons, we will typically end up with a superposition of states with different numbers of bosons at later times. The number of fermions, however, is conserved in this case.

In condensed matter physics, states with ill-defined particle numbers are particularly important for describing the various superfluids. Many of the defining characteristics of a superfluid arise from the notion that its quantum state is a superposition of states with different particle numbers. In addition, the concept of a coherent state (used to model the laser and the BCS ground state) refers to a state with an ill-defined particle number but a well-defined phase.

8.4.6 Axiomatic approaches

The preceding description of quantum field theory follows the spirit in which most physicists approach the subject. However, it is not mathematically rigorous. Over the past several decades, there have been many attempts to put quantum field theory on a firm mathematical footing by formulating a set of axioms for it. These attempts fall into two broad classes.

The first class of axioms, first proposed during the 1950s, include the Wightman, Osterwalder–Schrader, and Haag–Kastler systems. They attempted to formalize the physicists' notion of an "operator-valued field" within the context of functional analysis, and enjoyed limited success. It was possible to prove that any quantum field theory satisfying these axioms satisfied certain general theorems, such as the spin-statistics theorem and the CPT theorem. Unfortunately, it proved extraordinarily difficult to show that any realistic field theory, including the Standard Model, satisfied these axioms. Most of the theories that could be treated with these analytic axioms were physically trivial, being restricted to low-dimensions and lacking interesting dynamics. The construction of theories satisfying one of these sets of axioms falls in the field of constructive quantum field theory. Important work was done in this area in the 1970s by Segal, Glimm, Jaffe and others.

During the 1980s, a second set of axioms based on geometric ideas was proposed. This line of investigation, which restricts its attention to a particular class of quantum field theories known as topological quantum field theories, is associated most closely with Michael Atiyah and Graeme Segal, and was notably expanded upon by Edward Witten, Richard Borcherds, and Maxim Kontsevich. However, most of the physically relevant quantum field theories, such as the Standard Model, are not topological quantum field theories; the quantum field theory of the fractional quantum Hall effect is a notable exception. The main impact of axiomatic topological quantum field theory has been on mathematics, with important applications in representation theory, algebraic topology, and differential geometry.

Finding the proper axioms for quantum field theory is still an open and difficult problem in mathematics. One of the Millennium Prize Problems—proving the existence of a mass gap in Yang–Mills theory—is linked to this issue.

8.5 Associated phenomena

In the previous part of the article, we described the most general features of quantum field theories. Some of the quantum field theories studied in various fields of theoretical physics involve additional special ideas, such as renormalizability, gauge symmetry, and supersymmetry. These are described in the following sections.

8.5.1 Renormalization

Main article: Renormalization

Early in the history of quantum field theory, it was found that many seemingly innocuous calculations, such as the perturbative shift in the energy of an electron due to the presence of the electromagnetic field, give infinite results. The reason is that the perturbation theory for the shift in an energy involves a sum over all other energy levels, and there are infinitely many levels at short distances that each give a finite contribution which results in a divergent series.

Many of these problems are related to failures in classical electrodynamics that were identified but unsolved in the 19th century, and they basically stem from the fact that many of the supposedly "intrinsic" properties of an electron are tied to the electromagnetic field that it carries around with it. The energy carried by a single electron—its self energy—is not simply the bare value, but also includes the energy contained in its electromagnetic field, its attendant cloud of photons. The energy in a field of a spherical source diverges in both classical and quantum mechanics, but as discovered by Weisskopf with help from Furry, in quantum mechanics the divergence is much milder, going only as the logarithm of the radius of the sphere.

The solution to the problem, presciently suggested by Stueckelberg, independently by Bethe after the crucial experiment by Lamb, implemented at one loop by Schwinger, and systematically extended to all loops by Feynman and Dyson, with converging work by Tomonaga in isolated postwar Japan, comes from recognizing that all the infinities in the interactions of photons and electrons can be isolated into redefining a finite number of quantities in the equations by replacing them with the observed values: specifically the electron's mass and charge: this is called renormalization. The technique of renormalization recognizes that the problem is essentially purely mathematical, that extremely short distances are at fault. In order to define a theory on a continuum, first place a cutoff on the fields, by postulating that quanta cannot have energies above some extremely high value. This has the effect of replacing continuous space by a structure where very short wavelengths do not exist, as on a lattice. Lattices break rotational symmetry, and one of the crucial contributions made by Feynman, Pauli and Villars, and modernized by 't Hooft and Veltman, is a symmetry-preserving cutoff for perturbation theory (this process is called regularization). There is no known symmetrical cutoff outside of perturbation theory, so for rigorous or numerical work people often use an actual lattice.

On a lattice, every quantity is finite but depends on the spacing. When taking the limit of zero spacing, we make sure that the physically observable quantities like the observed electron mass stay fixed, which means that the constants in the Lagrangian defining the theory depend on the spacing. Hopefully, by allowing the constants to vary with the lattice spacing, all the results at long distances become insensitive to the lattice, defining a continuum limit.

The renormalization procedure only works for a certain class of quantum field theories, called **renormalizable quantum field theories**. A theory is **perturbatively renormalizable** when the constants in the Lagrangian only diverge at worst as logarithms of the lattice spacing for very short spacings. The continuum limit is then well defined in perturbation theory, and even if it is not fully well defined non-perturbatively, the problems only show up at distance scales that are exponentially small in the inverse coupling for weak couplings. The Standard Model of particle physics is perturbatively renormalizable, and so are its component theories (quantum electrodynamics/electroweak theory and quantum chromodynamics). Of the three components, quantum electrodynamics is believed to not have a continuum limit, while the asymptotically free $SU(2)$ and $SU(3)$ weak hypercharge and strong color interactions are nonperturbatively well defined.

The renormalization group describes how renormalizable theories emerge as the long distance low-energy effective field theory for any given high-energy theory. Because of this, renormalizable theories are insensitive to the precise nature of the underlying high-energy short-distance phenomena. This is a blessing because it allows physicists to formulate low energy theories without knowing the details of high energy phenomenon. It is also a curse, because once a renormalizable theory like the standard model is found to work, it gives very few clues to higher energy processes. The only way high energy processes can be seen in the standard model is when they allow otherwise forbidden events, or if they predict quantitative relations between the coupling constants.

8.5.2 Haag's theorem

See also: Haag's theorem

From a mathematically rigorous perspective, there exists no interaction picture in a Lorentz-covariant quantum field theory. This implies that the perturbative approach of Feynman diagrams in QFT is not strictly justified, despite producing vastly precise predictions validated by experiment. This is called Haag's theorem, but most particle physicists relying on QFT largely shrug it off.

8.5.3 Gauge freedom

A gauge theory is a theory that admits a symmetry with a local parameter. For example, in every quantum theory the global phase of the wave function is arbitrary and does not represent something physical. Consequently, the theory is invariant under a global change of phases (adding a constant to the phase of all wave functions, everywhere); this is a global symmetry. In quantum electrodynamics, the theory is also invariant under a *local* change of phase, that is — one may shift the phase of all wave functions so that the shift may be different at every point in space-time. This is a *local* symmetry. However, in order for a well-defined derivative operator to exist, one must introduce a new field, the gauge field, which also transforms in order for the local change of variables (the phase in our example) not to affect the derivative. In quantum electrodynamics this gauge field is the electromagnetic field. The change of local gauge of variables is termed gauge transformation. It is worth noting that by Noether's theorem, for every such symmetry there exists an associated conserved current. The aforementioned symmetry of the wavefunction under global phase changes implies the conservation of electric charge.

In quantum field theory the excitations of fields represent particles. The particle associated with excitations of the gauge field is the gauge boson, which is the photon in the case of quantum electrodynamics.

The degrees of freedom in quantum field theory are local fluctuations of the fields. The existence of a gauge symmetry reduces the number of degrees of freedom, simply because some fluctuations of the fields can be transformed to zero by gauge transformations, so they are equivalent to having no fluctuations at all, and they therefore have no physical meaning. Such fluctuations are usually called "non-physical degrees of freedom" or *gauge artifacts*; usually some of them have a negative norm, making them inadequate for a consistent theory. Therefore, if a classical field theory has a gauge symmetry, then its quantized version (i.e. the corresponding quantum field theory) will have this symmetry as well. In other words, a gauge symmetry cannot have a quantum anomaly. If a gauge symmetry is anomalous (i.e. not kept in the quantum theory) then the theory is non-consistent: for example, in quantum electrodynamics, had there been a gauge anomaly, this would require the appearance of photons with longitudinal polarization and polarization in the time direction, the latter having a negative norm, rendering the theory inconsistent; another possibility would be for these photons to appear only in intermediate processes but not in the final products of any interaction, making the theory non-unitary and again inconsistent (see optical theorem).

In general, the gauge transformations of a theory consist of several different transformations, which may not be commutative. These transformations are together described by a mathematical object known as a gauge group. Infinitesimal gauge transformations are the gauge group generators. Therefore, the number of gauge bosons is the group dimension (i.e. number of generators forming a basis).

All the fundamental interactions in nature are described by gauge theories. These are:

- Quantum chromodynamics, whose gauge group is $SU(3)$. The gauge bosons are eight gluons.

- The electroweak theory, whose gauge group is $U(1) \times SU(2)$, (a direct product of $U(1)$ and $SU(2)$).

- Gravity, whose classical theory is general relativity, admits the equivalence principle, which is a form of gauge symmetry. However, it is explicitly non-renormalizable.

8.5.4   Multivalued gauge transformations

The gauge transformations which leave the theory invariant involve, by definition, only single-valued gauge functions $\Lambda(x_i)$ which satisfy the Schwarz integrability criterion

$$\partial_{x_i x_j}\Lambda = \partial_{x_j x_i}\Lambda.$$

An interesting extension of gauge transformations arises if the gauge functions $\Lambda(x_i)$ are allowed to be multivalued functions which violate the integrability criterion. These are capable of changing the physical field strengths and are therefore not proper symmetry transformations. Nevertheless, the transformed field equations describe correctly the physical laws in the presence of the newly generated field strengths. See the textbook by H. Kleinert cited below for the applications to phenomena in physics.

8.5.5   Supersymmetry

Main article: Supersymmetry

Supersymmetry assumes that every fundamental fermion has a superpartner that is a boson and vice versa. It was introduced in order to solve the so-called Hierarchy Problem, that is, to explain why particles not protected by any symmetry (like the Higgs boson) do not receive radiative corrections to its mass driving it to the larger scales (GUT, Planck...). It was soon realized that supersymmetry has other interesting properties: its gauged version is an extension of general relativity (Supergravity), and it is a key ingredient for the consistency of string theory.

The way supersymmetry protects the hierarchies is the following: since for every particle there is a superpartner with the same mass, any loop in a radiative correction is cancelled by the loop corresponding to its superpartner, rendering the theory UV finite.

Since no superpartners have yet been observed, if supersymmetry exists it must be broken (through a so-called soft term, which breaks supersymmetry without ruining its helpful features). The simplest models of this breaking require that the energy of the superpartners not be too high; in these cases, supersymmetry is expected to be observed by experiments at the Large Hadron Collider. The Higgs particle has been detected at the LHC, and no such superparticles have been discovered.

8.6   See also

- Abraham–Lorentz force

- Basic concepts of quantum mechanics

- Common integrals in quantum field theory

- Einstein–Maxwell–Dirac equations

- Form factor (quantum field theory)

- Green–Kubo relations

- Green's function (many-body theory)

- Invariance mechanics

- List of quantum field theories

- Quantum electrodynamics

- Quantum field theory in curved spacetime

- Quantum flavordynamics

- Quantum hydrodynamics

- Quantum triviality

- Relation between Schrödinger's equation and the path integral formulation of quantum mechanics

- Relationship between string theory and quantum field theory

- Schwinger–Dyson equation

- Static forces and virtual-particle exchange

- Symmetry in quantum mechanics

- Theoretical and experimental justification for the Schrödinger equation

- Ward–Takahashi identity

- Wheeler–Feynman absorber theory

- Wigner's classification

- Wigner's theorem

8.7 Notes

8.8 References

[1] "Beautiful Minds, Vol. 20: Ed Witten". la Repubblica. 2010. Retrieved 22 June 2012. See here.

[2] J. J. Thorn et al. (2004). Observing the quantum behavior of light in an undergraduate laboratory. . J. J. Thorn, M. S. Neel, V. W. Donato, G. S. Bergreen, R. E. Davies, and M. Beck. American Association of Physics Teachers, 2004.DOI: 10.1119/1.1737397.

[3] David Tong, *Lectures on Quantum Field Theory*, chapter 1.

[4] Srednicki, Mark. *Quantum Field Theory* (1st ed.). p. 19.

[5] Srednicki, Mark. *Quantum Field Theory* (1st ed.). pp. 25–6.

[6] Zee, Anthony. *Quantum Field Theory in a Nutshell* (2nd ed.). p. 61.

[7] David Tong, *Lectures on Quantum Field Theory*, Introduction.

[8] Zee, Anthony. *Quantum Field Theory in a Nutshell* (2nd ed.). p. 3.

[9] Abraham Pais, *Inward Bound: Of Matter and Forces in the Physical World* ISBN 0-19-851997-4. Pais recounts how his astonishment at the rapidity with which Feynman could calculate using his method. Feynman's method is now part of the standard methods for physicists.

[10] Newton, T.D.; Wigner, E.P. (1949). "Localized states for elementary systems". *Reviews of Modern Physics* **21** (3): 400–406. Bibcode:1949RvMP...21..400N. doi:10.1103/RevModPhys.21.400.

8.9 Further reading

General readers

- Feynman, R.P. (2001) [1964]. *The Character of Physical Law*. MIT Press. ISBN 0-262-56003-8.

- Feynman, R.P. (2006) [1985]. *QED: The Strange Theory of Light and Matter*. Princeton University Press. ISBN 0-691-12575-9.

- Gribbin, J. (1998). *Q is for Quantum: Particle Physics from A to Z*. Weidenfeld & Nicolson. ISBN 0-297-81752-3.

- Schumm, Bruce A. (2004) *Deep Down Things*. Johns Hopkins Univ. Press. Chpt. 4.

Introductory texts

- McMahon, D. (2008). *Quantum Field Theory*. McGraw-Hill. ISBN 978-0-07-154382-8.

- Bogoliubov, N.; Shirkov, D. (1982). *Quantum Fields*. Benjamin-Cummings. ISBN 0-8053-0983-7.

- Frampton, P.H. (2000). *Gauge Field Theories. Frontiers in Physics (2nd ed.)*. Wiley.

- Greiner, W; Müller, B. (2000). *Gauge Theory of Weak Interactions*. Springer. ISBN 3-540-67672-4.

- Itzykson, C.; Zuber, J.-B. (1980). *Quantum Field Theory*. McGraw-Hill. ISBN 0-07-032071-3.

- Kane, G.L. (1987). *Modern Elementary Particle Physics*. Perseus Books. ISBN 0-201-11749-5.

- Kleinert, H.; Schulte-Frohlinde, Verena (2001). *Critical Properties of φ^4-Theories*. World Scientific. ISBN 981-02-4658-7.

- Kleinert, H. (2008). *Multivalued Fields in Condensed Matter, Electrodynamics, and Gravitation* (PDF). World Scientific. ISBN 978-981-279-170-2.

- Loudon, R (1983). *The Quantum Theory of Light*. Oxford University Press. ISBN 0-19-851155-8.

- Mandl, F.; Shaw, G. (1993). *Quantum Field Theory*. John Wiley & Sons. ISBN 978-0-471-94186-6.

- Peskin, M.; Schroeder, D. (1995). *An Introduction to Quantum Field Theory*. Westview Press. ISBN 0-201-50397-2.

- Ryder, L.H. (1985). *Quantum Field Theory*. Cambridge University Press. ISBN 0-521-33859-X.

- Schwartz, M.D. (2014). *Quantum Field Theory and the Standard Model*. Cambridge University Press. ISBN 978-1107034730.

- Srednicki, Mark (2007) *Quantum Field Theory*. Cambridge Univ. Press.

- Ynduráin, F.J. (1996). *Relativistic Quantum Mechanics and Introduction to Field Theory* (1st ed.). Springer. ISBN 978-3-540-60453-2.

- Zee, A. (2003). *Quantum Field Theory in a Nutshell*. Princeton University Press. ISBN 0-691-01019-6.

Advanced texts

- Brown, Lowell S. (1994). *Quantum Field Theory*. Cambridge University Press. ISBN 978-0-521-46946-3.

- Bogoliubov, N.; Logunov, A.A.; Oksak, A.I.; Todorov, I.T. (1990). *General Principles of Quantum Field Theory*. Kluwer Academic Publishers. ISBN 978-0-7923-0540-8.

- Weinberg, S. (1995). *The Quantum Theory of Fields* **1–3**. Cambridge University Press.

Articles:

- Gerard 't Hooft (2007) "The Conceptual Basis of Quantum Field Theory" in Butterfield, J., and John Earman, eds., *Philosophy of Physics, Part A*. Elsevier: 661–730.

- Frank Wilczek (1999) "Quantum field theory", *Reviews of Modern Physics* 71: S83–S95. Also doi=10.1103/Rev. Mod. Phys. 71.

8.10 External links

- Hazewinkel, Michiel, ed. (2001), "Quantum field theory", *Encyclopedia of Mathematics*, Springer, ISBN 978-1-55608-010-4

- Stanford Encyclopedia of Philosophy: "Quantum Field Theory", by Meinard Kuhlmann.

- Siegel, Warren, 2005. *Fields*. A free text, also available from arXiv:hep-th/9912205.

- Quantum Field Theory by P. J. Mulders

Chapter 9

Quantum gravity

Quantum gravity (**QG**) is a field of theoretical physics that seeks to describe the force of gravity according to the principles of quantum mechanics.

The current understanding of gravity is based on Albert Einstein's general theory of relativity, which is formulated within the framework of classical physics. On the other hand, the nongravitational forces are described within the framework of quantum mechanics, a radically different formalism for describing physical phenomena based on probability.[1] The necessity of a quantum mechanical description of gravity follows from the fact that one cannot consistently couple a classical system to a quantum one.[2]

Although a quantum theory of gravity is needed in order to reconcile general relativity with the principles of quantum mechanics, difficulties arise when one attempts to apply the usual prescriptions of quantum field theory to the force of gravity.[3] From a technical point of view, the problem is that the theory one gets in this way is not renormalizable and therefore cannot be used to make meaningful physical predictions. As a result, theorists have taken up more radical approaches to the problem of quantum gravity, the most popular approaches being string theory and loop quantum gravity.[4] A recent development is the theory of causal fermion systems which gives quantum mechanics, general relativity, and quantum field theory as limiting cases.[5][6][7][8][9][10]

Strictly speaking, the aim of quantum gravity is only to describe the quantum behavior of the gravitational field and should not be confused with the objective of unifying all fundamental interactions into a single mathematical framework. While any substantial improvement into the present understanding of gravity would aid further work towards unification, study of quantum gravity is a field in its own right with various branches having different approaches to unification. Although some quantum gravity theories, such as string theory, try to unify gravity with the other fundamental forces, others, such as loop quantum gravity, make no such attempt; instead, they make an effort to quantize the gravitational field while it is kept separate from the other forces. A theory of quantum gravity that is also a grand unification of all known interactions is sometimes referred to as a theory of everything (TOE).

One of the difficulties of quantum gravity is that quantum gravitational effects are only expected to become apparent near the Planck scale, a scale far smaller in distance (equivalently, far larger in energy) than what is currently accessible at high energy particle accelerators. As a result, quantum gravity is a mainly theoretical enterprise, although there are speculations about how quantum gravity effects might be observed in existing experiments.[11]

9.1   Overview

Much of the difficulty in meshing these theories at all energy scales comes from the different assumptions that these theories make on how the universe works. Quantum field theory depends on particle fields embedded in the flat space-time of special relativity. General relativity models gravity as a curvature within space-time that changes as a gravitational mass moves. Historically, the most obvious way of combining the two (such as treating gravity as simply another particle field) ran quickly into what is known as the renormalization problem. In the old-fashioned understanding of renormalization, gravity particles would attract each other and adding together all of the interactions results in many infinite values which

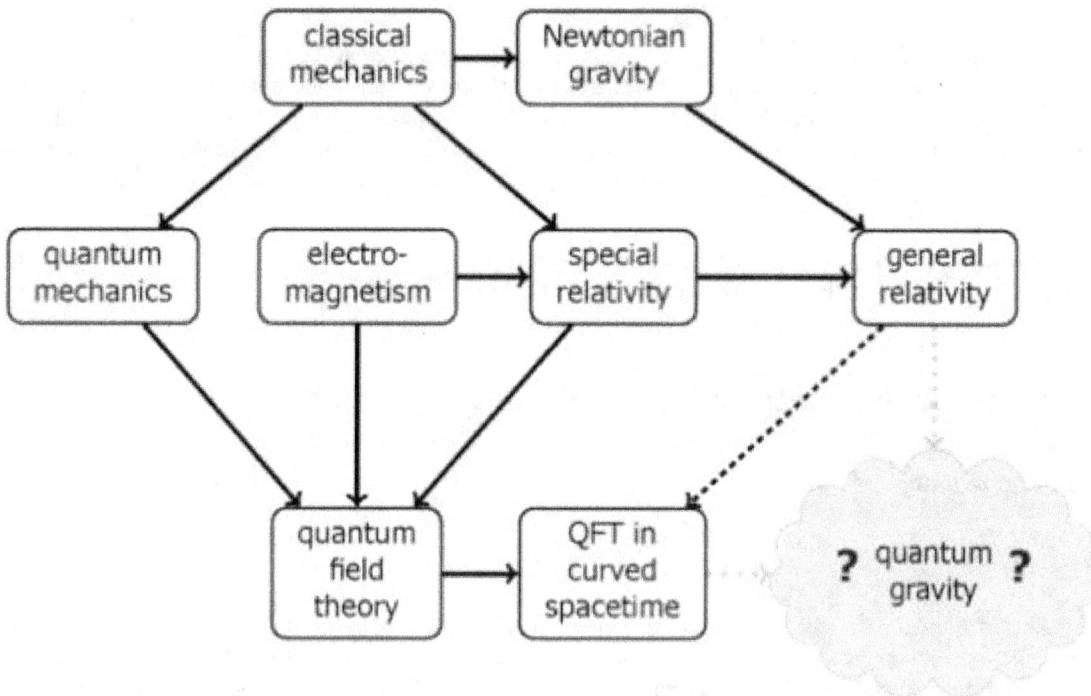

Diagram showing where quantum gravity sits in the hierarchy of physics theories

cannot easily be cancelled out mathematically to yield sensible, finite results. This is in contrast with quantum electrodynamics where, given that the series still do not converge, the interactions sometimes evaluate to infinite results, but those are few enough in number to be removable via renormalization.

9.1.1 Effective field theories

Quantum gravity can be treated as an effective field theory. Effective quantum field theories come with some high-energy cutoff, beyond which we do not expect that the theory provides a good description of nature. The "infinities" then become large but finite quantities depending on this finite cutoff scale, and correspond to processes that involve very high energies near the fundamental cutoff. These quantities can then be absorbed into an infinite collection of coupling constants, and at energies well below the fundamental cutoff of the theory, to any desired precision; only a finite number of these coupling constants need to be measured in order to make legitimate quantum-mechanical predictions. This same logic works just as well for the highly successful theory of low-energy pions as for quantum gravity. Indeed, the first quantum-mechanical corrections to graviton-scattering and Newton's law of gravitation have been explicitly computed[12] (although they are so astronomically small that we may never be able to measure them). In fact, gravity is in many ways a much better quantum field theory than the Standard Model, since it appears to be valid all the way up to its cutoff at the Planck scale.

While confirming that quantum mechanics and gravity are indeed consistent at reasonable energies, it is clear that near or above the fundamental cutoff of our effective quantum theory of gravity (the cutoff is generally assumed to be of the order of the Planck scale), a new model of nature will be needed. Specifically, the problem of combining quantum mechanics and gravity becomes an issue only at very high energies, and may well require a totally new kind of model.

9.1.2 Quantum gravity theory for the highest energy scales

The general approach to deriving a quantum gravity theory that is valid at even the highest energy scales is to assume that such a theory will be simple and elegant and, accordingly, to study symmetries and other clues offered by current theories that might suggest ways to combine them into a comprehensive, unified theory. One problem with this approach is that it is unknown whether quantum gravity will actually conform to a simple and elegant theory, as it should resolve the dual conundrums of special relativity with regard to the uniformity of acceleration and gravity, and general relativity with regard to spacetime curvature.

Such a theory is required in order to understand problems involving the combination of very high energy and very small dimensions of space, such as the behavior of black holes, and the origin of the universe.

9.2 Quantum mechanics and general relativity

9.2.1 The graviton

Main article: Graviton

At present, one of the deepest problems in theoretical physics is harmonizing the theory of general relativity, which describes gravitation, and applications to large-scale structures (stars, planets, galaxies), with quantum mechanics, which describes the other three fundamental forces acting on the atomic scale. This problem must be put in the proper context, however. In particular, contrary to the popular claim that quantum mechanics and general relativity are fundamentally incompatible, one can demonstrate that the structure of general relativity essentially follows inevitably from the quantum mechanics of interacting theoretical spin-2 massless particles (called gravitons).[13][14][15][16][17]

While there is no concrete proof of the existence of gravitons, quantized theories of matter may necessitate their existence. Supporting this theory is the observation that all fundamental forces except gravity have one or more known messenger particles, leading researchers to believe that at least one most likely does exist; they have dubbed this hypothetical particle the *graviton*. The predicted find would result in the classification of the graviton as a "force particle" similar to the photon of the electromagnetic field. Many of the accepted notions of a unified theory of physics since the 1970s assume, and to some degree depend upon, the existence of the graviton. These include string theory, superstring theory, M-theory, and loop quantum gravity. Detection of gravitons is thus vital to the validation of various lines of research to unify quantum mechanics and relativity theory.

9.2.2 The dilaton

Main article: Dilaton

The dilaton made its first appearance in Kaluza–Klein theory, a five-dimensional theory that combined gravitation and electromagnetism. Generally, it appears in string theory. More recently, it has appeared in the lower-dimensional many-bodied gravity problem[18] based on the field theoretic approach of Roman Jackiw. The impetus arose from the fact that complete analytical solutions for the metric of a covariant N-body system have proven elusive in general relativity. To simplify the problem, the number of dimensions was lowered to $(1+1)$, i.e. one spatial dimension and one temporal dimension. This model problem, known as $R=T$ theory[19] (as opposed to the general $G=T$ theory) was amenable to exact solutions in terms of a generalization of the Lambert W function. It was also found that the field equation governing the dilaton (derived from differential geometry) was the Schrödinger equation and consequently amenable to quantization.[20]

Thus, one had a theory which combined gravity, quantization and even the electromagnetic interaction, promising ingredients of a fundamental physical theory. It is worth noting that the outcome revealed a previously unknown and already existing *natural link* between general relativity and quantum mechanics. However, this theory needs to be generalized in $(2+1)$ or $(3+1)$ dimensions although, in principle, the field equations are amenable to such generalization as shown with the inclusion of a one-graviton process[21] and yielding the correct Newtonian limit in d dimensions if a dilaton is included. However, it is not yet clear what the fully generalized field equation governing the dilaton in (3+1) dimensions should be. This is further complicated by the fact that gravitons can propagate in *(3+1)* dimensions and consequently that would imply gravitons and dilatons exist in the real world. Moreover, detection of the dilaton is expected to be even more elusive than the graviton. However, since this approach allows for the combination of gravitational, electromagnetic and quantum effects, their coupling could potentially lead to a means of vindicating the theory, through cosmology and perhaps even *experimentally*.

9.2.3 Nonrenormalizability of gravity

Further information: Renormalization

General relativity, like electromagnetism, is a classical field theory. One might expect that, as with electromagnetism, there should be a corresponding quantum field theory.

However, gravity is perturbatively nonrenormalizable.[22][23] For a quantum field theory to be well-defined according to this understanding of the subject, it must be asymptotically free or asymptotically safe. The theory must be characterized by a choice of *finitely many* parameters, which could, in principle, be set by experiment. For example, in quantum electrodynamics, these parameters are the charge and mass of the electron, as measured at a particular energy scale.

On the other hand, in quantizing gravity, there are *infinitely many independent parameters* (counterterm coefficients) needed to define the theory. For a given choice of those parameters, one could make sense of the theory, but since we can never do infinitely many experiments to fix the values of every parameter, we do not have a meaningful physical theory:

- At low energies, the logic of the renormalization group tells us that, despite the unknown choices of these infinitely many parameters, quantum gravity will reduce to the usual Einstein theory of general relativity.

- On the other hand, if we could probe very high energies where quantum effects take over, then *every one* of the infinitely many unknown parameters would begin to matter, and we could make no predictions at all.

As explained below, there is a way around this problem by treating QG as an effective field theory.

Any meaningful theory of quantum gravity that makes sense and is predictive at all energy scales must have some deep principle that reduces the infinitely many unknown parameters to a finite number that can then be measured.

- One possibility is that normal perturbation theory is not a reliable guide to the renormalizability of the theory, and that there really *is* a UV fixed point for gravity. Since this is a question of non-perturbative quantum field theory, it is difficult to find a reliable answer, but some people still pursue this option.

- Another possibility is that there are new symmetry principles that constrain the parameters and reduce them to a finite set. This is the route taken by string theory, where all of the excitations of the string essentially manifest themselves as new symmetries.

9.2.4 QG as an effective field theory

Main article: Effective field theory

In an effective field theory, all but the first few of the infinite set of parameters in a non-renormalizable theory are suppressed by huge energy scales and hence can be neglected when computing low-energy effects. Thus, at least in the low-energy regime, the model is indeed a predictive quantum field theory.[12] (A very similar situation occurs for the very similar effective field theory of low-energy pions.) Furthermore, many theorists agree that even the Standard Model should really be regarded as an effective field theory as well, with "nonrenormalizable" interactions suppressed by large energy scales and whose effects have consequently not been observed experimentally.

Recent work[112] has shown that by treating general relativity as an effective field theory, one can actually make legitimate predictions for quantum gravity, at least for low-energy phenomena. An example is the well-known calculation of the tiny first-order quantum-mechanical correction to the classical Newtonian gravitational potential between two masses.

9.2.5 Spacetime background dependence

Main article: Background independence

A fundamental lesson of general relativity is that there is no fixed spacetime background, as found in Newtonian mechanics and special relativity; the spacetime geometry is dynamic. While easy to grasp in principle, this is the hardest idea to understand about general relativity, and its consequences are profound and not fully explored, even at the classical level. To a certain extent, general relativity can be seen to be a relational theory,[24] in which the only physically relevant information is the relationship between different events in space-time.

On the other hand, quantum mechanics has depended since its inception on a fixed background (non-dynamic) structure. In the case of quantum mechanics, it is time that is given and not dynamic, just as in Newtonian classical mechanics. In relativistic quantum field theory, just as in classical field theory, Minkowski spacetime is the fixed background of the theory.

String theory

String theory can be seen as a generalization of quantum field theory where instead of point particles, string-like objects propagate in a fixed spacetime background, although the interactions among closed strings give rise to space-time in a dynamical way. Although string theory had its origins in the study of quark confinement and not of quantum gravity, it was soon discovered that the string spectrum contains the graviton, and that "condensation" of certain vibration modes of strings is equivalent to a modification of the original background. In this sense, string perturbation theory exhibits exactly the features one would expect of a perturbation theory that may exhibit a strong dependence on asymptotics (as seen, for example, in the AdS/CFT correspondence) which is a weak form of background dependence.

Background independent theories

Loop quantum gravity is the fruit of an effort to formulate a background-independent quantum theory.

Topological quantum field theory provided an example of background-independent quantum theory, but with no local degrees of freedom, and only finitely many degrees of freedom globally. This is inadequate to describe gravity in 3+1 dimensions, which has local degrees of freedom according to general relativity. In 2+1 dimensions, however, gravity is a topological field theory, and it has been successfully quantized in several different ways, including spin networks.

9.2.6 Semi-classical quantum gravity

Quantum field theory on curved (non-Minkowskian) backgrounds, while not a full quantum theory of gravity, has shown many promising early results. In an analogous way to the development of quantum electrodynamics in the early part of the 20th century (when physicists considered quantum mechanics in classical electromagnetic fields), the consideration of quantum field theory on a curved background has led to predictions such as black hole radiation.

Phenomena such as the Unruh effect, in which particles exist in certain accelerating frames but not in stationary ones, do not pose any difficulty when considered on a curved background (the Unruh effect occurs even in flat Minkowskian backgrounds). The vacuum state is the state with the least energy (and may or may not contain particles). See Quantum field theory in curved spacetime for a more complete discussion.

9.2.7   Points of tension

There are other points of tension between quantum mechanics and general relativity.

- First, classical general relativity breaks down at singularities, and quantum mechanics becomes inconsistent with general relativity in the neighborhood of singularities (however, no one is certain that classical general relativity applies near singularities in the first place).

- Second, it is not clear how to determine the gravitational field of a particle, since under the Heisenberg uncertainty principle of quantum mechanics its location and velocity cannot be known with certainty. The resolution of these points may come from a better understanding of general relativity.[25]

- Third, there is the problem of time in quantum gravity. Time has a different meaning in quantum mechanics and general relativity and hence there are subtle issues to resolve when trying to formulate a theory which combines the two.[26]

9.3   Candidate theories

There are a number of proposed quantum gravity theories.[27] Currently, there is still no complete and consistent quantum theory of gravity, and the candidate models still need to overcome major formal and conceptual problems. They also face the common problem that, as yet, there is no way to put quantum gravity predictions to experimental tests, although there is hope for this to change as future data from cosmological observations and particle physics experiments becomes available.[28][29]

9.3.1   String theory

Main article: String theory
One suggested starting point is ordinary quantum field theories which, after all, are successful in describing the other three basic fundamental forces in the context of the standard model of elementary particle physics. However, while this leads to an acceptable effective (quantum) field theory of gravity at low energies,[30] gravity turns out to be much more problematic at higher energies. Where, for ordinary field theories such as quantum electrodynamics, a technique known as renormalization is an integral part of deriving predictions which take into account higher-energy contributions,[31] gravity turns out to be nonrenormalizable: at high energies, applying the recipes of ordinary quantum field theory yields models that are devoid of all predictive power.[32]

One attempt to overcome these limitations is to replace ordinary quantum field theory, which is based on the classical concept of a point particle, with a quantum theory of one-dimensional extended objects: string theory.[33] At the energies reached in current experiments, these strings are indistinguishable from point-like particles, but, crucially, different modes of oscillation of one and the same type of fundamental string appear as particles with different (electric and other) charges. In this way, string theory promises to be a unified description of all particles and interactions.[34] The theory is successful in that one mode will always correspond to a graviton, the messenger particle of gravity; however, the price to pay are unusual features such as six extra dimensions of space in addition to the usual three for space and one for time.[35]

In what is called the second superstring revolution, it was conjectured that both string theory and a unification of general relativity and supersymmetry known as supergravity[36] form part of a hypothesized eleven-dimensional model known as M-theory, which would constitute a uniquely defined and consistent theory of quantum gravity.[37][38] As presently understood, however, string theory admits a very large number (10^{500} by some estimates) of consistent vacua, comprising the so-called "string landscape". Sorting through this large family of solutions remains a major challenge.

9.3.2   Loop quantum gravity

Main article: Loop quantum gravity
Loop quantum gravity is based first of all on the idea to take seriously the insight of general relativity that spacetime is

a dynamical field and therefore is a quantum object. The second idea is that the quantum discreteness that determines the particle-like behavior of other field theories (for instance, the photons of the electromagnetic field) also affects the structure of space.

The main result of loop quantum gravity is the derivation of a granular structure of space at the Planck length. This is derived as follows. In the case of electromagnetism, the quantum operator representing the energy of each frequency of the field has discrete spectrum. Therefore the energy of each frequency is quantized, and the quanta are the photons. In the case of gravity, the operators representing the area and the volume of each surface or space region have discrete spectrum. Therefore area and volume of any portion of space are quantized, and the quanta are elementary quanta of space. It follows that spacetime has an elementary quantum granular structure at the Planck scale, which cuts-off the ultraviolet infinities of quantum field theory.

The quantum state of spacetime is described in the theory by means of a mathematical structure called spin networks. Spin networks were initially introduced by Roger Penrose in abstract form, and later shown by Carlo Rovelli and Lee Smolin to derive naturally from a non perturbative quantization of general relativity. Spin networks do not represent quantum states of a field in spacetime: they represent directly quantum states of spacetime.

The theory is based on the reformulation of general relativity known as Ashtekar variables, which represent geometric gravity using mathematical analogues of electric and magnetic fields.[39][40] In the quantum theory space is represented by a network structure called a spin network, evolving over time in discrete steps.[41][42][43][44]

The dynamics of the theory is today constructed in several versions. One version starts with the canonical quantization of general relativity. The analogue of the Schrödinger equation is a Wheeler–DeWitt equation, which can be defined in the theory.[45] In the covariant, or spinfoam formulation of the theory, the quantum dynamics is obtained via a sum over discrete versions of spacetime, called spinfoams. These represent histories of spin networks.

9.3.3 Scale Relativity

Main article: Scale relativity

Most quantum gravity theories assume quantum laws as a starting point. However, in the framework of scale relativity, this is not needed.[46] The theory is an extension of special and general relativity, including the relativity of scale transformations. It thus takes a geometrical approach to the problem, where quantum phenomena became a manifestation of the fractality of spacetime. This is similar to the geometrical interpretation of gravitation in general relativity, where gravitation become a manifestation of spacetime curvature instead of a force. Although much remains to be developed, validated predictions have already been obtained in physics, astrophysics and cosmology.

9.3.4 Other approaches

There are a number of other approaches to quantum gravity. The approaches differ depending on which features of general relativity and quantum theory are accepted unchanged, and which features are modified.[47][48] Examples include:

- Acoustic metric and other analog models of gravity

- Asymptotic safety in quantum gravity

- Euclidean quantum gravity

- Causal dynamical triangulation[49]

- Causal fermion systems,[5][6][7][8][9][10] giving quantum mechanics, general relativity and quantum field theory as limiting cases.

- Causal sets[50]

- Covariant Feynman path integral approach

- Group field theory[51]

- Wheeler-DeWitt equation

- Geometrodynamics

- Hořava–Lifshitz gravity

- MacDowell–Mansouri action

- Noncommutative geometry.

- Path-integral based models of quantum cosmology[52]

- Regge calculus

- String-nets giving rise to gapless helicity ±2 excitations with no other gapless excitations[53]

- Superfluid vacuum theory a.k.a. theory of BEC vacuum

- Supergravity

- Twistor theory[54]

- Canonical quantum gravity

- E8 Theory

9.4 Weinberg–Witten theorem

In quantum field theory, the Weinberg–Witten theorem places some constraints on theories of composite gravity/emergent gravity. However, recent developments attempt to show that if locality is only approximate and the holographic principle is correct, the Weinberg–Witten theorem would not be valid.

9.5 Experimental tests

As was emphasized above, quantum gravitational effects are extremely weak and therefore difficult to test. For this reason, the possibility of experimentally testing quantum gravity had not received much attention prior to the late 1990s. However, in the past decade, physicists have realized that evidence for quantum gravitational effects can guide the development of the theory. Since theoretical development has been slow, the field of phenomenological quantum gravity, which studies the possibility of experimental tests, has obtained increased attention.[55][56]

The most widely pursued possibilities for quantum gravity phenomenology include violations of Lorentz invariance, imprints of quantum gravitational effects in the cosmic microwave background (in particular its polarization), and decoherence induced by fluctuations in the space-time foam.

The BICEP2 experiment detected what was initially thought to be primordial B-mode polarization caused by gravitational waves in the early universe. If truly primordial, these waves were born as quantum fluctuations in gravity itself. Cosmologist Ken Olum (Tufts University) stated: "I think this is the only observational evidence that we have that actually shows that gravity is quantized....It's probably the only evidence of this that we will ever have."[57]

9.6 See also

9.7 References

[1] Griffiths, David J. (2004). *Introduction to Quantum Mechanics*. Pearson Prentice Hall. OCLC 803860989.

[2] Wald, Robert M. (1984). *General Relativity.* University of Chicago Press. p. 382. OCLC 471881415.

[3] Zee, Anthony (2010). *Quantum Field Theory in a Nutshell* (2nd ed.). Princeton University Press. p. 172. OCLC 659549695.

[4] Penrose, Roger (2007). *The road to reality : a complete guide to the laws of the universe.* Vintage. p. 1017. OCLC 716437154.

[5] F. Finster, J. Kleiner, Causal Fermion Systems as a Candidate for a Unified Physical Theory, http://arxiv.org/abs/1502.03587

[6] F. Finster, The Principle of the Fermionic Projector, hep-th/0001048, hep-th/0202059, hep- th/0210121, AMS/IP Studies in

Advanced Mathematics, vol. **35**, American Mathematical Society, Providence, RI, 2006.

[7] F. Finster, A formulation of quantum field theory realizing a sea of interacting Dirac particles, arXiv:0911.2102 [hep-th], Lett.

Math. Phys. **97** (2011), no. 2, 165–183.

[8] F. Finster, An action principle for an interacting fermion system and its analysis in the continuum limit, arXiv:0908.1542

[math-ph] (2009).

[9] F. Finster, The continuum limit of a fermion system involving neutrinos: Weak and gravitational interactions, arXiv:1211.3351

[math-ph] (2012).

[10] F.Finster,Perturbative quantumfield theory in the framework of the fermionic projector,arXiv:1310.4121[math-ph],J.Math. Phys. **55** (2014), no. 4, 042301.

[11] Quantum effects in the early universe might have an observable effect on the structure of the present universe, for example, or

gravity might play a role in the unification of the other forces. Cf. the text by Wald cited above.

[12] Donoghue (1995). "Introduction to the Effective Field Theory Description of Gravity". arXiv:gr-qc/9512024. (verify against

ISBN 9789810229085)

[13] Kraichnan, R. H. (1955). "Special-Relativistic Derivation of Generally Covariant Gravitation Theory". *Physical Review* **98** (4):

1118–1122. Bibcode:1955PhRv...98.1118K. doi:10.1103/PhysRev.98.1118.

[14] Gupta,S.N.(1954). "Gravitation and Electromagnetism". *Physical Review***96**(6): 1683–1685. Bibcode:1954PhRv...96.1683G

doi:10.1103/PhysRev.96.1683.

[15] Gupta,S.N.(1957). "Einstein's and Other Theories of Gravitation". *Reviews of Modern Physics***29**(3): 334–336. Bibcode:1957 RvMP...29..334G. doi:10.1103/RevModPhys.29.334.

[16] Gupta, S. N. (1962). "Quantum Theory of Gravitation". *Recent Developments in General Relativity.* Pergamon Press. pp.

251–258.

[17] Deser, S. (1970). "Self-Interaction and Gauge Invariance". *General Relativity and Gravitation* **1**: 9–18. arXiv:gr-qc/0411023.

Bibcode:1970GReGr...1....9D. doi:10.1007/BF00759198.

[18] Ohta, Tadayuki; Mann, Robert (1996). "Canonical reduction of two-dimensional gravity for particle dynamics". *Classical and*

*Quantum Gravity***13**(9): 2585–2602. arXiv:gr-qc/9605004. Bibcode:1996CQGra..13.2585O.doi:10.1088/0264-9381/13/9/022.

[19] Sikkema, A E; Mann, R B (1991). "Gravitation and cosmology in (1+1) dimensions". *Classical and Quantum Gravity* **8**:

[26] Edward Anderson (2010). "The Problem of Time in Quantum Gravity". arXiv:1009.2157 [gr-qc]. (also published as chapter
4 of ISBN 9781611229578)

[27] A timeline and overview can be found in Rovelli, Carlo (2000). "Notes for a brief history of quantum gravity". arXiv: gr-
qc/0006061. (verify against ISBN 9789812777386)

[28] Ashtekar, Abhay (2007). "Loop Quantum Gravity: Four Recent Advances and a Dozen Frequently Asked Questions". *11th Marcel Grossmann Meeting on Recent Developments in Theoretical and Experimental General Relativity*. p. 126. arXiv :0705.2222. Bibcode:2008mgm..conf..126A. doi:10.1142/9789812834300_0008.

[29] Schwarz, John H. (2007). "String Theory: Progress and Problems". *Progress of Theoretical Physics Supplement* **170**: 214–226. arXiv:hep-th/0702219. Bibcode:2007PThPS.170..214S. doi:10.1143/PTPS.170.214.

[30] Donoghue, John F. (editor) (1995). "Introduction to the Effective Field Theory Description of Gravity". In Cornet, Fernando.
Effective Theories: Proceedings of the Advanced School, Almunecar, Spain, 26 June–1 July 1995. Singapore: World Scientific. arXiv:gr-qc/9512024. ISBN 981-02-2908-9.

[31] Weinberg, Steven (1996). "Chapters 17–18". *The Quantum Theory of Fields II: Modern Applications*. Cambridge University Press. ISBN 0-521-55002-5.

[32] Goroff, Marc H.; Sagnotti, Augusto; Sagnotti, Augusto (1985). "Quantum gravity at two loops". *Physics Letters B* **160**: 81–86. Bibcode:1985PhLB..160...81G. doi:10.1016/0370-2693(85)91470-4.

[33] An accessible introduction at the undergraduate level can be found inZwiebach ,Barton(2004). *A First Course in String Theory*. Cambridge University Press. ISBN 0-521-83143-1., and more complete overviews in Polchinski, Joseph (1998). *String Theory Vol. I: An Introduction to the Bosonic String*. Cambridge University Press. ISBN 0-521-63303-6. and Polchinski, Joseph (1998b). *String Theory Vol. II: Superstring Theory and Beyond*. Cambridge University Press. ISBN 0-521-63304-4.

[34] Ibanez, L. E. (2000). "The second string (phenomenology) revolution". *Classical & Quantum Gravity* **17** (5): 1117 –1128. arXiv:hep-ph/9911499. Bibcode:2000CQGra..17.1117I. doi:10.1088/0264-9381/17/5/321.

[35] For the graviton as part of the string spectrum, e.g. Green, Schwarz & Witten 1987, sec. 2.3 and 5.3; for the extra dimensions, ibid sec. 4.2.

[36] Weinberg, Steven (2000). "Chapter 31". *The Quantum Theory of Fields II: Modern Applications*. Cambridge University Press. ISBN 0-521-55002-5.

[37] Townsend, Paul K. (1996). *Four Lectures on M-Theory*. ICTP Series in Theoretical Physics. p. 385. arXiv:hep-th/ 9612121. Bibcode:1997hepcbconf..385T.

[38] Duff, Michael (1996). "M-Theory (the Theory Formerly Known as Strings)". *International Journal of Modern Physics A* **11** (32): 5623–5642. arXiv:hep-th/9608117. Bibcode:1996IJMPA..11.5623D. doi:10.1142/S0217751X96002583.

[39] Ashtekar, Abhay (1986). "New variables for classical and quantum gravity". *Physical Review Letters* **57** (18): 2244 –2247. Bibcode:1986PhRvL..57.2244A. doi:10.1103/PhysRevLett.57.2244. PMID 10033673.

[40] Ashtekar, Abhay (1987). "New Hamiltonian formulation of general relativity". *Physical Review D* **36** (6): 1587 –1602. Bibcode:1987PhRvD..36.1587A. doi:10.1103/PhysRevD.36.1587.

[41] Thiemann, Thomas (2006). "Loop Quantum Gravity: An Inside View". *Approaches to Fundamental Physics*. Lecture Notes in Physics**721**: 185. arXiv:hep-th/0608210. Bibcode:2007LNP...721..185T.doi:10.1007/978-3-540-71117-9_ 10. ISBN978-3- 540-71115-5.

[42] Rovelli, Carlo (1998). "Loop Quantum Gravity". *Living Reviews in Relativity* **1**. Retrieved 2008-03-13.

[43] Ashtekar, Abhay; Lewandowski, Jerzy (2004). "Background Independent Quantum Gravity: A Status Report". *Classical & Quantum Gravity***21**(15): R53–R152. arXiv:gr-qc/0404018. Bibcode:2004CQGra..21R..53A.doi:10.1088/0264-9381/21

[46] Nottale, L. (2011). *Scale Relativity and Fractal Space-Time: A New Approach to Unifying Relativity and Quantum Mechanics.* World Scientific Publishing Company. ISBN 1848166508.:p. 458

[47] Isham, Christopher J. (1994). "Prima facie questions in quantum gravity". In Ehlers, Jürgen; Friedrich, Helmut. *Canonical Gravity: From Classical to Quantum.* Springer. arXiv:gr-qc/9310031. ISBN 3-540-58339-4.

[48] Sorkin, Rafael D. (1997). "Forks in the Road, on the Way to Quantum Gravity". *International Journal of Theoretical Physics* **36** (12): 2759–2781. arXiv:gr-qc/9706002. Bibcode:1997IJTP...36.2759S. doi:10.1007/BF02435709.

[49] Loll, Renate (1998). "Discrete Approaches to Quantum Gravity in Four Dimensions". *Living Reviews in Relativity* **1**: 13. arXiv:gr-qc/9805049. Bibcode:1998LRR.....1...13L. doi:10.12942/lrr-1998-13. Retrieved 2008-03-09.

[50] Sorkin, Rafael D. (2005). "Causal Sets: Discrete Gravity". In Gomberoff, Andres; Marolf, Donald. *Lectures on Quantum Gravity.* Springer. arXiv:gr-qc/0309009. ISBN 0-387-23995-2.

[51] See Daniele Oriti and references therein.

[52] Hawking, Stephen W. (1987). "Quantum cosmology". In Hawking, Stephen W.; Israel, Werner. *300 Years of Gravitation.* Cambridge University Press. pp. 631–651. ISBN 0-521-37976-8.

[53] Wen 2006

[54] See ch. 33 in Penrose 2004 and references therein.

[55] Hossenfelder, Sabine (2011). "Experimental Search for Quantum Gravity". In V. R. Frignanni. *Classical and Quantum Gravity: Theory, Analysis and Applications.* Chapter 5: Nova Publishers. ISBN 978-1-61122-957-8.

[56] Hossenfelder, Sabine (2010-10-17). V. R. Frignanni, ed. "Experimental Search for Quantum Gravity". *Classical and Quantum Gravity: Theory, Analysis and Applications* (Nova Publishers) **5** (2011). arXiv:1010.3420. Bibcode:2010arXiv1010.3420H. |chapter= ignored (help)

[57] Camille Carlisle. "First Direct Evidence of Big Bang Inflation". SkyandTelescope.com. Retrieved March 18, 2014.

9.8 Further reading

- Ahluwalia, D. V. (2002). "Interface of Gravitational and Quantum Realms". *Modern Physics Letters A* **17** (15–17): 1135. arXiv:gr-qc/0205121. Bibcode:2002MPLA...17.1135A. doi:10.1142/S021773230200765X.

- Ashtekar, Abhay (2005). "The winding road to quantum gravity" (PDF). *Current Science* **89**: 2064–2074.

- Carlip, Steven (2001). "Quantum Gravity: a Progress Report". *Reports on Progress in Physics* **64** (8): 885–942. arXiv:gr-qc/0108040. Bibcode:2001RPPh...64..885C. doi:10.1088/0034-4885/64/8/301.

- Herbert W. Hamber (2009). *Quantum Gravitation.* Springer Publishing. doi:10.1007/978-3-540-85293-3. ISBN 978-3-540-85292-6.

- Kiefer, Claus (2007). *Quantum Gravity.* Oxford University Press. ISBN 0-19-921252-X.

- Kiefer, Claus (2005). "Quantum Gravity: General Introduction and Recent Developments". *Annalen der Physik* **15**: 129–148. arXiv:gr-qc/0508120. Bibcode:2006AnP...518..129K. doi:10.1002/andp.200510175.

- Lämmerzahl, Claus, ed. (2003). *Quantum Gravity: From Theory to Experimental Search.* Lecture Notes in Physics. Springer. ISBN 3-540-40810-X.

- Rovelli, Carlo (2004). *Quantum Gravity.* Cambridge University Press. ISBN 0-521-83733-2.

- Trifonov, Vladimir (2008). "GR-friendly description of quantum systems". *International Journal of Theoretical Physics* **47** (2): 492–510. arXiv:math-ph/0702095. Bibcode:2008IJTP...47..492T. doi:10.1007/s10773-007-9474-3.

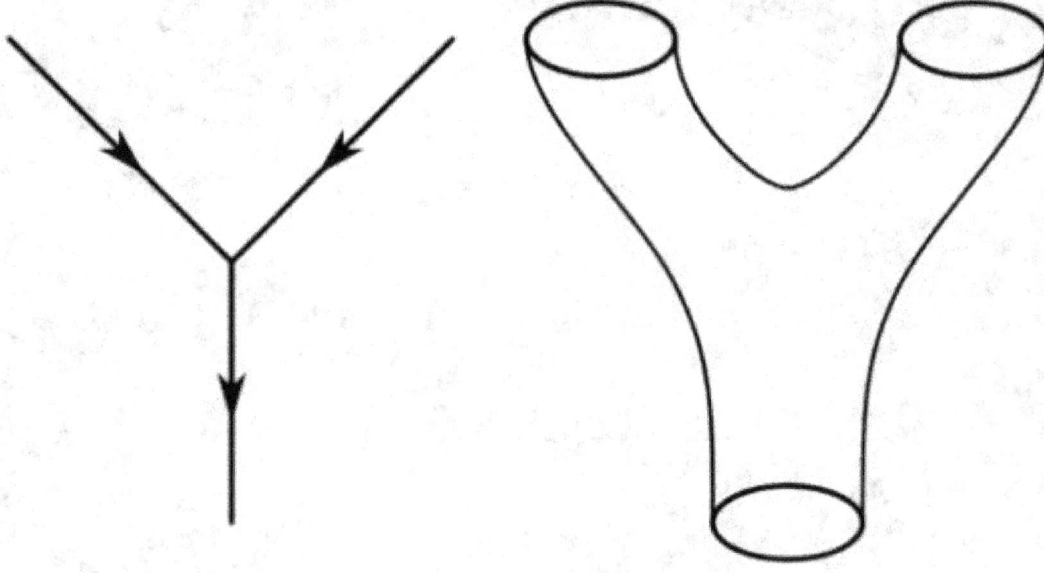

Interaction in the subatomic world: world lines of point-like particles in the Standard Model or a world sheet swept up by closed strings in string theory

Projection of a Calabi–Yau manifold, one of the ways of compactifying the extra dimensions posited by string theory

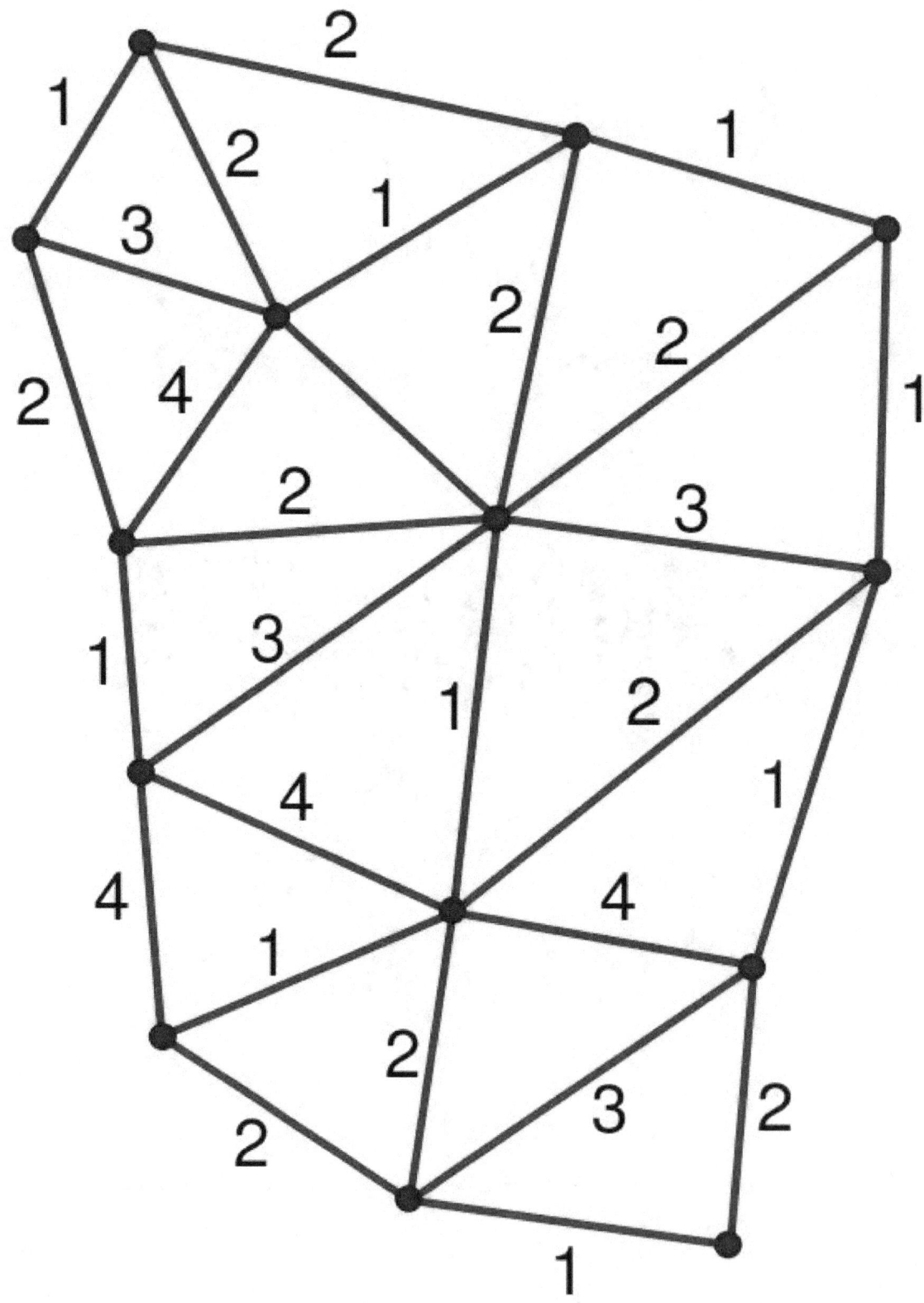

Simple spin network of the type used in loop quantum gravity

Schrödinger's flower. Morphogenesis of a flower-like structure, solution of a growth process equation that takes the form of a Schrödinger equation under fractal conditions.

Chapter 10

Loop quantum gravity

Loop quantum gravity (**LQG**) is a theory that attempts to describe the quantum properties of the universe and gravity. It is also a theory of quantumspacetimebecause,according togeneral relativity,
gravity is a manifestation of the geometry of spacetime. LQG is an attempt to merge quantum mechanics and general relativity. The main output of the theory is a physical picture of space where space is granular. The granularity is a direct consequence of the quantization. It has the same nature as the granularity of the photons in the quantum theory of electromagnetism and the discrete levels of the energy of the atoms. Here, it is space itself that is discrete. In other words, there is a minimum distance possible to travel through it.

More precisely , space can be viewed as an extremely fine fabric or network "woven" of finite loops . These networks of loops are called spin networks. The evolution of a spin network over time is called a spin foam. The predicted size of this structure is the Planck length, which is approximately 10-35 meters. According to the theory , there is no meaning to distance at scales smaller than the Planck scale . Therefore , LQG predicts that not just matter, but space itself, has an atomic structure.

Today LQG is a vast area of research,developing in several directions,which involves about30research groups worldwide .[1] They all share the basic physical assumptions and the mathematical description of quantum space. The full development of the theory is being pursued in two directions: the more traditional canonical loop quantum gravity, and the newer covariant loop quantum gravity, more commonly called spin foam theory.

Research into the physical consequences of the theory is proceeding in several directions. Among these, the most well- developed is the application of LQG to cosmology , called loop quantum cosmology (LQC). LQC applies LQG ideas to the study of the early universe and the physics of the Big Bang. Its most spectacular consequence is that the evolution of the universe can be continued beyond the Big Bang. The Big Bang appears thus to be replaced by a sort of cosmic Big Bounce.

10.1 History

Main article: History of loop quantum gravity

In 1986 , Abhay Ashtekar reformulated Einstein 's general relativity in a language closer to that of the rest of fundamental physics. Shortly after, Ted Jacobson and Lee Smolin realized that the formal equation of quantum gravity , called the Wheeler –DeWitt equation , admitted solutions labelled by loops, when rewritten in the new Ashtekar variables , and Carlo Rovelli and Lee Smolin defined a nonperturbative and background -independent quantum theory of gravity in terms of these loop solutions. Jorge Pullin and Jerzy Lewandowski understood that the intersections of the loops are essential for the consistency of the theory, and the theory should be formulated in terms of intersecting loops, or graphs.

In 1994, Rovelli and Smolin showed that the quantum operators of the theory associated to area and volume have a discrete spectrum . That is, geometry is quantized . This result defines an explicit basis of states of quantum geometry.

which turned out to be labelled by Roger Penrose's spin networks, which are graphs labelled by spins.

The canonical version of the dynamics was put on firm ground by Thomas Thiemann, who defined an anomaly-free Hamiltonian operator, showing the existence of a mathematically consistent background-independent theory. The covariant or spinfoam version of the dynamics developed during several decades, and crystallized in 2008, from the joint work of research groups in France, Canada, UK, Poland, and Germany, lead to the definition of a family of transition amplitudes, which in the classical limit can be shown to be related to a family of truncations of general relativity.[2] The finiteness of these amplitudes was proven in 2011.[3][4] It requires the existence of a positive cosmological constant, and this is consistent with observed acceleration in the expansion of the Universe.

10.2 General covariance and background independence

In theoretical physics, general covariance is the invariance of the form of physical laws under arbitrary differentiable coordinate transformations. The essential idea is that coordinates are only artifices used in describing nature, and hence should play no role in the formulation of fundamental physical laws. A more significant requirement is the principle of general relativity that states that the laws of physics take the same form in all reference systems. This is a generalization of the principle of special relativity which states that the laws of physics take the same form in all inertial frames.

In mathematics, a diffeomorphism is an isomorphism in the category of smooth manifolds. It is an invertible function that maps one differentiable manifold to another, such that both the function and its inverse are smooth. These are the defining symmetry transformations of General Relativity since the theory is formulated only in terms of a differentiable manifold.

In general relativity, general covariance is intimately related to "diffeomorphism invariance". This symmetry is one of the defining features of the theory. However, it is a common misunderstanding that "diffeomorphism invariance" refers to the invariance of the physical predictions of a theory under arbitrary coordinate transformations; this is untrue and in fact every physical theory is invariant under coordinate transformations this way. Diffeomorphisms, as mathematicians define them, correspond to something much more radical; intuitively a way they can be envisaged is as simultaneously dragging all the physical fields (including the gravitational field) over the bare differentiable manifold while staying in the same coordinate system. Diffeomorphisms are the true symmetry transformations of general relativity, and come about from the assertion that the formulation of the theory is based on a bare differentiable manifold, but not on any prior geometry — the theory is background-independent (this is a profound shift, as all physical theories before general relativity had as part of their formulation a prior geometry). What is preserved under such transformations are the coincidences between the values the gravitational field take at such and such a "place" and the values the matter fields take there. From these relationships one can form a notion of matter being located with respect to the gravitational field, or vice versa. This is what Einstein discovered: that physical entities are located with respect to one another only and not with respect to the spacetime manifold. As Carlo Rovelli puts it: "No more fields on spacetime: just fields on fields.".[5] This is the true meaning of the saying "The stage disappears and becomes one of the actors"; space-time as a "container" over which physics takes place has no objective physical meaning and instead the gravitational interaction is represented as just one of the fields forming the world. This is known as the relationalist interpretation of space-time. The realization by Einstein that general relativity should be interpreted this way is the origin of his remark "Beyond my wildest expectations".

In LQG this aspect of general relativity is taken seriously and this symmetry is preserved by requiring that the physical states remain invariant under the generators of diffeomorphisms. The interpretation of this condition is well understood for purely spatial diffeomorphisms. However, the understanding of diffeomorphisms involving time (the Hamiltonian constraint) is more subtle because it is related to dynamics and the so-called "problem of time" in general relativity.[6] A generally accepted calculational framework to account for this constraint has yet to be found.[7][8] A plausible candidate for the quantum hamiltonian constraint is the operator introduced by Thiemann.[9]

LQG is formally background independent. The equations of LQG are not embedded in, or dependent on, space and time (except for its invariant topology). Instead, they are expected to give rise to space and time at distances which are large compared to the Planck length. The issue of background independence in LQG still has some unresolved subtleties. For example, some derivations require a fixed choice of the topology, while any consistent quantum theory of gravity should include topology change as a dynamical process.

10.3   Constraints and their Poisson bracket algebra

Main articles: Poisson bracket and Hamiltonian constraint

10.3.1   The constraints of classical canonical general relativity

Main article: Lie derivative

In the Hamiltonian formulation of ordinary classical mechanics the Poisson bracket is an important concept. A "canonical coordinate system" consists of canonical position and momentum variables that satisfy canonical Poisson-bracket relations,

$$\{q_i, p_j\} = \delta_{ij}$$

where the Poisson bracket is given by

$$\{f, g\} = \sum_{i=1}^{N} \left(\frac{\partial f}{\partial q_i} \frac{\partial g}{\partial p_i} - \frac{\partial f}{\partial p_i} \frac{\partial g}{\partial q_i} \right).$$

for arbitrary phase space functions $f(q_i, p_j)$ and $g(q_i, p_j)$. With the use of Poisson brackets, the Hamilton's equations can be rewritten as,

$$\dot{q}_i = \{q_i, H\},$$

$$\dot{p}_i = \{p_i, H\}.$$

These equations describe a "flow" or orbit in phase space generated by the Hamiltonian H. Given any phase space function $F(q, p)$, we have

$$\frac{d}{dt} F(q_i, p_i) = \{F, H\}.$$

Let us consider constrained systems, of which General relativity is an example. In a similar way the Poisson bracket between a constraint and the phase space variables generates a flow along an orbit in (the unconstrained) phase space generated by the constraint. There are three types of constraints in Ashtekar's reformulation of classical general relativity:

$SU(2)$ Gauss gauge constraints

The Gauss constraints

$$G_j(x) = 0.$$

This represents an infinite number of constraints one for each value of x. These come about from re-expressing General relativity as an $SU(2)$ Yang–Mills type gauge theory (Yang–Mills is a generalization of Maxwell's theory where the gauge field transforms as a vector under Gauss transformations, that is, the Gauge field is of the form $A_a^i(x)$ where i is an internal index. See Ashtekar variables). These infinite number of Gauss gauge constraints can be smeared with test fields with internal indices, $\lambda^j(x)$,

$$G(\lambda) = \int d^3 x G_j(x) \lambda^j(x).$$

which we demand vanish for any such function. These smeared constraints defined with respect to a suitable space of smearing functions give an equivalent description to the original constraints.

In fact Ashtekar's formulation may be thought of as ordinary $SU(2)$ Yang–Mills theory together with the following special constraints, resulting from diffeomorphism invariance, and a Hamiltonian that vanishes. The dynamics of such a theory are thus very different from that of ordinary Yang–Mills theory.

Spatial diffeomorphisms constraints

The spatial diffeomorphism constraints

$$C_a(x) = 0$$

can be smeared by the so-called shift functions $\vec{N}(x)$ to give an equivalent set of smeared spatial diffeomorphism constraints,

$$C(\vec{N}) = \int d^3x C_a(x) N^a(x) .$$

These generate spatial diffeomorphisms along orbits defined by the shift function $N^a(x)$.

Hamiltonian constraints

The Hamiltonian

$$H(x) = 0$$

can be smeared by the so-called lapse functions $N(x)$ to give an equivalent set of smeared Hamiltonian constraints,

$$H(N) = \int d^3x H(x) N(x) .$$

These generate time diffeomorphisms along orbits defined by the lapse function $N(x)$.

In Ashtekar formulation the gauge field $A_a^i(x)$ is the configuration variable (the configuration variable being analogous to q in ordinary mechanics) and its conjugate momentum is the (densitized) triad (electrical field) $\tilde{E}_i^a(x)$. The constraints are certain functions of these phase space variables.

We consider the action of the constraints on arbitrary phase space functions. An important notion here is the Lie derivative, $\mathcal{L}_V$, which is basically a derivative operation that infinitesimally "shifts" functions along some orbit with tangent vector V .

10.3.2 The Poisson bracket algebra

Of particular importance is the Poisson bracket algebra formed between the (smeared) constraints themselves as it completely determines the theory. In terms of the above smeared constraints the constraint algebra amongst the Gauss' law reads,

$$\{G(\lambda), G(\mu)\} = G([\lambda, \mu])$$

where $[\lambda, \mu]^k = \lambda_i \mu_j \epsilon^{ijk}$. And so we see that the Poisson bracket of two Gauss' law is equivalent to a single Gauss' law evaluated on the commutator of the smearings. The Poisson bracket amongst spatial diffeomorphisms constraints reads

$$\{C(\vec{N}), C(\vec{M})\} = C(\mathcal{L}_{\vec{N}}\vec{M})$$

and we see that its effect is to "shift the smearing". The reason for this is that the smearing functions are not functions of the canonical variables and so the spatial diffeomorphism does not generate diffeomorphims on them. They do however generate diffeomorphims on everything else. This is equivalent to leaving everything else fixed while shifting the smearing .The action of the spatial diffeomorphism on the Gauss law is

$$\{C(\vec{N}), G(\lambda)\} = G(\mathcal{L}_{\vec{N}}\lambda) .$$

again, it shifts the test field λ . The Gauss law has vanishing Poisson bracket with the Hamiltonian constraint. The spatial diffeomorphism constraint with a Hamiltonian gives a Hamiltonian with its smearing shifted,

$$\{C(\vec{N}), H(M)\} = H(\mathcal{L}_{\vec{N}}M) .$$

Finally, the poisson bracket of two Hamiltonians is a spatial diffeomorphism,

$$\{H(N), H(M)\} = C(K)$$

where K is some phase space function. That is, it is a sum over infinitesimal spatial diffeomorphisms constraints where the coefficients of proportionality are not constants but have non-trivial phase space dependence.

A (Poisson bracket) Lie algebra, with constraints C_I , is of the form

$$\{C_I, C_J\} = f_{IJ}^K C_K$$

where f_{IJ}^K are constants (the so-called structure constants). The above Poisson bracket algebra for General relativity does not form a true Lie algebra as we have structure functions rather than structure constants for the Poisson bracket between two Hamiltonians. This leads to difficulties.

10.3.3   Dirac observables

The constraints define a constraint surface in the original phase space. The gauge motions of the constraints apply to all phase space but have the feature that they leave the constraint surface where it is, and thus the orbit of a point in the hypersurface under gauge transformations will be an orbit entirely within it. Dirac observables are defined as phase space functions, O , that Poisson commute with all the constraints when the constraint equations are imposed,

$$\{G_j, O\}_{G_j = C_a = H = 0} = \{C_a, O\}_{G_j = C_a = H = 0} = \{H, O\}_{G_j = C_a = H = 0} = 0 \,,$$

that is, they are quantities defined on the constraint surface that are invariant under the gauge transformations of the theory.

Then, solving only the constraint $G_j = 0$ and determining the Dirac observables with respect to it leads us back to the ADM phase space with constraints H, C_a . The dynamics of general relativity is generated by the constraints, it can be shown that six Einstein equations describing time evolution (really a gauge transformation) can be obtained by calculating the Poisson brackets of the three-metric and its conjugate momentum with a linear combination of the spatial diffeomorphism and Hamiltonian constraint. The vanishing of the constraints, giving the physical phase space, are the four other Einstein equations.[10]

10.4   Quantization of the constraints – the equations of quantum general relativity

10.4.1   Pre-history and Ashtekar new variables

Main articles: Frame fields in general relativity, Ashtekar variables and Self-dual Palatini action

Many of the technical problems in canonical quantum gravity revolve around the constraints. Canonical general relativity was originally formulated in terms of metric variables, but there seemed to be insurmountable mathematical difficulties in promoting the constraints to quantum operators because of their highly non-linear dependence on the canonical variables. The equations were much simplified with the introduction of Ashtekars new variables. Ashtekar variables describe canonical general relativity in terms of a new pair canonical variables closer to that of gauge theories. The first step consists of using densitized triads $\tilde{E}_i^a$ (a triad E_i^a is simply three orthogonal vector fields labeled by $i = 1, 2, 3$ and the densitized triad is defined by $\tilde{E}_i^a = \sqrt{\det(q)} E_i^a$) to encode information about the spatial metric,

$$\det(q) q^{ab} = \tilde{E}_i^a \tilde{E}_j^b \delta^{ij} \,.$$

(where δ^{ij} is the flat space metric, and the above equation expresses that q^{ab} , when written in terms of the basis E_i^a , is locally flat). (Formulating general relativity with triads instead of metrics was not new.) The densitized triads are not unique, and in fact one can perform a local in space rotation with respect to the internal indices i . The canonically conjugate variable is related to the extrinsic curvature by $K_a^i = K_{ab} \tilde{E}^{ai} / \sqrt{\det(q)}$. But problems similar to using the metric formulation arise when one tries to quantize the theory. Ashtekar's new insight was to introduce a new configuration variable,

$$A_a^i = \Gamma_a^i - i K_a^i$$

that behaves as a complex SU(2) connection where Γ_a^i is related to the so-called spin connection via $\Gamma_a^i = \Gamma_{ajk} \epsilon^{jki}$. Here A_a^i is called the chiral spin connection. It defines a covariant derivative $\mathcal{D}_a$. It turns out that $\tilde{E}_i^a$ is the conjugate momentum of A_a^i , and together these form Ashtekar's new variables.

The expressions for the constraints in Ashtekar variables; the Gauss's law, the spatial diffeomorphism constraint and the (densitized) Hamiltonian constraint then read:

$$G^i = \mathcal{D}_a \tilde{E}^a_i = 0$$
$$C_a = \tilde{E}^b_i F^i_{ab} - A^i_a (D_b \tilde{E}^b_i) = V_a - A^i_a G^i = 0\,,$$
$$\tilde{H} = \epsilon_{ijk} \tilde{E}^a_i \tilde{E}^b_j F^i_{ab} = 0$$

respectively, where F^i_{ab} is the field strength tensor of the connection A^i_a and where V_a is referred to as the vector constraint. The above-mentioned local in space rotational invariance is the original of the SU(2) gauge invariance here expressed by the Gauss law. Note that these constraints are polynomial in the fundamental variables, unlike as with the constraints in the metric formulation. This dramatic simplification seemed to open up the way to quantizing the constraints. (See the article Self-dual Palatini action for a derivation of Ashtekar's formulism).

With Ashtekar's new variables, given the configuration variable A^i_a, it is natural to consider wavefunctions $\Psi(A^i_a)$. This is the connection representation. It is analogous to ordinary quantum mechanics with configuration variable q and wavefunctions $\psi(q)$. The configuration variable gets promoted to a quantum operator via:

$$\hat{A}^i_a \Psi(A) = A^i_a \Psi(A)\,,$$

(analogous to $\hat{q}\psi(q) = q\psi(q)$) and the triads are (functional) derivatives,

$$\hat{\tilde{E}}^a_i \Psi(A) = -i \frac{\delta \Psi(A)}{\delta A^i_a}\,.$$

(analogous to $\hat{p}\psi(q) = -i\hbar d\psi(q)/dq$). In passing over to the quantum theory the constraints become operators on a kinematic Hilbert space (the unconstrained SU(2) Yang–Mills Hilbert space). Note that different ordering of the A 's and $\tilde{E}$'s when replacing the $\tilde{E}$'s with derivatives give rise to different operators - the choice made is called the factor ordering and should be chosen via physical reasoning. Formally they read

$$\hat{G}_j |\psi\rangle = 0$$
$$\hat{C}_a |\psi\rangle = 0$$
$$\hat{\tilde{H}} |\psi\rangle = 0\,.$$

There are still problems in properly defining all these equations and solving them. For example the Hamiltonian constraint Ashtekar worked with was the densitized version instead of the original Hamiltonian, that is, he worked with $\tilde{H} = \sqrt{\det(q)}H$. There were serious difficulties in promoting this quantity to a quantum operator. Moreover, although Ashtekar variables had the virtue of simplifying the Hamiltonian, they are complex. When one quantizes the theory, it is difficult to ensure that one recovers real general relativity as opposed to complex general relativity.

10.4.2   Quantum constraints as the equations of quantum general relativity

We now move on to demonstrate an important aspect of the quantum constraints. We consider Gauss' law only. First we state the classical result that the Poisson bracket of the smeared Gauss' law $G(\lambda) = \int d^3x \lambda^j (D_a E^a)^j$ with the connections is

$$\{G(\lambda), A^i_a\} = \partial_a \lambda^i + g \epsilon^{ijk} A^j_a \lambda^k = (D_a \lambda)^i.$$

The quantum Gauss' law reads

$$\hat{G}_j \Psi(A) = -i D_a \frac{\delta \lambda \Psi[A]}{\delta A^j_a} = 0.$$

If one smears the quantum Gauss' law and study its action on the quantum state one finds that the action of the constraint on the quantum state is equivalent to shifting the argument of Ψ by an infinitesimal (in the sense of the parameter λ small) gauge transformation,

$$\left[1 + \int d^3x \lambda^j(x) \hat{G}_j\right] \Psi(A) = \Psi[A + D\lambda] = \Psi[A].$$

and the last identity comes from the fact that the constraint annihilates the state. So the constraint, as a quantum operator, is imposing the same symmetry that its vanishing imposed classically: it is telling us that the functions $\Psi[A]$ have to be gauge invariant functions of the connection. The same idea is true for the other constraints.

Therefore the two step process in the classical theory of solving the constraints $C_I = 0$ (equivalent to solving the admissibility conditions for the initial data) and looking for the gauge orbits (solving the 'evolution' equations) is replaced by a one step process in the quantum theory, namely looking for solutions Ψ of the quantum equations $\hat{C}_I \Psi = 0$. This is because it obviously solves the constraint at the quantum level and it simultaneously looks for states that are gauge invariant because $\hat{C}_I$ is the quantum generator of gauge transformations (gauge invariant functions are constant along the gauge orbits and thus characterize them).[11] Recall that, at the classical level, solving the admissibility conditions and evolution equations was equivalent to solving all of Einstein's field equations, this underlines the central role of the quantum constraint equations in canonical quantum gravity.

10.4.3   Introduction of the loop representation

Main articles: Holonomy, Wilson loop and Knot invariant

It was in particular the inability to have good control over the space of solutions to the Gauss' law and spacial diffeomorphism constraints that led Rovelli and Smolin to consider a new representation - the loop representation in gauge theories and quantum gravity.[12]

We need the notion of a holonomy. A holonomy is a measure of how much the initial and final values of a spinor or vector differ after parallel transport around a closed loop; it is denoted

$h_\gamma[A]$.

Knowledge of the holonomies is equivalent to knowledge of the connection, up to gauge equivalence. Holonomies can also be associated with an edge; under a Gauss Law these transform as

$(h'_e)_{\alpha\beta} = U^{-1}_{\alpha\gamma}(x)(h_e)_{\gamma\sigma}U_{\sigma\beta}(y)$.

For a closed loop $x = y$ if we take the trace of this, that is, putting $\alpha = \beta$ and summing we obtain

$(h'_e)_{\alpha\alpha} = U^{-1}_{\alpha\gamma}(x)(h_e)_{\gamma\sigma}U_{\sigma\alpha}(x) = [U_{\sigma\alpha}(x)U^{-1}_{\alpha\gamma}(x)](h_e)_{\gamma\sigma} = \delta_{\sigma\gamma}(h_e)_{\gamma\sigma} = (h_e)_{\gamma\gamma}$

or

$\text{Tr}\, h'_\gamma = \text{Tr}\, h_\gamma$.

The trace of an holonomy around a closed loop is written

$W_\gamma[A]$

and is called a Wilson loop. Thus Wilson loops are gauge invariant. The explicit form of the Holonomy is

$h_\gamma[A] = \mathcal{P} \exp\left\{ - \int_{\gamma_0}^{\gamma_1} ds \dot{\gamma}^a A_a^i(\gamma(s))T_i \right\}$

where γ is the curve along which the holonomy is evaluated, and s is a parameter along the curve, $\mathcal{P}$ denotes path ordering meaning factors for smaller values of s appear to the left, and T_i are matrices that satisfy the SU(2) algebra

$[T^i, T^j] = 2i\epsilon^{ijk}T^k$.

The Pauli matrices satisfy the above relation. It turns out that there are infinitely many more examples of sets of matrices that satisfy these relations, where each set comprises $(N + 1) \times (N + 1)$ matrices with $N = 1, 2, 3, \dots$, and where none of these can be thought to 'decompose' into two or more examples of lower dimension. They are called different irreducible representations of the SU(2) algebra. The most fundamental representation being the Pauli matrices. The holonomy is labelled by a half integer $N/2$ according to the irreducible representation used.

The use of Wilson loops explicitly solves the Gauss gauge constraint. To handle the spatial diffeomorphism constraint we need to go over to the loop representation. As Wilson loops form a basis we can formally expand any Gauss gauge invariant function as,

$\Psi[A] = \sum_\gamma \Psi[\gamma]W_\gamma[A]$.

This is called the loop transform. We can see the analogy with going to the momentum representation in quantum mechanics(see Position and momentum space). There one has a basis of states $\exp(ikx)$ labelled by a number k and one expands

$$\psi[x] = \int dk \psi(k) \exp(ikx).$$

and works with the coefficients of the expansion $\psi(k)$.

The inverse loop transform is defined by

$$\Psi[\gamma] = \int [dA] \Psi[A] W_\gamma[A].$$

This defines the loop representation. Given an operator $\hat{O}$ in the connection representation,

$$\Phi[A] = \hat{O}\Psi[A] \qquad Eq\ 1.$$

one should define the corresponding operator $\hat{O}'$ on $\Psi[\gamma]$ in the loop representation via,

$$\Phi[\gamma] = \hat{O}'\Psi[\gamma] \qquad Eq\ 2,$$

where $\Phi[\gamma]$ is defined by the usual inverse loop transform,

$$\Phi[\gamma] = \int [dA]\Phi[A]W_\gamma[A] \qquad Eq\ 3..$$

A transformation formula giving the action of the operator $\hat{O}'$ on $\Psi[\gamma]$ in terms of the action of the operator $\hat{O}$ on $\Psi[A]$ is then obtained by equating the R.H.S. of $Eq\ 2$ with the R.H.S. of $Eq\ 3$ with $Eq\ 1$ substituted into $Eq\ 3$, namely

$$\hat{O}'\Psi[\gamma] = \int [dA]W_\gamma[A]\hat{O}\Psi[A],$$

or

$$\hat{O}'\Psi[\gamma] = \int [dA](\hat{O}^\dagger W_\gamma[A])\Psi[A],$$

where by $\hat{O}^\dagger$ we mean the operator $\hat{O}$ but with the reverse factor ordering (remember from simple quantum mechanics where the product of operators is reversed under conjugation). We evaluate the action of this operator on the Wilson loop as a calculation in the connection representation and rearranging the result as a manipulation purely in terms of loops (one should remember that when considering the action on the Wilson loop one should choose the operator one wishes to transform with the opposite factor ordering to the one chosen for its action on wavefunctions $\Psi[A]$). This gives the physical meaning of the operator $\hat{O}'$. For example if $\hat{O}^\dagger$ corresponded to a spatial diffeomorphism, then this can be thought of as keeping the connection field A of $W_\gamma[A]$ where it is while performing a spatial diffeomorphism on γ instead. Therefore the meaning of $\hat{O}'$ is a spatial diffeomorphism on γ, the argument of $\Psi[\gamma]$.

In the loop representation we can then solve the spatial diffeomorphism constraint by considering functions of loops $\Psi[\gamma]$ that are invariant under spatial diffeomorphisms of the loop γ. That is, we construct what mathematicians call knot invariants. This opened up an unexpected connection between knot theory and quantum gravity.

What about the Hamiltonian constraint? Let us go back to the connection representation. Any collection of non-intersecting Wilson loops satisfy Ashtekar's quantum Hamiltonian constraint. This can be seen from the following. With a particular ordering of terms and replacing $\tilde{E}_i^a$ by a derivative, the action of the quantum Hamiltonian constraint on a Wilson loop is

$$\hat{\tilde{H}}^\dagger W_\gamma[A] = -\epsilon_{ijk}\hat{F}_{ab}^k \frac{\delta}{\delta A_a^i} \frac{\delta}{\delta A_b^j} W_\gamma[A].$$

When a derivative is taken it brings down the tangent vector, $\dot{\gamma}^a$, of the loop, γ. So we have something like

$$\hat{F}_{ab}^i \dot{\gamma}^a \dot{\gamma}^b.$$

However, as F_{ab}^i is anti-symmetric in the indices a and b this vanishes (this assumes that γ is not discontinuous anywhere and so the tangent vector is unique). Now let us go back to the loop representation.

We consider wavefunctions $\Psi[\gamma]$ that vanish if the loop has discontinuities and that are knot invariants. Such functions solve the Gauss law, the spatial diffeomorphism constraint and (formally) the Hamiltonian constraint. Thus we have identified an infinite set of exact (if only formal) solutions to all the equations of quantum general relativity![12] This generated a lot of interest in the approach and eventually led to LQG.

10.4.4   Geometric operators, the need for intersecting Wilson loops and spin network states

The easiest geometric quantity is the area. Let us choose coordinates so that the surface Σ is characterized by $x^3 = 0$. The area of small parallelogram of the surface Σ is the product of length of each side times $\sin \theta$ where θ is the angle between the sides. Say one edge is given by the vector $\vec{u}$ and the other by $\vec{v}$ then,

$$A = \|\vec{u}\|\|\vec{v}\| \sin \theta = \sqrt{\|\vec{u}\|^2\|\vec{v}\|^2(1 - \cos^2 \theta)} \quad = \sqrt{\|\vec{u}\|^2\|\vec{v}\|^2 - (\vec{u} \cdot \vec{v})^2}$$

From this we get the area of the surface Σ to be given by

$$A_\Sigma = \int_\Sigma dx^1 dx^2 \sqrt{\det(q^{(2)})}$$

where $\det(q^{(2)}) = q_{11}q_{22} - q_{12}^2$ and is the determinant of the metric induced on Σ. This can be rewritten as

$$\det(q^{(2)}) = \frac{e^{3ab}e^{3cd}q_{ac}q_{bc}}{2}.$$

The standard formula for an inverse matrix is

$$q^{ab} = \frac{e^{acd}e^{bef}q_{ce}q_{df}}{3!\det(q)}$$

Note the similarity between this and the expression for $\det(q^{(2)})$. But in Ashtekar variables we have $\tilde{E}_i^a \tilde{E}^{bi} = \det(q)q^{ab}$. Therefore

$$A_\Sigma = \int_\Sigma dx^1 dx^2 \sqrt{\tilde{E}_i^3 \tilde{E}^{3i}}.$$

According to the rules of canonical quantization we should promote the triads $\tilde{E}_i^3$ to quantum operators,

$$\hat{\tilde{E}}_i^3 \sim \frac{\delta}{\delta A_3^i}.$$

It turns out that the area A_Σ can be promoted to a well defined quantum operator despite the fact that we are dealing with product of two functional derivatives and worse we have a square-root to contend with as well.[13] Putting $N = 2J$, we talk of being in the J-th representation. We note that $\sum_i T^i T^i = J(J + 1)1$. This quantity is important in the final formula for the area spectrum. We simply state the result below,

$$\hat{A}_\Sigma W_\gamma[A] = 8\pi \ell_{\text{Planck}}^2 \beta \sum_I \sqrt{j_I(j_I + 1)} W_\gamma[A]$$

where the sum is over all edges I of the Wilson loop that pierce the surface Σ.

The formula for the volume of a region R is given by

$$V = \int_R d^3 x \sqrt{\det(q)} = \frac{1}{6} \int_R dx^3 \sqrt{\epsilon_{abc}\epsilon^{ijk} \tilde{E}_i^a \tilde{E}_j^b \tilde{E}_k^c}.$$

The quantization of the volume proceeds the same way as with the area. As we take the derivative, and each time we do so we bring down the tangent vector $\dot{\gamma}^a$, when the volume operator acts on non-intersecting Wilson loops the result vanishes. Quantum states with non-zero volume must therefore involve intersections. Given that the anti-symmetric summation is taken over in the formula for the volume we would need at least intersections with three non-coplanar lines. Actually it turns out that one needs at least four-valent vertices for the volume operator to be non-vanishing.

We now consider Wilson loops with intersections. We assume the real representation where the gauge group is $SU(2)$. Wilson loops are an over complete basis as there are identities relating different Wilson loops. These come about from the fact that Wilson loops are based on matrices (the holonomy) and these matrices satisfy identities. Given any two $SU(2)$ matrices $\mathbb{A}$ and $\mathbb{B}$ it is easy to check that,

$$\text{Tr}(\mathbb{A}) \text{Tr}(\mathbb{B}) = \text{Tr}(\mathbb{A}\mathbb{B}) + \text{Tr}(\mathbb{A}\mathbb{B}^{-1}).$$

This implies that given two loops γ and η that intersect, we will have,

$$W_\gamma[A]W_\eta[A] = W_{\gamma \circ \eta}[A] + W_{\gamma \circ \eta^{-1}}[A]$$

where by η^{-1} we mean the loop η traversed in the opposite direction and $\gamma \circ \eta$ means the loop obtained by going around the loop γ and then along η. See figure below. Given that the matrices are unitary one has that $W_\gamma[A] = W_{\gamma^{-1}}[A]$. Also given the cyclic property of the matrix traces (i.e. $Tr(\mathbb{A}\mathbb{B}) = Tr(\mathbb{B}\mathbb{A})$) one has that $W_{\gamma \circ \eta}[A] = W_{\eta \circ \gamma}[A]$. These identities can be combined with each other into further identities of increasing complexity adding more loops. These identities are the so-called Mandelstam identities. Spin networks certain are linear combinations of intersecting Wilson loops designed to address the over completeness introduced by the Mandelstam identities (for trivalent intersections they eliminate the over-completeness entirely) and actually constitute a basis for all gauge invariant functions.

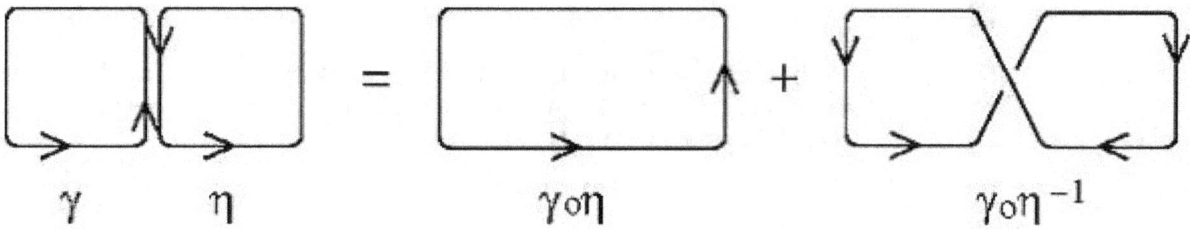

Graphical representation of the simplest non-trivial Mandestam identity relating different Wilson loops.

As mentioned above the holonomy tells you how to propagate test spin half particles. A spin network state assigns an amplitude to a set of spin half particles tracing out a path in space, merging and splitting. These are described by spin networks γ : the edges are labelled by spins together with 'intertwiners' at the vertices which are prescription for how to sum over different ways the spins are rerouted. The sum over rerouting are chosen as such to make the form of the intertwiner invariant under Gauss gauge transformations.

10.4.5 Real variables, modern analysis and LQG

Main article: Hamiltonian constraint of LQG

Let us go into more detail about the technical difficulties associated with using Ashtekar's variables:

With Ashtekar's variables one uses a complex connection and so the relevant gauge group as actually $SL(2, \mathbb{C})$ and not $SU(2)$. As $SL(2, \mathbb{C})$ is non-compact it creates serious problems for the rigorous construction of the necessary mathematical machinery. The group $SU(2)$ is on the other hand is compact and the relevant constructions needed have been developed.

As mentioned above, because Ashtekar's variables are complex it results in complex general relativity. To recover the real theory one has to impose what are known as the reality conditions. These require that the densitized triad be real and that the real part of the Ashtekar connection equals the compatible spin connection (the compatibility condition being $\nabla_a e_b^i = 0$) determined by the desitized triad. The expression for compatible connection Γ_a^i is rather complicated and as such non-polynomial formula enters through the back door.

Before we state the next difficulty we should give a definition; a tensor density of weight W transforms like an ordinary tensor, except that in additional the W th power of the Jacobian,

$$J = \left| \frac{\partial x^a}{\partial x'^b} \right|$$

appears as a factor, i.e.

$$T'^{a...}_{b...} = J^W \frac{\partial x'^a}{\partial x^c} \ldots \frac{\partial x^d}{\partial x'^b} T^{c...}_{d...} .$$

It turns out that it is impossible, on general grounds, to construct a UV-finite, diffeomorphism non-violating operator corresponding to $\sqrt{\det(q)} H$. The reason is that the rescaled Hamiltonian constraint is a scalar density of weight two while it can be shown that only scalar densities of weight one have a chance to result in a well defined operator. Thus, one is forced to work with the original unrescaled, density one-valued, Hamiltonian constraint. However, this is non-polynomial and the whole virtue of the complex variables is questioned. In fact, all the solutions constructed for Ashtekar's Hamiltonian constraint only vanished for finite regularization (physics), however, this violates spatial diffeomorphism invariance.

Without the implementation and solution of the Hamiltonian constraint no progress can be made and no reliable predictions are possible!

To overcome the first problem one works with the configuration variable

$$A_a^i = \Gamma_a^i + \beta K_a^i$$

where β is real (as pointed out by Barbero, who introduced real variables some time after Ashtekar's variables[14][15]).

The Guass law and the spatial diffeomorphism constraints are the same. In real Ashtekar variables the Hamiltonian is

$$H = \frac{\epsilon_{ijk} F^k_{ab} \tilde{E}^a_i \tilde{E}^b_j}{\sqrt{\det(q)}} + 2 \frac{\beta^2+1}{\beta^2} \frac{(\tilde{E}^a_i \tilde{E}^b_j - \tilde{E}^a_j \tilde{E}^b_i)}{\sqrt{\det(q)}} (A^i_a - \Gamma^i_a)(A^j_b - \Gamma^j_b) = H_E + H' \ .$$

The complicated relationship between Γ^i_a and the desitized triads causes serious problems upon quantization. It is with the choice $\beta = \pm i$ that the second more complicated term is made to vanish. However, as mentioned above Γ^i_a reappears in the reality conditions. Also we still have the problem of the $1/\sqrt{\det(q)}$ factor.

Thiemann was able to make it work for real β. First he could simplify the troublesome $1/\sqrt{\det(q)}$ by using the identity

$$\{A^k_c, V\} = \frac{\epsilon_{abc} \epsilon^{ijk} \tilde{E}^a_i \tilde{E}^b_j}{\sqrt{\det(q)}}$$

where V is the volume. The A^k_c and V can be promoted to well defined operators in the loop representation and the Poisson bracket is replaced by a commutator upon quantization; this takes care of the first term. It turns out that a similar trick can be used to treat the second term. One introduces the quantity

$$K = \int d^3x K^i_a \tilde{E}^a_i$$

and notes that

$$K^i_a = \{A^i_a, K\} \ .$$

We are then able to write

$$A^i_a - \Gamma^i_a = \beta K^i_a = \beta \{A^i_a, K\} \ .$$

The reason the quantity K is easier to work with at the time of quantization is that it can be written as

$$K = -\{V, \int d^3x H_E\}$$

where we have used that the integrated densitized trace of the extrinsic curvature, K, is the "time derivative of the volume".

In the long history of canonical quantum gravity formulating the Hamiltonian constraint as a quantum operator (Wheeler–DeWitt equation) in a mathematically rigorous manner has been a formidable problem. It was in the loop representation that a mathematically well defined Hamiltonian constraint was finally formulated in 1996.[9] We leave more details of its construction to the article Hamiltonian constraint of LQG. This together with the quantum versions of the Gauss law and spatial diffeomorphism constrains written in the loop representation are the central equations of LQG (modern canonical quantum General relativity).

Finding the states that are annihilated by these constraints (the physical states), and finding the corresponding physical inner product, and observables is the main goal of the technical side of LQG.

A very important aspect of the Hamiltonian operator is that it only acts at vertices (a consequence of this is that Thiemann's Hamiltonian operator, like Ashtekar's operator, annihilates non-intersecting loops except now it is not just formal and has rigorous mathematical meaning). More precisely, its action is non-zero on at least vertices of valence three and greater and results in a linear combination of new spin networks where the original graph has been modified by the addition of lines at each vertex together and a change in the labels of the adjacent links of the vertex.

10.4.6   Solving the quantum constraints

Main articles: spectrum, dual space and Rigged Hilbert space

We solve, at least approximately, all the quantum constraint equations and for the physical inner product to make physical predictions.

Before we move on to the constraints of LQG, lets us consider certain cases. We start with a kinematic Hilbert space $\mathcal{H}_{\text{Kin}}$ as so is equipped with an inner product—the kinematic inner product $\langle \phi, \psi \rangle_{\text{Kin}}$.

i) Say we have constraints $\hat{C}_I$ whose zero eigenvalues lie in their discrete spectrum. Solutions of the first constraint, $\hat{C}_1$, correspond to a subspace of the kinematic Hilbert space, $\mathcal{H}_1 \subset \mathcal{H}_{\text{Kin}}$. There will be a projection operator P_1 mapping

$\mathcal{H}_{Kin}$ onto $\mathcal{H}_1$. The kinematic inner product structure is easily employed to provide the inner product structure after solving this first constraint; the new inner product $\langle \phi, \psi \rangle_1$ is simply

$$\langle \phi, \psi \rangle_1 = \langle P\phi, P\psi \rangle_{Kin}$$

They are based on the same inner product and are states normalizable with respect to it.

ii) The zero point is not contained in the point spectrum of all the $\hat{C}_I$, there is then no non-trivial solution $\Psi \in \mathcal{H}_{Kin}$ to the system of quantum constraint equations $\hat{C}_I \Psi = 0$ for all I .

For example the zero eigenvalue of the operator

$$\hat{C} = \left(i\tfrac{d}{dx} - k \right)$$

on $L_2(\mathbb{R}, dx)$ lies in the continuous spectrum $\mathbb{R}$ but the formal "eigenstate" $\exp(-ikx)$ is not normalizable in the kinematic inner product,

$$\int_{-\infty}^{\infty} dx \psi^*(x)\psi(x) = \int_{-\infty}^{\infty} dx e^{ikx} e^{-ikx} = \int_{-\infty}^{\infty} dx = \infty$$

and so does not belong to the kinematic Hilbert space $\mathcal{H}_{Kin}$. In these cases we take a dense subset S of $\mathcal{H}_{Kin}$ (intuitively this means either any point in S is either in $\mathcal{H}_{Kin}$ or arbitrarily close to a point in $\mathcal{H}_{Kin}$) with very good convergence properties and consider its dual space S' (intuitively these map elements of S onto finite complex numbers in a linear manner), then $S \subset \mathcal{H}_{Kin} \subset S'$ (as S' contains distributional functions). The constraint operator is then implemented on this larger dual space, which contains distributional functions, under the adjoint action on the operator. One looks for solutions on this larger space. This comes at the price that the solutions must be given a new Hilbert space inner product with respect to which they are normalizable (see article on rigged Hilbert space). In this case we have a generalized projection operator on the new space of states. We cannot use the above formula for the new inner product as it diverges, instead the new inner product is given by the simply modification of the above,

$$\langle \phi, \psi \rangle_1 = \langle P\phi, \psi \rangle_{Kin}.$$

The generalized projector P is known as a rigging map.

Let us move to LQG, additional complications will arise from the fact the constraint algebra is not a Lie algebra due to the bracket between two Hamiltonian constraints.

The Gauss law is solved by the use of spin network states. They provide a basis for the Kinematic Hilbert space $\mathcal{H}_{Kin}$. The spatial diffeomorphism constraint has been solved. The induced inner product on $\mathcal{H}_{Diff}$ (we do not pursue the details) has a very simple description in terms of spin network states; given two spin networks s and s' , with associated spin network states ψ_s and $\psi_{s'}$, the inner product is 1 if s and s' are related to each other by a spatial diffeomorphism and zero otherwise.

The Hamiltonian constraint maps diffeomorphism invariant states onto non-diffeomorphism invaiant states as so does not preserve the diffeomorphism Hilbert space $\mathcal{H}_{Diff}$. This is an unavoidable consequence of the operator algebra, in particular the commutator:

$$[\hat{C}(\vec{N}), \hat{H}(M)] \propto \hat{H}(\mathcal{L}_{\vec{N}} M)$$

as can be seen by applying this to $\psi_s \in \mathcal{H}_{Diff}$,

$$(\vec{C}(\vec{N})\hat{H}(M) - \hat{H}(M)\vec{C}(\vec{N}))\psi_s \propto \hat{H}(\mathcal{L}_{\vec{N}} M)\psi_s$$

and using $\vec{C}(\vec{N})\psi_s = 0$ to obtain

$$\vec{C}(\vec{N})[\hat{H}(M)\psi_s] \propto \hat{H}(\mathcal{L}_{\vec{N}} M)\psi_s \neq 0$$

and so $\hat{H}(M)\psi_s$ is not in $\mathcal{H}_{Diff}$.

This means that you can't just solve the diffeomorphism constraint and then the Hamiltonian constraint. This problem can be circumvented by the introduction of the master constraint, with its trivial operator algebra, one is then able in principle to construct the physical inner product from $\mathcal{H}_{Diff}$.

10.5   Spin foams

Main articles: spin network, spin foam, BF model and Barrett–Crane model

In loop quantum gravity (LQG), a spin network represents a "quantum state" of the gravitational field on a 3-dimensional hypersurface. The set of all possible spin networks (or, more accurately, "s-knots" - that is, equivalence classes of spin networks under diffeomorphisms) is countable; it constitutes a basis of LQG Hilbert space.

In physics, a spin foam is a topological structure made out of two-dimensional faces that represents one of the configurations that must be summed to obtain a Feynman's path integral (functional integration) description of quantum gravity. It is closely related to loop quantum gravity.

10.5.1   Spin foam derived from the Hamiltonian constraint operator

The Hamiltonian constraint generates 'time' evolution. Solving the Hamiltonian constraint should tell us how quantum states evolve in 'time' from an initial spin network state to a final spin network state. One approach to solving the Hamiltonian constraint starts with what is called the Dirac delta function. This is a rather singular function of the real line, denoted $\delta(x)$, that is zero everywhere except at $x = 0$ but whose integral is finite and nonzero. It can be represented as a Fourier integral,

$$\delta(x) = \int e^{ikx} dk \ .$$

One can employ the idea of the delta function to impose the condition that the Hamiltonian constraint should vanish. It is obvious that

$$\prod_{x \in \Sigma} \delta(\hat{H}(x))$$

is non-zero only when $\hat{H}(x) = 0$ for all x in Σ. Using this we can 'project' out solutions to the Hamiltonian constraint. With analogy to the Fourier integral given above, this (generalized) projector can formally be written as

$$\int [dN] e^{i \int d^3 x N(x) \hat{H}(x)} \ .$$

Interestingly, this is formally spatially diffeomorphism-invariant. As such it can be applied at the spatially diffeomorphism-invariant level. Using this the physical inner product is formally given by

$$\left\langle \int [dN] e^{i \int d^3 x N(x) \hat{H}(x)} s_{\text{int}} s_{\text{fin}} \right\rangle_{\text{Diff}}$$

where s_{int} are the initial spin network and s_{fin} is the final spin network.

The exponential can be expanded

$$\left\langle \int [dN] (1 + i \int d^3 x N(x) \hat{H}(x) + \tfrac{i^2}{2!} [\int d^3 x N(x) \hat{H}(x)][\int d^3 x' N(x') \hat{H}(x')] + \dots) s_{\text{int}}, s_{\text{fin}} \right\rangle_{\text{Diff}}$$

and each time a Hamiltonian operator acts it does so by adding a new edge at the vertex. The summation over different sequences of actions of $\hat{H}$ can be visualized as a summation over different histories of 'interaction vertices' in the 'time' evolution sending the initial spin network to the final spin network. This then naturally gives rise to the two-complex (a combinatorial set of faces that join along edges, which in turn join on vertices) underlying the spin foam description; we evolve forward an initial spin network sweeping out a surface, the action of the Hamiltonian constraint operator is to produce a new planar surface starting at the vertex. We are able to use the action of the Hamiltonian constraint on the vertex of a spin network state to associate an amplitude to each "interaction" (in analogy to Feynman diagrams). See figure below. This opens up a way of trying to directly link canonical LQG to a path integral description. Now just as a spin networks describe quantum space, each configuration contributing to these path integrals, or sums over history, describe 'quantum space-time'. Because of their resemblance to soap foams and the way they are labeled John Baez gave these 'quantum space-times' the name 'spin foams'.

There are however severe difficulties with this particular approach, for example the Hamiltonian operator is not self-adjoint, in fact it is not even a normal operator (i.e. the operator does not commute with its adjoint) and so the spectral theorem cannot be used to define the exponential in general. The most serious problem is that the $\hat{H}(x)$'s are not mutually

The action of the Hamiltonian constraint translated to the path integral or so-called spin foam description. A single node splits into three nodes, creating a spin foam vertex. $N(x_n)$ is the value of N at the vertex and H_{nop} are the matrix elements of the Hamiltonian constraint $\hat{H}$.

commuting, it can then be shown the formal quantity $\int [dN] e^{i \int d^3 x N(x) \hat{H}(x)}$ cannot even define a (generalized) projector. The master constraint (see below) does not suffer from these problems and as such offers a way of connecting the canonical theory to the path integral formulation.

10.5.2 Spin foams from BF theory

It turns out there are alternative routes to formulating the path integral, however their connection to the Hamiltonian formalism is less clear. One way is to start with the BF theory. This is a simpler theory to general relativity. It has no local degrees of freedom and as such depends only on topological aspects of the fields. BF theory is what is known as a topological field theory. Surprisingly, it turns out that general relativity can be obtained from BF theory by imposing a constraint,[16] BF theory involves a field B_{ab}^{IJ} and if one chooses the field B to be the (anti-symmetric) product of two tetrads

$$B_{ab}^{IJ} = \frac{1}{2}(E_a^I E_b^J - E_b^I E_a^J)$$

(tetrads are like triads but in four spacetime dimensions), one recovers general relativity. The condition that the B field be given by the product of two tetrads is called the simplicity constraint. The spin foam dynamics of the topological field theory is well understood. Given the spin foam 'interaction' amplitudes for this simple theory, one then tries to implement the simplicity conditions to obtain a path integral for general relativity. The non-trivial task of constructing a spin foam model is then reduced to the question of how this simplicity constraint should be imposed in the quantum theory. The first attempt at this was the famous Barrett–Crane model.[17] However this model was shown to be problematic, for example there did not seem to be enough degrees of freedom to ensure the correct classical limit.[18] It has been argued that the simplicity constraint was imposed too strongly at the quantum level and should only be imposed in the sense of expectation values just as with the Lorenz gauge condition $\partial_\mu \hat{A}^\mu$ in the Gupta–Bleuler formalism of quantum electrodynamics. New models have now been put forward, sometimes motivated by imposing the simplicity conditions in a weaker sense.

Another difficulty here is that spin foams are defined on a discretization of spacetime. While this presents no problems for a topological field theory as it has no local degrees of freedom, it presents problems for GR. This is known as the problem triangularization dependence.

10.5.3 Modern formulation of spin foams

Just as imposing the classical simplicity constraint recovers general relativity from BF theory, one expects an appropriate quantum simplicity constraint will recover quantum gravity from quantum BF theory.

Much progress has been made with regard to this issue by Engle, Pereira, and Rovelli[19] and Freidal and Krasnov[20] in defining spin foam interaction amplitudes with much better behaviour.

An attempt to make contact between EPRL-FK spin foam and the canonical formulation of LQG has been made.[21]

10.5.4 Spin foam derived from the master constraint operator

See below.

10.6 The semi-classical limit

10.6.1 What is the semiclassical limit?

Main articles: Correspondence principle and classical limit

The **classical limit** or **correspondence limit** is the ability of a physical theory to approximate or "recover" classical mechanics when considered over special values of its parameters.[22] The classical limit is used with physical theories that predict non-classical behavior.

In physics, the **correspondence principle** states that the behavior of systems described by the theory of quantum mechanics (or by the old quantum theory) reproduces classical physics in the limit of large quantum numbers. In other words, it says that for large orbits and for large energies, quantum calculations must agree with classical calculations.[23]

The principle was formulated by Niels Bohr in 1920,[24] though he had previously made use of it as early as 1913 in developing his model of the atom.[25]

There are two basic requirements in establishing the semi-classical limit of any quantum theory:

i) reproduction of the Poisson brackets (of the diffeomorphism constraints in the case of general relativity). This is extremely important because, as noted above, the Poisson bracket algebra formed between the (smeared) constraints themselves completely determines the classical theory. This is analogous to establishing Ehrenfest's theorem;

ii) the specification of a complete set of classical observables whose corresponding operators (see complete set of commuting observables for the quantum mechanical definition of a complete set of observables) when acted on by appropriate semi-classical states reproduce the same classical variables with small quantum corrections (a subtle point is that states that are semi-classical for one class of observables may not be semi-classical for a different class of observables[26]).

This may be easily done, for example, in ordinary quantum mechanics for a particle but in general relativity this becomes a highly non-trivial problem as we will see below.

10.6.2 Why might LQG not have general relativity as its semiclassical limit?

Any candidate theory of quantum gravity must be able to reproduce Einstein's theory of general relativity as a classical limit of a quantum theory. This is not guaranteed because of a feature of quantum field theories which is that they have different sectors, these are analogous to the different phases that come about in the thermodynamical limit of statistical systems. Just as different phases are physically different, so are different sectors of a quantum field theory. It may turn out that LQG belongs to an unphysical sector - one in which you do not recover general relativity in the semi classical limit (in fact there might not be any physical sector at all).

Theorems establishing the uniqueness of the loop representation as defined by Ashtekar et al. (i.e. a certain concrete realization of a Hilbert space and associated operators reproducing the correct loop algebra - the realization that everybody was using) have been given by two groups (Lewandowski, Okolow, Sahlmann and Thiemann;[27] and Christian Fleischhack[28]). Before this result was established it was not known whether there could be other examples of Hilbert spaces with operators invoking the same loop algebra, other realizations, not equivalent to the one that had been used so far. These uniqueness theorems imply no others exist and so if LQG does not have the correct semiclassical limit then this would mean the end of the loop representation of quantum gravity altogether.

10.6.3 Difficulties checking the semiclassical limit of LQG

There are difficulties in trying to establish LQG gives Einstein's theory of general relativity in the semi classical limit. There are a number of particular difficulties in establishing the semi-classical limit

1. There is no operator corresponding to infinitesimal spacial diffeomorphisms (it is not surprising that the theory has no generator of infinitesimal spatial 'translations' as it predicts spatial geometry has a discrete nature, compare to the situation in condensed matter). Instead it must be approximated by finite spatial diffeomorphisms and so the Poisson bracket structure of the classical theory is not exactly reproduced. This problem can be circumvented with the introduction of the so-called master constraint (see below)[29]

2. There is the problem of reconciling the discrete combinatorial nature of the quantum states with the continuous nature of the fields of the classical theory.

3. There are serious difficulties arising from the structure of the Poisson brackets involving the spatial diffeomorphism and Hamiltonian constraints. In particular, the algebra of (smeared) Hamiltonian constraints does not close, it is proportional to a sum over infinitesimal spatial diffeomorphisms (which, as we have just noted, does not exist in the quantum theory) where the coefficients of proportionality are not constants but have non-trivial phase space dependence - as such it does not form a Lie algebra. However, the situation is much improved by the introduction of the master constraint.[29]

4. The semi-classical machinery developed so far is only appropriate to non-graph-changing operators, however, Thiemann's Hamiltonian constraint is a graph-changing operator - the new graph it generates has degrees of freedom upon which the coherent state does not depend and so their quantum fluctuations are not suppressed. There is also the restriction, so far, that these coherent states are only defined at the Kinematic level, and now one has to lift them to the level of $\mathcal{H}_{Diff}$ and $\mathcal{H}_{Phys}$. It can be shown that Thiemann's Hamiltonian constraint is required to be graph changing in order to resolve problem 3 in some sense. The master constraint algebra however is trivial and so the requirement that it be graph changing can be lifted and indeed non-graph changing master constraint operators have been defined.

5. Formulating observables for classical general relativity is a formidable problem by itself because of its non-linear nature and space-time diffeomorphism invariance. In fact a systematic approximation scheme to calculate observables has only been recently developed.[30][31]

Difficulties in trying to examine the semi classical limit of the theory should not be confused with it having the wrong semi classical limit.

10.6.4 Progress in demonstrating LQG has the correct semiclassical limit

Much details here to be written up...

Concerning issue number 2 above one can consider so-called weave states. Ordinary measurements of geometric quantities are macroscopic, and planckian discreteness is smoothed out. The fabric of a T-shirt is analogous. At a distance it is a smooth curved two-dimensional surface. But a closer inspection we see that it is actually composed of thousands of one-dimensional linked threads. The image of space given in LQG is similar, consider a very large spin network formed by a very large number of nodes and links, each of Planck scale. But probed at a macroscopic scale, it appears as a three-dimensional continuous metric geometry.

As far as the editor knows problem 4 of having semi-classical machinery for non-graph changing operators is as the moment still out of reach.

To make contact with familiar low energy physics it is mandatory to have to develop approximation schemes both for the physical inner product and for Dirac observables.

The spin foam models have been intensively studied can be viewed as avenues toward approximation schemes for the physical inner product.

Markopoulou et al. adopted the idea of noiseless subsystems in an attempt to solve the problem of the low energy limit in background independent quantum gravity theories[32][33][34] The idea has even led to the intriguing possibility of matter of the standard model being identified with emergent degrees of freedom from some versions of LQG (see section below: *LQG and related research programs*).

As Wightman emphasized in the 1950s, in Minkowski QFTs the $n-$ point functions

$$W(x_1, \ldots, x_n) = \langle 0|\phi(x_n) \ldots \phi(x_1)|0 \rangle ,$$

completely determine the theory. In particular, one can calculate the scattering amplitudes from these quantities. As explained below in the section on the *Background independent scattering amplitudes*, in the background-independent context, the $n-$ point functions refer to a state and in gravity that state can naturally encode information about a specific geometry which can then appear in the expressions of these quantities. To leading order LQG calculations have been shown to agree in an appropriate sense with the $n-$ point functions calculated in the effective low energy quantum general relativity.

10.7 Improved dynamics and the master constraint

Main articles: Hamiltonian (quantum mechanics), Hamiltonian constraint of LQG and Friedrichs extension

10.7.1 The master constraint

Thiemann's master constraint should not be confused with the master equation which has to do with random processes. The Master Constraint Programme for Loop Quantum Gravity (LQG) was proposed as a classically equivalent way to impose the infinite number of Hamiltonian constraint equations

$$H(x) = 0$$

(x being a continuous index) in terms of a single master constraint,

$$M = \int d^3x \frac{[H(x)]^2}{\sqrt{\det(q(x))}} .$$

which involves the square of the constraints in question. Note that $H(x)$ were infinitely many whereas the master constraint is only one. It is clear that if M vanishes then so do the infinitely many $H(x)$'s. Conversely, if all the $H(x)$'s vanish then so does M, therefore they are equivalent. The master constraint M involves an appropriate averaging over all space and so is invariant under spatial diffeomorphisms (it is invariant under spatial "shifts" as it is a summation over all such spatial "shifts" of a quantity that transforms as a scalar). Hence its Poisson bracket with the (smeared) spacial diffeomorphism constraint, $C(\vec{N})$, is simple:

$$\{M, C(\vec{N})\} = 0 .$$

(it is $su(2)$ invariant as well). Also, obviously as any quantity Poisson commutes with itself, and the master constraint being a single constraint, it satisfies

$$\{M, M\} = 0 .$$

We also have the usual algebra between spatial diffeomorphisms. This represents a dramatic simplification of the Poisson bracket structure, and raises new hope in understanding the dynamics and establishing the semi-classical limit.[35]

An initial objection to the use of the master constraint was that on first sight it did not seem to encode information about the observables; because the Mater constraint is quadratic in the constraint, when you compute its Poisson bracket with any quantity, the result is proportional to the constraint, therefore it always vanishes when the constraints are imposed and as such does not select out particular phase space functions. However, it was realized that the condition

$$\{\{M, O\}, O\}_{M=0} = 0$$

is equivalent to O being a Dirac observable. So the master constraint does capture information about the observables. Because of its significance this is known as the Master equation.[35]

That the master constraint Poisson algebra is an honest Lie algebra opens up the possibility of using a certain method, known as group averaging, in order to construct solutions of the infinite number of Hamiltonian constraints, a physical inner product thereon and Dirac observables via what is known as refined algebraic quantization RAQ[36]

10.7.2   The quantum master constraint

Define the quantum master constraint (regularisation issues aside) as

$$\hat{M} := \int d^3x \left(\widehat{\frac{H}{\det(q(x))^{1/4}}} \right)^{\dagger}(x) \left(\widehat{\frac{H}{\det(q(x))^{1/4}}} \right)(x) .$$

Obviously,

$$\left(\widehat{\frac{H}{\det(q(x))^{1/4}}} \right)(x)\Psi = 0$$

for all x implies $\hat{M}\Psi = 0$. Conversely, if $\hat{M}\Psi = 0$ then

$$0 = < \Psi, \hat{M}\Psi > = \int d^3x \left\| \left(\widehat{\frac{H}{\det(q(x))^{1/4}}} \right)(x)\Psi \right\|^2 \quad Eq\ 4$$

implies

$$\left(\widehat{\frac{H}{\det(q(x))^{1/4}}} \right)(x)\Psi = 0 .$$

What is done first is, we are able to compute the matrix elements of the would-be operator $\hat{M}$, that is, we compute the quadratic form Q_M . It turns out that as Q_M is a graph changing, diffeomorphism invariant quadratic form it cannot exist on the kinematic Hilbert space H_{Kin} , and must be defined on H_{Diff} . The fact that the master constraint operator $\hat{M}$ is densely defined on H_{Diff} , it is obvious that $\hat{M}$ is a positive and symmetric operator in H_{Diff} . Therefore, the quadratic form Q_M associated with $\hat{M}$ is closable. The closure of Q_M is the quadratic form of a unique self-adjoint operator $\overline{M}$, called the Friedrichs extension of $\hat{M}$. We relabel $\overline{M}$ as $\hat{M}$ for simplicity. (Note that the presence of an inner product, viz Eq 4, means there are no superfluous solutions i.e. there are no Ψ such that $\left(\widehat{\frac{H}{\det(q(x))^{1/4}}} \right)(x)\Psi \neq 0$ but for which $\hat{M}\Psi = 0$).

It is also possible to construct a quadratic form Q_{M_E} for what is called the extended master constraint (discussed below) on H_{Kin} which also involves the weighted integral of the square of the spatial diffeomorphism constraint (this is possible because Q_{M_E} is not graph changing).

The spectrum of the master constraint may not contain zero due to normal or factor ordering effects which are finite but similar in nature to the infinite vacuum energies of background-dependent quantum field theories. In this case it turns out to be physically correct to replace $\hat{M}$ with $\hat{M}' := \hat{M} - min(spec(\hat{M}))\hat{1}$ provided that the "normal ordering constant" vanishes in the classical limit, that is, $\lim_{\hbar \to 0} min(spec(\hat{M})) = 0$, so that $\hat{M}'$ is a valid quantisation of M .

10.7.3   Testing the master constraint

The constraints in their primitive form are rather singular, this was the reason for integrating them over test functions to obtain smeared constraints. However, it would appear that the equation for the master constraint, given above, is even more singular involving the product of two primitive constraints (although integrated over space). Squaring the constraint is dangerous as it could lead to worsened ultraviolent behaviour of the corresponding operator and hence the master constraint programme must be approached with due care.

In doing so the master constraint programme has been satisfactorily tested in a number of model systems with non-trivial constraint algebras, free and interacting field theories.[37][38][39][40][41] The master constraint for LQG was established as a genuine positive self-adjoint operator and the physical Hilbert space of LQG was shown to be non-empty.[42] an obvious consistency test LQG must pass to be a viable theory of quantum General relativity.

10.7.4   Applications of the master constraint

The master constraint has been employed in attempts to approximate the physical inner product and define more rigorous path integrals.[43][44][45][46]

The Consistent Discretizations approach to LQG,[47][48] is an application of the master constraint program to construct the physical Hilbert space of the canonical theory.

10.7.5   Spin foam from the master constraint

It turns out that the master constraint is easily generalized to incorporate the other constraints. It is then referred to as the extended master constraint, denoted M_E. We can define the extended master constraint which imposes both the Hamiltonian constraint and spatial diffeomorphism constraint as a single operator,

$$M_E = \int_\Sigma d^3 x \, \frac{H(x)^2 - q^{ab} V_a(x) V_b(x)}{\sqrt{det(q)}} \, .$$

Setting this single constraint to zero is equivalent to $H(x) = 0$ and $V_a(x) = 0$ for all x in Σ. This constraint implements the spatial diffeomorphism and Hamiltonian constraint at the same time on the Kinematic Hilbert space. The physical inner product is then defined as

$$\langle \phi, \psi \rangle_{\text{Phys}} = \lim_{T \to \infty} \left\langle \phi, \int_{-T}^{T} dt \, e^{it\hat{M}_E} \psi \right\rangle$$

(as $\delta(\hat{M}_E) = \lim_{T \to \infty} \int_{-T}^{T} dt \, e^{it\hat{M}_E}$). A spin foam representation of this expression is obtained by splitting the t-parameter in discrete steps and writing

$$e^{it\hat{M}_E} = \lim_{n \to \infty} [e^{it\hat{M}_E/n}]^n = \lim_{n \to \infty} [1 + it\hat{M}_E/n]^n.$$

The spin foam description then follows from the application of $[1 + it\hat{M}_E/n]$ on a spin network resulting in a linear combination of new spin networks whose graph and labels have been modified. Obviously an approximation is made by truncating the value of n to some finite integer. An advantage of the extended master constraint is that we are working at the kinematic level and so far it is only here we have access semi-classical coherent states. Moreover, one can find none graph changing versions of this master constraint operator, which are the only type of operators appropriate for these coherent states.

10.7.6   Algebraic quantum gravity

The master constraint programme has evolved into a fully combinatorial treatment of gravity known as Algebraic Quantum Gravity (AQG).[49] The non-graph changing master constraint operator is adapted in the framework of algebraic quantum gravity. While AQG is inspired by LQG, it differs drastically from it because in AQG there is fundamentally no topology or differential structure - it is background independent in a more generalized sense and could possibly have something to say about topology change. In this new formulation of quantum gravity AQG semiclassical states always control the fluctuations of all present degrees of freedom. This makes the AQG semiclassical analysis superior over that of LQG, and progress has been made in establishing it has the correct semiclassical limit and providing contact with familiar low energy physics.[50][51] See Thiemann's book for details.

10.8   Physical applications of LQG

10.8.1   Black hole entropy

Main articles: Black hole thermodynamics, Isolated horizon and Immirzi parameter

The Immirzi parameter (also known as the Barbero-Immirzi parameter) is a numerical coefficient appearing in loop quantum gravity. It may take real or imaginary values.

An artist depiction of two black holes merging, a process in which the laws of thermodynamics are upheld.

Black hole thermodynamics is the area of study that seeks to reconcile the laws of thermodynamics with the existence of black hole event horizons. The no hair conjecture of general relativity states that a black hole is characterized only by its mass, its charge, and its angular momentum; hence, it has no entropy. It appears, then, that one can violate the second law of thermodynamics by dropping an object with nonzero entropy into a black hole.[52] Work by Stephen Hawking and Jacob Bekenstein showed that one can preserve the second law of thermodynamics by assigning to each black hole a *black-hole entropy*

$$S_{\mathrm{BH}} = \frac{k_{\mathrm{B}} A}{4 \ell_{\mathrm{P}}^2},$$

where A is the area of the hole's event horizon, k_{B} is the Boltzmann constant, and $\ell_{\mathrm{P}} = \sqrt{G\hbar/c^3}$ is the Planck length.[53] The fact that the black hole entropy is also the maximal entropy that can be obtained by the Bekenstein bound (wherein the Bekenstein bound becomes an equality) was the main observation that led to the holographic principle.[52]

An oversight in the application of the no-hair theorem is the assumption that the relevant degrees of freedom accounting for the entropy of the black hole must be classical in nature; what if they were purely quantum mechanical instead and had non-zero entropy? Actually, this is what is realized in the LQG derivation of black hole entropy, and can be seen as a consequence of its background-independence – the classical black hole spacetime comes about from the semi-classical limit of the quantum state of the gravitational field, but there are many quantum states that have the same semiclassical limit. Specifically, in LQG[54] it is possible to associate a quantum geometrical interpretation to the microstates: These are the quantum geometries of the horizon which are consistent with the area, A, of the black hole and the topology of the horizon (i.e. spherical). LQG offers a geometric explanation of the finiteness of the entropy and of the proportionality of the area of the horizon.[55][56] These calculations have been generalized to rotating black holes.[57]

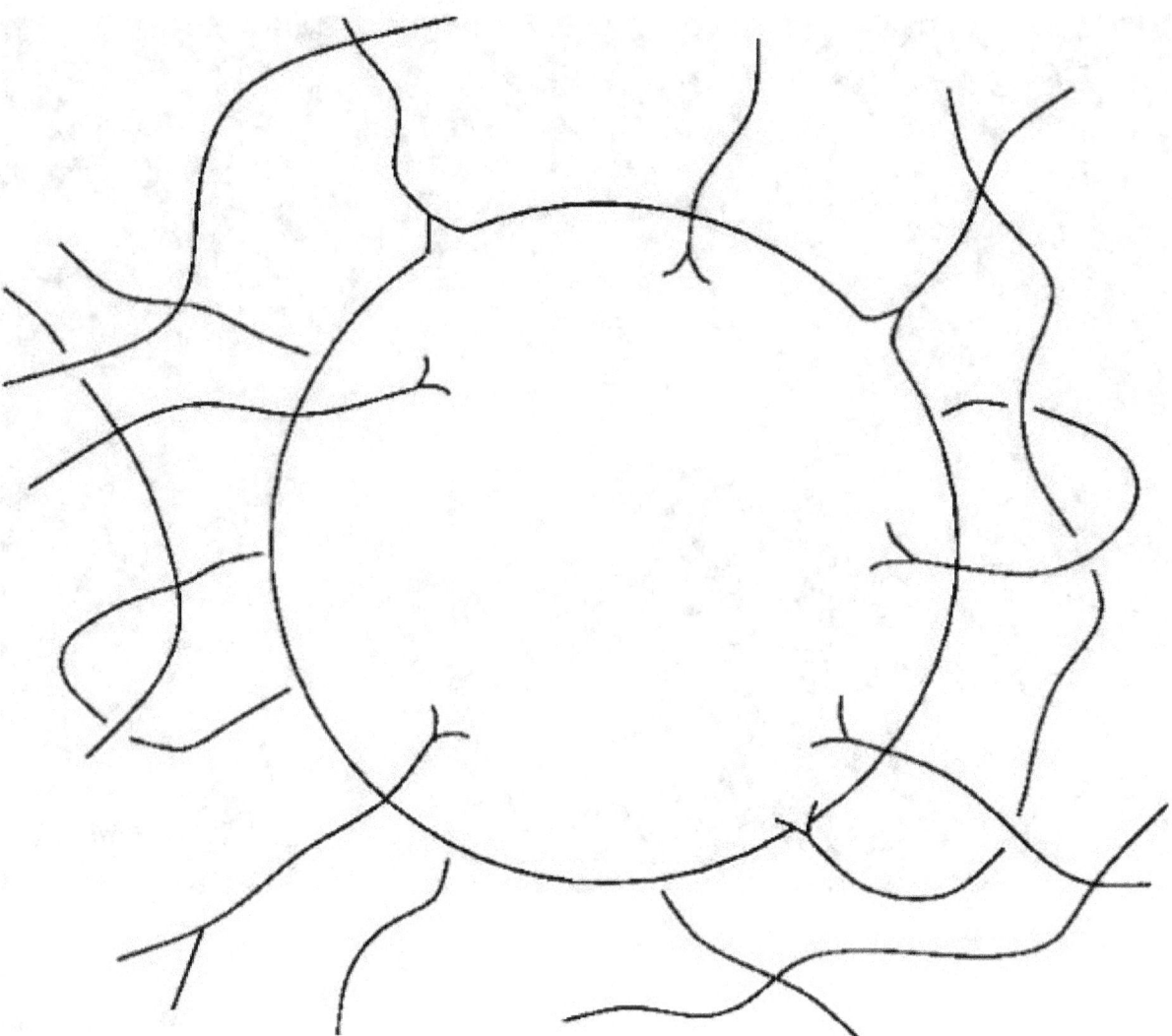

Representation of quantum geometries of the horizon. Polymer excitations in the bulk puncture the horizon, endowing it with quantized area. Intrinsically the horizon is flat except at punctures where it acquires a quantized deficit angle or quantized amount of curvature. These deficit angles add up to 4π .

It is possible to derive, from the covariant formulation of full quantum theory (Spinfoam) the correct relation between energy and area (1st law), the Unruh temperature and the distribution that yields Hawking entropy.[58] The calculation makes use of the notion of dynamical horizon and is done for non-extremal black holes.

A recent success of the theory in this direction is the computation of the entropy of all non singular black holes directly from theory and independent of Immirzi parameter.[59] The result is the expected formula $S = A/4$, where S is the entropy and A the area of the black hole, derived by Bekenstein and Hawking on heuristic grounds. This is the only known derivation of this formula from a fundamental theory, for the case of generic non singular black holes. Older attempts at this calculation had difficulties. The problem was that although Loop quantum gravity predicted that the entropy of a black hole is proportional to the area of the event horizon, the result depended on a crucial free parameter in the theory, the above-mentioned Immirzi parameter. However, there is no known computation of the Immirzi parameter, so it had to be fixed by demanding agreement with Bekenstein and Hawking's calculation of the black hole entropy.

10.8.2 Loop quantum cosmology

Main articles: loop quantum cosmology, Big bounce and inflation (cosmology)

The popular and technical literature makes extensive references to LQG-related topic of loop quantum cosmology. LQC was mainly developed by Martin Bojowald, it was popularized Loop quantum cosmology in *Scientific American* for predicting a Big Bounce prior to the Big Bang. Loop quantum cosmology (LQC) is a symmetry-reduced model of classical general relativity quantized using methods that mimic those of loop quantum gravity (LQG) that predicts a "quantum bridge" between contracting and expanding cosmological branches.

Achievements of LQC have been the resolution of the big bang singularity, the prediction of a Big Bounce, and a natural mechanism for inflation (cosmology).

LQC models share features of LQG and so is a useful toy model. However, the results obtained are subject to the usual restriction that a truncated classical theory, then quantized, might not display the true behaviour of the full theory due to artificial suppression of degrees of freedom that might have large quantum fluctuations in the full theory. It has been argued that singularity avoidance in LQC are by mechanisms only available in these restrictive models and that singularity avoidance in the full theory can still be obtained but by a more subtle feature of LQG.[60][61]

10.8.3 Loop quantum gravity phenomenology

Quantum gravity effects are notoriously difficult to measure because the Planck length is so incredibly small. However recently physicists have started to consider the possibility of measuring quantum gravity effects, mostly from astrophysical observations and gravitational wave detectors.The energy of those fluctuations at scales this small cause space-perturbations which are visible at higher scales.

10.8.4 Background independent scattering amplitudes

Loop quantum gravity is formulated in a background-independent language. No spacetime is assumed a priori, but rather it is built up by the states of theory themselves - however scattering amplitudes are derived from n -point functions (Correlation function (quantum field theory)) and these, formulated in conventional quantum field theory, are functions of points of a background space-time. The relation between the background-independent formalism and the conventional formalism of quantum field theory on a given spacetime is far from obvious, and it is far from obvious how to recover low-energy quantities from the full background-independent theory. One would like to derive the n -point functions of the theory from the background-independent formalism, in order to compare them with the standard perturbative expansion of quantum general relativity and therefore check that loop quantum gravity yields the correct low-energy limit.

A strategy for addressing this problem has been suggested;[62] the idea is to study the boundary amplitude, namely a path integral over a finite space-time region, seen as a function of the boundary value of the field.[63] In conventional quantum field theory, this boundary amplitude is well-defined[64][65] and codes the physical information of the theory; it does so in quantum gravity as well, but in a fully background-independent manner.[66] A generally covariant definition of n -point functions can then be based on the idea that the distance between physical points –arguments of the n -point function is determined by the state of the gravitational field on the boundary of the spacetime region considered.

Progress has been made in calculating background independent scattering amplitudes this way with the use of spin foams. This is a way to extract physical information from the theory. Claims to have reproduced the correct behaviour for graviton scattering amplitudes and to have recovered classical gravity have been made. "We have calculated Newton's law starting from a world with no space and no time." - Carlo Rovelli.

10.9 Gravitons, string theory, supersymmetry, extra dimensions in LQG

Main articles: graviton, string theory, supersymmetry, Kaluza–Klein theory and supergravity

Some quantum theories of gravity posit a spin-2 quantum field that is quantized, giving rise to gravitons. In string theory one generally starts with quantized excitations on top of a classically fixed background. This theory is thus described as background dependent. Particles like photons as well as changes in the spacetime geometry (gravitons) are both described as excitations on the string worldsheet. The background dependence of string theory can have important physical consequences, such as determining the number of quark generations. In contrast, loop quantum gravity, like general relativity, is manifestly background independent, eliminating the background required in string theory. Loop quantum gravity, like string theory, also aims to overcome the nonrenormalizable divergences of quantum field theories.

LQG never introduces a background and excitations living on this background, so LQG does not use gravitons as building blocks. Instead one expects that one may recover a kind of semiclassical limit or weak field limit where something like "gravitons" will show up again. In contrast, gravitons play a key role in string theory where they are among the first (massless) level of excitations of a superstring.

LQG differs from string theory in that it is formulated in 3 and 4 dimensions and without supersymmetry or Kaluza-Klein extra dimensions, while the latter requires both to be true. There is no experimental evidence to date that confirms string theory's predictions of supersymmetry and Kaluza–Klein extra dimensions. In a 2003 paper A dialog on quantum gravity,[67] Carlo Rovelli regards the fact LQG is formulated in 4 dimensions and without supersymmetry as a strength of the theory as it represents the most parsimonious explanation, consistent with current experimental results, over its rival string/M-theory. Proponents of string theory will often point to the fact that, among other things, it demonstrably reproduces the established theories of general relativity and quantum field theory in the appropriate limits, which Loop Quantum Gravity has struggled to do. In that sense string theory's connection to established physics may be considered more reliable and less speculative, at the mathematical level. Peter Woit in Not Even Wrong and Lee Smolin in The Trouble with Physics regard string/M-theory to be in conflict with current known experimental results.

Since LQG has been formulated in 4 dimensions (with and without supersymmetry), and M-theory requires supersymmetry and 11 dimensions, a direct comparison between the two has not been possible. It is possible to extend mainstream LQG formalism to higher-dimensional supergravity, general relativity with supersymmetry and Kaluza–Klein extra dimensions should experimental evidence establish their existence. It would therefore be desirable to have higher-dimensional Supergravity loop quantizations at one's disposal in order to compare these approaches. In fact a series of recent papers have been published attempting just this.[68][69][70][71][72][73][74][75] Most recently, Thiemann (and alumni) have made progress toward calculating black hole entropy for supergravity in higher dimensions. It will be interesting to compare these results to the corresponding super string calculations.[76][77]

As of April 2013 LHC has failed to find evidence of supersymmetry or Kaluza–Klein extra dimensions, which has encouraged LQG researchers. Shaposhnikov in his paper "Is there a new physics between electroweak and Planck scales?" has proposed the neutrino minimal standard model,[78] which claims the most parsimonious theory is a standard model extended with neutrinos, plus gravity, and that extra dimensions, GUT physics, and supersymmetry, string/M-theory physics are unrealized in nature, and that any theory of quantum gravity must be four dimensional, like loop quantum gravity.

10.10   LQG and related research programs

Main articles: noncommutative geometry, twistor theory, entropic gravity, Sundance Bilson-Thompson, Asymptotic safety in quantum gravity, Causal dynamical triangulation, group field theory and consistent discretizations

Several research groups have attempted to combine LQG with other research programs: Johannes Aastrup, Jesper M. Grimstrup et al. research combines noncommutative geometry with loop quantum gravity,[79] Laurent Freidel, Simone Speziale, et al., spinors and twistor theory with loop quantum gravity,[80] and Lee Smolin et al. with Verlinde entropic gravity and loop gravity.[81] Stephon Alexander, Antonino Marciano and Lee Smolin have attempted to explain the origins of weak force chirality in terms of Ashketar's variables, which describe gravity as chiral,[82] and LQG with Yang–Mills theory fields[83] in four dimensions. Sundance Bilson-Thompson, Hackett et al.,[84][85] has attempted to introduce standard model via LQG"s degrees of freedom as an emergent property (by employing the idea noiseless subsystems a useful notion introduced in more general situation for constrained systems by Fotini Markopoulou-Kalamara et al.[86]) LQG has also drawn philosophical comparisons with causal dynamical triangulation[87] and asymptotically safe gravity,[88] and the spinfoam with group field theory and AdS/CFT correspondence.[89] Smolin and Wen have suggested combining LQG with String-net liquid, tensors, and Smolin and Fotini Markopoulou-Kalamara Quantum Graphity. There is the consistent discretizations approach. In addition to what has already mentioned above, Pullin and Gambini provide a framework to connect the path integral and canonical approaches to quantum gravity. They may help reconcile the spin foam and canonical loop representation approaches. Recent research by Chris Duston and Matilde Marcolli introduces topology change via topspin networks.[90]

10.11 Problems and comparisons with alternative approaches

Main article: List of unsolved problems in physics

Some of the major unsolved problems in physics are theoretical, meaning that existing theories seem incapable of explaining a certain observed phenomenon or experimental result. The others are experimental, meaning that there is a difficulty in creating an experiment to test a proposed theory or investigate a phenomenon in greater detail.

Can quantum mechanics and general relativity be realized as a fully consistent theory (perhaps as a quantum field theory)? Is spacetime fundamentally continuous or discrete? Would a consistent theory involve a force mediated by a hypothetical graviton, or be a product of a discrete structure of spacetime itself (as in loop quantum gravity)? Are there deviations from the predictions of general relativity at very small or very large scales or in other extreme circumstances that flow from a quantum gravity theory?

The theory of LQG is one possible solution to the problem of quantum gravity, as is string theory. There are substantial differences however. For example, string theory also addresses unification, the understanding of all known forces and particles as manifestations of a single entity, by postulating extra dimensions and so-far unobserved additional particles and symmetries. Contrary to this, LQG is based only on quantum theory and general relativity and its scope is limited to understanding the quantum aspects of the gravitational interaction. On the other hand, the consequences of LQG are radical, because they fundamentally change the nature of space and time and provide a tentative but detailed physical and mathematical picture of quantum spacetime.

Presently, no semiclassical limit recovering general relativity has been shown to exist. This means it remains unproven that LQG's description of spacetime at the Planck scale has the right continuum limit (described by general relativity with possible quantum corrections). Specifically, the dynamics of the theory is encoded in the Hamiltonian constraint, but there is no candidate Hamiltonian.[91] Other technical problems include finding off-shell closure of the constraint algebra and physical inner product vector space, coupling to matter fields of Quantum field theory, fate of the renormalization of the graviton in perturbation theory that lead to ultraviolet divergence beyond 2-loops (see One-loop Feynman diagram in Feynman diagram).[91]

While there has been a recent proposal relating to observation of naked singularities,[92] and doubly special relativity as a part of a program called loop quantum cosmology, there is no experimental observation for which loop quantum gravity makes a prediction not made by the Standard Model or general relativity (a problem that plagues all current theories of quantum gravity). Because of the above-mentioned lack of a semiclassical limit, LQG has not yet even reproduced the predictions made by general relativity.

An alternative criticism is that general relativity may be an effective field theory, and therefore quantization ignores the fundamental degrees of freedom.

10.12 See also

10.13 Notes

[1] Rovelli, Carlo (August 2008). "Loop Quantum Gravity" (PDF). *CERN*. Retrieved 14 September 2014.

[2] Rovelli, C. (2011). "Zakopane lectures on loop gravity". arXiv:1102.3660 [gr-qc].

[3] Muxin, H. (2011). "Cosmological constant in loop quantum gravity vertex amplitude". *Physical Review D* **84** (6): 064010. arXiv:1105.2212. Bibcode:2011PhRvD..84f4010H. doi:10.1103/PhysRevD.84.064010.

[4] Fairbairn, W. J.; Meusburger, C. (2011). "q-Deformation of Lorentzian spin foam models". arXiv:1112.2511 [gr-qc].

[5] Rovelli, C. (2004). *Quantum Gravity*. Cambridge Monographs on Mathematical Physics. p. 71. ISBN 978-0-521-83733-0.

[6] Kauffman, S.; Smolin, L. (7 April 1997). "A Possible Solution For The Problem Of Time In Quantum Cosmology". *Edge.org*. Retrieved 2014-08-20.

[7] Smolin, L. (2006). "The Case for Background Independence". In Rickles, D.; French, S.; Saatsi, J. T. *The Structural Foundations of Quantum Gravity*. Clarendon Press. pp. 196*ff*. arXiv:hep-th/0507235. ISBN 978-0-19-926969-3.

[8] Rovelli, C. (2004). *Quantum Gravity*. Cambridge Monographs on Mathematical Physics. p. 13ff. ISBN 978-0-521-83733-0.

[9] Thiemann, T. (1996). "Anomaly-free formulation of non-perturbative, four-dimensional Lorentzian quantum gravity". *Physics Letters B* **380**: 257–264. arXiv:gr-qc/9606088. Bibcode:1996PhLB..380..257T. doi:10.1016/0370-2693(96)00532-1.

[10] Baez, J.; de Muniain, J. P. (1994). *Gauge Fields, Knots and Quantum Gravity*. Series on Knots and Everything. Vol. 4. World Scientific. Part III, chapter 4. ISBN 978-981-02-1729-7.

[11] Thiemann, T. (2003). "Lectures on Loop Quantum Gravity". *Lecture Notes in Physics* **631**: 41–135. arXiv:gr-qc/0210094. Bibcode:2003LNP...631...41T. doi:10.1007/978-3-540-45230-0_3.

[12] Rovelli,C.;Smolin,L.(1988). "Knot Theory and Quantum Gravity". *Physical Review Letters* **61**(10): 1155–1958. doi:10.1103/PhysRevLett.61.1155.

[13] Gambini, R.; Pullin, J. (2011). *A First Course in Loop Quantum Gravity*. Oxford University Press. Section 8.2. ISBN 978-0-19-959075-9.

[14] Fernando, J.; Barbero, G. (1995). "Reality Conditions and Ashtekar Variables: A Different Perspective". *Physical Review D* **51**: 5498–5506. arXiv:gr-qc/9410013. Bibcode:1995PhRvD..51.5498B. doi:10.1103/PhysRevD.51.5498.

[15] Fernando, J.; Barbero, G. (1995). "Real Ashtekar Variables for Lorentzian Signature Space-times". *Physical Review D* **51**: 5507–5520. arXiv:gr-qc/9410014. Bibcode:1995PhRvD..51.5507B. doi:10.1103/PhysRevD.51.5507.

[16] Bojowald, M.; Alejandro. P. "Spin Foam Quantization and Anomalies". arXiv:gr-qc/0303026 [gr-qc].

[17] Barrett, J.; Crane, L. (2000). "A Lorentzian signature model for quantum general relativity". *Classical and Quantum Gravity* **17**: 3101–3118. arXiv:gr-qc/9904025. Bibcode:2000CQGra..17.3101B. doi:10.1088/0264-9381/17/16/302..

[18] Rovelli, C.; Alesci, E. (2007). "The complete LQG propagator I. Difficulties with the Barrett–Crane vertex". *Physical Review D* **76**: 104012. arXiv:hep-th/0703074. Bibcode:2007PhRvD..76b4012B. doi:10.1103/PhysRevD.76.024012.

[19] Engle, J.; Pereira, R.; Rovelli, C. (2009). "Loop-Quantum-Gravity Vertex Amplitude". *Physical Review Letters* **99**: 161301. arXiv:0705.2388. Bibcode:2007PhRvL..99p1301E. doi:10.1103/physrevlett.99.161301.

[20] Freidal, L.; Krasnov, K. (2008). "A new spin foam model for 4D gravity". *Classical and Quantum Gravity* **25**: 125018. arXiv:0708.1595. Bibcode:2008CQGra..25l5018F. doi:10.1088/0264-9381/25/12/125018.

[21] Alesci, E.; Thiemann, T.; Zipfel, A. (2011). "Linking covariant and canonical LQG: new solutions to the Euclidean Scalar Constraint". arXiv:1109.1290.

[22] Bohm, D. (1989). *Quantum Theory*. Dover Publications. ISBN 978-0-486-65969-5.

[23] Tipler, P.; Llewellyn, R. (2008). *Modern Physics* (5th ed.). W. H. Freeman and Co. pp. 160–161. ISBN 978-0-7167-7550-8.

[24] Bohr, N. (1920). "Über die Serienspektra der Element". *Zeitschrift für Physik* **2** (5): 423–478. Bibcode:1920ZPhy....2..423B. doi:10.1007/BF01329978. (English translation in Bohr 1976, pp. 241–282)

[25] Jammer, M. (1989). *The Conceptual Development of Quantum Mechanics* (2nd ed.). Tomash Publishers. Section 3.2. ISBN 978-0-88318-617-6.

[26] Ashtekar, A.; Bombelli, L.; Corichi, A. (2005). "Semiclassical States for Constrained Systems". *Physical Review D* **72**: 025008. arXiv:hep-ph/0504114. Bibcode:2005PhRvD..72a5008C. doi:10.1103/PhysRevD.72.015008.

[27] Lewandowski, J.; Okołów, A.; Sahlmann, H.; Thiemann, T. (2005). "Uniqueness of Diffeomorphism Invariant States on Holonomy-Flux Algebras". *Communications in Mathematical Physics***267**: 703–733. arXiv:gr-qc/0504147. Bibcode:2006CM doi:10.1007/s00220-006-0100-7.

[28] Fleischhack, C. (2006). "Irreducibility of the Weyl algebra in loop quantum gravity". *Physical Review Letters* **97**: 061302. Bibcode:2006PhRvL..97f1302F. doi:10.1103/physrevlett.97.061302.

[29] Thiemann, T. (2008). *Modern Canonical General Relativity*. Cambridge Monographs on Mathematical Physics. Cambridge University Press. Section 10.6. ISBN 978-0-521-74187-3.

[30] "Partial and Complete Observables for Hamiltonian Constrained Systems". *General Relativity and Gravitation* **39**: 1891–1927. 2007. arXiv:gr-qc/0411013. Bibcode:2007GReGr..39.1891D. doi:10.1007/s10714-007-0495-2.

[31] "Partial and Complete Observables for Canonical General Relativity". *Classical and Quantum Gravity* **23**: 6155–6184. arXiv:gr-qc/0507106. Bibcode:2006CQGra..23.6155D. doi:10.1088/0264-9381/23/22/006.

[32] Dreyer, O.; Markopoulou, f.; Smolin, L. (2006). "Symmetry and entropy of black hole horizons". *Nuclear Physics B* **774**: 1–13. arXiv:hep-th/0409056. Bibcode:2006NuPhB.744....1D. doi:10.1016/j.nuclphysb.2006.02.045.

[33] Kribs, D. W.; Markopoulou, F. "Geometry from quantum particles". arXiv:gr-qc/0510052.

[34] Markopoulou, F.; Poulin, D. "Noiseless subsystems and the low energy limit of spin foam models" (unpublished).

[35] *The Phoenix Project: Master Constraint Programme for Loop Quantum Gravity*. Class.Quant.Grav.23:2211-2248,2006 or http://fr.arxiv.org/pdf/gr-qc/0305080

[36] *Modern Canonical Quantum General Relativity* by Thomas Thiemann

[37] *Testing the Master Constraint Programme for Loop Quantum Gravity I. General Framework*, Bianca Dittrich, Thomas Thiemann, Class.Quant.Grav. 23 (2006) 1025-1066.

[38] *Testing the Master Constraint Programme for Loop Quantum Gravity II. Finite Dimensional Systems*, Bianca Dittrich, Thomas Thiemann, Class.Quant.Grav. 23 (2006) 1067-1088.

[39] *Testing the Master Constraint Programme for Loop Quantum Gravity III. SL(2,R) Models*, Bianca Dittrich, Thomas Thiemann, Class.Quant.Grav. 23 (2006) 1089-1120.

[40] *Testing the Master Constraint Programme for Loop Quantum Gravity IV. Free Field Theories*, Bianca Dittrich, Thomas Thiemann, Class.Quant.Grav. 23 (2006) 1121-1142.

[41] *Testing the Master Constraint Programme for Loop Quantum Gravity V. Interacting Field Theories*, Bianca Dittrich, Thomas Thiemann, Class.Quant.Grav. 23 (2006) 1143-1162.

[42] *Quantum Spin Dynamics VIII. The Master Constraint*, Thomas Thiemann, Class.Quant.Grav. 23 (2006) 2249-2266.

[43] *Approximating the physical inner product of Loop Quantum Cosmology*, Benjamin Bahr, Thomas Thiemann, Class.Quant.Grav 2138.2007.

[44] *On the Relation between Operator Constraint --, Master Constraint --, Reduced Phase Space --, and Path Integral Quantisation*, Muxin Han, Thomas Thiemann, Class.Quant.Grav.27:225019,2010.

[45] *On the Relation between Rigging Inner Product and Master Constraint Direct Integral Decomposition*, Muxin Han, Thomas Thiemann, J.Math.Phys.51:092501,2010.

[46] *A Path-integral for the Master Constraint of Loop Quantum Gravity*, Muxin Han, Class.Quant.Grav.27:215009,2010

[47] *Emergent diffeomorphism invariance in a discrete loop quantum gravity model*, Rodolfo Gambini, Jorge Pullin, Class.Quant.Gra

[48] Section 10.2.2 *A First Course in Loop quantum Gravity*, Rodolfo Gambinni, Jorge Pullin, Oxford University Press, first published 2011.

[49] *Algebraic Quantum Gravity (AQG) I. Conceptual Setup*, K. Giesel, T. Thiemann, Class.Quant.Grav.24:2465-2498,2007.

[50] *Algebraic Quantum Gravity (AQG) II. Semiclassical Analysis*, K. Giesel, T. Thiemann, Class.Quant.Grav.24:2499-2564,2007.

[51] *Algebraic Quantum Gravity (AQG) III. Semiclassical Perturbation Theory*, K. Giesel, T. Thiemann, Class.Quant.Grav.24:2565-2588,2007.

[52] Bousso, Raphael (2002). "The Holographic Principle". *Reviews of Modern Physics* **74** (3): 825–874. arXiv:hep-th/0203101. Bibcode:2002RvMP...74..825B. doi:10.1103/RevModPhys.74.825.

[53] Majumdar,Parthasarathi(1998). "Black Hole Entropy and Quantum Gravity"**73**. p. 147. arXiv:gr-qc/9807045. Bibcode:

[54] See List of loop quantum gravity researchers

[55] Rovelli, Carlo (1996). "Black Hole Entropy from Loop Quantum Gravity". *Physical Review Letters* **77** (16): 3288–3291. arXiv:gr-qc/9603063. Bibcode:1996PhRvL..77.3288R. doi:10.1103/PhysRevLett.77.3288.

[56] Ashtekar, Abhay; Baez, John; Corichi, Alejandro; Krasnov, Kirill (1998). "Quantum Geometry and Black Hole Entropy". *Physical Review Letters* **80** (5): 904–907. arXiv:gr-qc/9710007. Bibcode:1998PhRvL..80..904A. doi:10.1103/PhysRevLett.80.904.

[57] *Quantum horizons and black hole entropy: Inclusion of distortion and rotation*, Abhay Ashtekar, Jonathan Engle, Chris Van Den Broeck, Class.Quant.Grav.22:L27-L34, 2005.

[58] Bianchi, Eugenio (2012). "Entropy of Non-Extremal Black Holes from Loop Gravity". arXiv:1204.5122.

[59] http://inspirehep.net/record/940357?ln=en. http://inspirehep.net/record/1111991.

[60] *On (Cosmological) Singularity Avoidance in Loop Quantum Gravity*, Johannes Brunnemann, Thomas Thiemann, Class.Quant.Grav. 23 (2006) 1395-1428.

[61] *Unboundedness of Triad -- Like Operators in Loop Quantum Gravity*, Johannes Brunnemann, Thomas Thiemann, Class.Quant.Grav. 23 (2006) 1429-1484.

[62] L. Modesto, C. Rovelli:*Particle scattering in loop quantum gravity*, Phys Rev Lett 95 (2005) 191301

[63] R Oeckl, *A 'general boundary' formulation for quantum mechanics and quantum gravity*, Phys Lett B575 (2003) 318-324 ; *Schrodinger's cat and the clock: lessons for quantum gravity*, Class Quant Grav 20 (2003) 5371-5380l

[64] F. Conrady, C. Rovelli *Generalized Schrodinger equation in Euclidean field theory"*, Int J Mod Phys A 19, (2004) 1-32.

[65] L. Doplicher, *Generalized Tomonaga-Schwinger equation from the Hadamard formula*, Phys Rev D70 (2004) 064037

[66] F. Conrady, L. Doplicher, R. Oeckl, C. Rovelli, M. Testa, *Minkowski vacuum in background independent quantum gravity*, Phys Rev D69 (2004) 064019.

[67] http://arxiv.org/abs/arXiv:hep-th/0310077

[68] *New Variables for Classical and Quantum Gravity in all Dimensions I. Hamiltonian Analysis*, Norbert Bodendorfer, Thomas Thiemann, Andreas Thurn, Class. Quantum Grav. 30 (2013) 045001

[69] *New Variables for Classical and Quantum Gravity in all Dimensions II. Lagrangian Analysis*, Norbert Bodendorfer, Thomas Thiemann, Andreas Thurn, Quantum Grav. 30 (2013) 045002

[70] *New Variables for Classical and Quantum Gravity in all Dimensions III. Quantum Theory*, Norbert Bodendorfer, Thomas Thiemann, Andreas Thurn, Class. Quantum Grav. 30 (2013) 045003

[71] *New Variables for Classical and Quantum Gravity in all Dimensions IV. Matter Coupling*, Norbert Bodendorfer, Thomas Thiemann, Andreas Thurn, Class. Quantum Grav. 30 (2013) 045004

[72] *On the Implementation of the Canonical Quantum Simplicity Constraint*, Norbert Bodendorfer, Thomas Thiemann, Andreas Thurn, Class. Quantum Grav. 30 (2013) 045005

[73] *Towards Loop Quantum Supergravity (LQSG) I. Rarita-Schwinger Sector*, Norbert Bodendorfer, Thomas Thiemann, Andreas Thurn, Class. Quantum Grav. 30 (2013) 045006

[74] *Towards Loop Quantum Supergravity (LQSG) II. p-Form Sector*, Norbert Bodendorfer, Thomas Thiemann, Andreas Thurn, Class. Quantum Grav. 30 (2013) 045007

[75] *Towards Loop Quantum Supergravity (LQSG)*, Norbert Bodendorfer, Thomas Thiemann, Andreas Thurn, Phys. Lett. B 711: 205-211 (2012)

[76] *New Variables for Classical and Quantum Gravity in all Dimensions V. Isolated Horizon Boundary Degrees of Freedom*, Norbert Bodendorfer, Thomas Thiemann, Andreas Thurn, http://uk.arxiv.org/pdf/1304.2679.

[77] *Black hole entropy from loop quantum gravity in higher dimensions*, Norbert Bodendorfer http://uk.arxiv.org/pdf/1307.5029

[78] http://arxiv.org/abs/0708.3550

[79] http://arxiv.org/abs/1203.6164

[80] http://arxiv.org/abs/1006.0199

[81] http://arxiv.org/abs/1001.3668

[82] http://arxiv.org/abs/1212.5246

[83] http://arxiv.org/abs/1105.3480

[84] *Quantum gravity and the standard model*,Sundance O.Bilson-Thompson,Fotini Markopoulou,Lee Smolin,Class.Quant.Grav. 3994,2007.

[85] For a precise review and outlook of this research see: *Emergent Braided Matter of Quantum Geometry*, Sundance Bilson-Thompson, Jonathan Hackett, Louis Kauffman, Yidun Wan, SIGMA 8 (2012), 014, 43 pages.

[86] *Constrained Mechanics and Noiseless Subsystems*, Tomasz Konopka, Fotini Markopoulou, arXiv:gr-qc/0601028.

[87] http://www.perimeterinstitute.ca/people/renate-loll

[88] wwnpqft.inln.cnrs.fr/pdf/Bianchi.pdf

[89] http://arxiv.org/abs/0804.0632

[90] http://arxiv.org/abs/1308.2934

[91] Nicolai, Hermann; Peeters, Kasper; Zamaklar, Marija (2005). "Loop quantum gravity: an outside view". *Classical and Quantum Gravity* **22** (19): R193–R247. arXiv:hep-th/0501114. Bibcode:2005CQGra..22R.193N. doi:10.1088/0264-9381/22/19/R01.

[92] Goswami; Joshi, Pankaj S.; Singh, Parampreet; et al. (2006). "Quantum evaporation of a naked singularity". *Physical Review Letters* **96** (3): 31302. arXiv:gr-qc/0506129. Bibcode:2006PhRvL..96c1302G. doi:10.1103/PhysRevLett.96.031302.

10.14 References

- Topical Reviews

  - Rovelli, Carlo (2011). "Zakopane lectures on loop gravity". arXiv:1102.3660.

  - Rovelli, Carlo (1998). "Loop Quantum Gravity". *Living Reviews in Relativity* 1. Retrieved 2008-03-13.

  - Thiemann, Thomas (2003). "Lectures on Loop Quantum Gravity". *Lectures Notes in Physics.* Lecture Notes in Physics **631**: 41–135. arXiv:gr-qc/0210094. Bibcode:2003LNP...631...41T. doi:10.1007/978-3-540-45230-0_3. ISBN 978-3-540-40810-9.

  - Ashtekar, Abhay; Lewandowski, Jerzy (2004). "Background Independent Quantum Gravity: A Status Report". *Classical and Quantum Gravity*21(15): R53–R152. arXiv:gr-qc/0404018. Bibcode:2004CQGra..21R doi:10.1088/0264-9381/21/15/R01.

  - Carlo Rovelli and Marcus Gaul, *Loop Quantum Gravity and the Meaning of Diffeomorphism Invariance*, e-print available as gr-qc/9910079.

  - Lee Smolin, *The case for background independence*, e-print available as hep-th/0507235.

  - Alejandro Corichi, *Loop Quantum Geometry: A primer*, e-print available as .

  - Alejandro Perez, *Introduction to loop quantum gravity and spin foams*, e-print available as .

  - Hermann Nicolai and Kasper Peeters *Loop and spin foam quantum gravity: A Brief guide for beginners.*, e-print available as .

- Popular books:

  - Lee Smolin, *Three Roads to Quantum Gravity*

  - Carlo Rovelli, *Che cos'è il tempo? Che cos'è lo spazio?*, Di Renzo Editore, Roma, 2004. French translation: *Qu'est ce que le temps? Qu'est ce que l'espace?*, Bernard Gilson ed, Brussel, 2006. English translation: *What is Time? What is space?*, Di Renzo Editore, Roma, 2006.

  - Julian Barbour, *The End of Time: The Next Revolution in Our Understanding of the Universe*

  - Musser, George (2008). "The Complete Idiot's Guide to String Theory". *The Physics Teacher* (Indianapolis: Alpha) **47** (2): 368. Bibcode:2009PhTea..47Q.128H. doi:10.1119/1.3072469. ISBN 978-1-59257-702-6. – Focuses on string theory but has an extended discussion of loop gravity as well.

- Magazine articles:

  - Lee Smolin, "Atoms of Space and Time", *Scientific American*, January 2004

  - Martin Bojowald, "Following the Bouncing Universe", *Scientific American*, October 2008

- Easier introductory, expository or critical works:

  - Abhay Ashtekar, *Gravity and the quantum*, e-print available as gr-qc/0410054 (2004)

  - John C. Baez and Javier Perez de Muniain, *Gauge Fields, Knots and Quantum Gravity*, World Scientific (1994)

  - Carlo Rovelli, *A Dialog on Quantum Gravity*, e-print available as hep-th/0310077 (2003)

  - Rodolfo Gambini and Jorge Pullin, *A First Course in Loop Quantum Gravity*, Oxford (2011)

  - Carlo Rovelli and Francesca Vidotto, *Covariant Loop Quantum Gravity*, Cambridge (2014); draft available online

- More advanced introductory/expository works:

  - Carlo Rovelli, *Quantum Gravity*, Cambridge University Press (2004); draft available online

  - Thomas Thiemann, *Introduction to modern canonical quantum general relativity*, e-print available as gr-qc/0110034

  - Thomas Thiemann, *Introduction to Modern Canonical Quantum General Relativity*, Cambridge University Press (2007)

  - Abhay Ashtekar, *New Perspectives in Canonical Gravity*, Bibliopolis (1988).

  - Abhay Ashtekar, *Lectures on Non-Perturbative Canonical Gravity*, World Scientific (1991)

  - Rodolfo Gambini and Jorge Pullin, *Loops, Knots, Gauge Theories and Quantum Gravity*, Cambridge University Press (1996)

  - Hermann Nicolai, Kasper Peeters, Marija Zamaklar, *Loop quantum gravity: an outside view*, e-print available as hep-th/0501114

  - H. Nicolai and K. Peeters, *Loop and Spin Foam Quantum Gravity: A Brief Guide for Beginners*, e-print available as hep-th/0601129

  - T. Thiemann The LQG – String: Loop Quantum Gravity Quantization of String Theory (2004)

- Conference proceedings:

  - John C. Baez (ed.), *Knots and Quantum Gravity*

- Fundamental research papers:

  - Ashtekar, Abhay (1986). "New variables for classical and quantum gravity". *Physical Review Letters* **57** (18): 2244–2247. Bibcode:1986PhRvL..57.2244A. doi:10.1103/PhysRevLett.57.2244. PMID 10033673

  - Ashtekar, Abhay (1987). "New Hamiltonian formulation of general relativity". *Physical Review D* **36** (6): 1587–1602. Bibcode:1987PhRvD..36.1587A. doi:10.1103/PhysRevD.36.1587

- Roger Penrose, *Angular momentum: an approach to combinatorial space-time* in *Quantum Theory and Beyond*, ed. Ted Bastin, Cambridge University Press, 1971

- Rovelli, Carlo; Smolin, Lee (1988). "Knot theory and quantum gravity". *Physical Review Letters* **61** (10): 1155–1158. Bibcode:1988PhRvL..61.1155R. doi:10.1103/PhysRevLett.61.1155.

- Rovelli, Carlo; Smolin, Lee (1990). "Loop space representation of quantum general relativity". *Nuclear Physics* **B331**: 80–152.

- Carlo Rovelli and Lee Smolin, *Discreteness of area and volume in quantum gravity*, Nucl. Phys., **B442** (1995) 593-622, e-print available as gr-qc/9411005

- Kuchař, Karel (1973). "Canonical Quantization of Gravity". In Israel, Werner. *Relativity, Astrophysics and Cosmology*. D. Reidel. pp. 237–288. ISBN 90-277-0369-8.

- Thiemann, Thomas (2006). "Loop Quantum Gravity: An Inside View". *Approaches to Fundamental Physics*. Lecture Notes in Physics **721**: 185–263. arXiv:hep-th/0608210. Bibcode:2007LNP...721..185T.doi:10.1007 3-540-71117-9_10. ISBN 978-3-540-71115-5.

10.15 External links

- "Loop Quantum Gravity" by Carlo Rovelli Physics World, November 2003

- Quantum Foam and Loop Quantum Gravity

- Abhay Ashtekar: Semi-Popular Articles . Some excellent popular articles suitable for beginners about space, time, GR, and LQG.

- Loop Quantum Gravity: Lee Smolin.

- Loop Quantum Gravity on arxiv.org

- A list of LQG references catered to fresh graduates

- Loop Quantum Gravity Lectures Online by Lee Smolin

- Spin networks, spin foams and loop quantum gravity

- Wired magazine, News: *Moving Beyond String Theory*

- April 2006 Scientific American Special Issue, *A Matter of Time*, has Lee Smolin LQG Article *Atoms of Space and Time*

- September 2006, The Economist, article *Looping the loop*

- Gamma-ray Large Area Space Telescope: http://glast.gsfc.nasa.gov/

- Zeno meets modern science. Article from Acta Physica Polonica B by Z.K. Silagadze.

- Did pre-big bang universe leave its mark on the sky? - According to a model based on "loop quantum gravity" theory, a parent universe that existed before ours may have left an imprint (*New Scientist*, 10 April 2008)

Chapter 11

Isomorphism

This article is about mathematics.

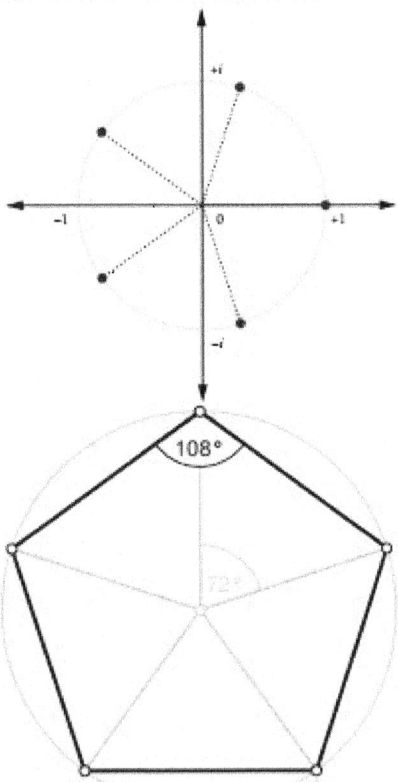

The group of fifth roots of unity under multiplication is isomorphic to the group of rotations of the regular pentagon under composition.

In mathematics, an **isomorphism** (from the Ancient Greek: ἴσος *isos* "equal", and μορφή *morphe* "form" or "shape") is a homomorphism (or more generally a morphism) that admits an inverse.[note 1] Two mathematical objects are **isomorphic** if an isomorphism exists between them. An *automorphism* is an isomorphism whose source and target coincide. The interest of isomorphisms lies in the fact that two isomorphic objects cannot be distinguished by using only the properties used to define morphisms; thus isomorphic objects may be considered the same as long as one considers only these properties and their consequences.

For most algebraic structures, including groups and rings, a homomorphism is an isomorphism if and only if it is bijective.

In topology, where the morphisms are continuous functions, isomorphisms are also called *homeomorphisms* or *bicontinuous functions*. In mathematical analysis, where the morphisms are differentiable functions, isomorphisms are also called *diffeomorphisms*.

A **canonical isomorphism** is a canonical map that is an isomorphism. Two objects are said to be **canonically isomorphic** if there is a canonical isomorphism between them. For example, the canonical map from a finite-dimensional vector space V to its second dual space is a canonical isomorphism; on the other hand, V is isomorphic to its dual space but not canonically in general.

Isomorphisms are formalized using category theory. A morphism $f : X \to Y$ in a category is an isomorphism if it admits a two-sided inverse, meaning that there is another morphism $g : Y \to X$ in that category such that $gf = 1X$ and $fg = 1Y$, where $1X$ and $1Y$ are the identity morphisms of X and Y, respectively.[1]

11.1 Examples

11.1.1 Logarithm and exponential

Let $\mathbb{R}^+$ be the multiplicative group of positive real numbers, and let $\mathbb{R}$ be the additive group of real numbers.

The logarithm function $\log: \mathbb{R}^+ \to \mathbb{R}$ satisfies $\log(xy) = \log x + \log y$ for all $x, y \in \mathbb{R}^+$, so it is a group homomorphism. The exponential function $\exp: \mathbb{R} \to \mathbb{R}^+$ satisfies $\exp(x + y) = (\exp x)(\exp y)$ for all $x, y \in \mathbb{R}$, so it too is a homomorphism.

The identities $\log \exp x = x$ and $\exp \log y = y$ show that log and exp are inverses of each other. Since log is a homomorphism that has an inverse that is also a homomorphism, log is an isomorphism of groups.

Because log is an isomorphism, it translates multiplication of positive real numbers into addition of real numbers. This facility makes it possible to multiply real numbers using a ruler and a table of logarithms, or using a slide rule with a logarithmic scale.

11.1.2 Integers modulo 6

Consider the group $(\mathbb{Z}_6, +)$, the integers from 0 to 5 with addition modulo 6. Also consider the group $(\mathbb{Z}_2 \times \mathbb{Z}_3, +)$, the ordered pairs where the x coordinates can be 0 or 1, and the y coordinates can be 0, 1, or 2, where addition in the x-coordinate is modulo 2 and addition in the y-coordinate is modulo 3.

These structures are isomorphic under addition, if you identify them using the following scheme:

$(0,0) \to 0$

$(1,1) \to 1$

$(0,2) \to 2$

$(1,0) \to 3$

$(0,1) \to 4$

$(1,2) \to 5$

or in general $(a,b) \to (3a + 4b)$ mod 6.

For example note that $(1,1) + (1,0) = (0,1)$, which translates in the other system as $1 + 3 = 4$.

Even though these two groups "look" different in that the sets contain different elements, they are indeed **isomorphic**: their structures are exactly the same. More generally, the direct product of two cyclic groups $\mathbb{Z}_m$ and $\mathbb{Z}_n$ is isomorphic to $(\mathbb{Z}_{mn}, +)$ if and only if m and n are coprime.

11.1.3   Relation-preserving isomorphism

If one object consists of a set X with a binary relation R and the other object consists of a set Y with a binary relation S then an isomorphism from X to Y is a bijective function $f: X \to Y$ such that:[2]

$$S(f(u), f(v)) \iff R(u, v)$$

S is reflexive, irreflexive, symmetric, antisymmetric, asymmetric, transitive, total, trichotomous, a partial order, total order, strict weak order, total preorder (weak order), an equivalence relation, or a relation with any other special properties, if and only if R is.

For example, R is an ordering ≤ and S an ordering ⊑ , then an isomorphism from X to Y is a bijective function $f: X \to Y$ such that

$$f(u) \sqsubseteq f(v) \iff u \le v.$$

Such an isomorphism is called an *order isomorphism* or (less commonly) an *isotone isomorphism*.

If $X = Y$, then this is a relation-preserving automorphism.

11.2   Isomorphism vs. bijective morphism

In a concrete category (that is, roughly speaking, a category whose objects are sets and morphisms are mappings between sets), such as the category of topological spaces or categories of algebraic objects like groups, rings, and modules, an isomorphism must be bijective on the underlying sets. In algebraic categories (specifically, categories of varieties in the sense of universal algebra), an isomorphism is the same as a homomorphism which is bijective on underlying sets. However, there are concrete categories in which bijective morphisms are not necessarily isomorphisms (such as the category of topological spaces), and there are categories in which each object admits an underlying set but in which isomorphisms need not be bijective (such as the homotopy category of CW-complexes).

11.3   Applications

In abstract algebra, two basic isomorphisms are defined:

- Group isomorphism, an isomorphism between groups

- Ring isomorphism, an isomorphism between rings. (Note that isomorphisms between fields are actually ring isomorphisms)

Just as the automorphisms of an algebraic structure form a group, the isomorphisms between two algebras sharing a common structure form a heap. Letting a particular isomorphism identify the two structures turns this heap into a group.

In mathematical analysis, the Laplace transform is an isomorphism mapping hard differential equations into easier algebraic equations.

In category theory, let the category C consist of two classes, one of *objects* and the other of morphisms. Then a general definition of isomorphism that covers the previous and many other cases is: an isomorphism is a morphism $f: a \to b$ that has an inverse, i.e. there exists a morphism $g: b \to a$ with $fg = 1b$ and $gf = 1a$. For example, a bijective linear map is an isomorphism between vector spaces, and a bijective continuous function whose inverse is also continuous is an isomorphism between topological spaces, called a homeomorphism.

In graph theory, an isomorphism between two graphs G and H is a bijective map f from the vertices of G to the vertices of H that preserves the "edge structure" in the sense that there is an edge from vertex u to vertex v in G if and only if there is an edge from $f(u)$ to $f(v)$ in H. See graph isomorphism.

In mathematical analysis, an isomorphism between two Hilbert spaces is a bijection preserving addition, scalar multiplication, and inner product.

In early theories of logical atomism, the formal relationship between facts and true propositions was theorized by Bertrand Russell and Ludwig Wittgenstein to be isomorphic. An example of this line of thinking can be found in Russell's Introduction to Mathematical Philosophy.

In cybernetics, the Good Regulator or Conant-Ashby theorem is stated "Every Good Regulator of a system must be a model of that system". Whether regulated or self-regulating an isomorphism is required between regulator part and the processing part of the system.

11.4 Relation with equality

See also: Equality (mathematics)

In certain areas of mathematics, notably category theory, it is valuable to distinguish between *equality* on the one hand and *isomorphism* on the other.[3] Equality is when two objects are exactly the same, and everything that's true about one object is true about the other, while an isomorphism implies everything that's true about a designated part of one object's structure is true about the other's. For example, the sets

$$A = \{x \in \mathbb{Z} \mid x^2 < 2\} \text{ and } B = \{-1, 0, 1\}$$

are *equal*; they are merely different presentations—the first an intensional one (in set builder notation), and the second extensional (by explicit enumeration)—of the same subset of the integers. By contrast, the sets $\{A,B,C\}$ and $\{1,2,3\}$ are not *equal*—the first has elements that are letters, while the second has elements that are numbers. These are isomorphic as sets, since finite sets are determined up to isomorphism by their cardinality (number of elements) and these both have three elements, but there are many choices of isomorphism—one isomorphism is

$$A \mapsto 1, B \mapsto 2, C \mapsto 3, \text{ while another is } A \mapsto 3, B \mapsto 2, C \mapsto 1,$$

and no one isomorphism is intrinsically better than any other.[note 2][note 3] On this view and in this sense, these two sets are not equal because one cannot consider them *identical*: one can choose an isomorphism between them, but that is a weaker claim than identity—and valid only in the context of the chosen isomorphism.

Sometimes the isomorphisms can seem obvious and compelling, but are still not equalities. As a simple example, the genealogical relationships among Joe, John, and Bobby Kennedy are, in a real sense, the same as those among the American football quarterbacks in the Manning family: Archie, Peyton, and Eli. The father-son pairings and the elder-brother-younger-brother pairings correspond perfectly. That similarity between the two family structures illustrates the origin of the word *isomorphism* (Greek *iso-*, "same," and *-morph*, "form" or "shape"). But because the Kennedys are not the same people as the Mannings, the two genealogical structures are merely isomorphic and not equal.

Another example is more formal and more directly illustrates the motivation for distinguishing equality from isomorphism: the distinction between a finite-dimensional vector space V and its dual space $V^* = \{ \varphi: V \rightarrow \mathbf{K}\}$ of linear maps from V to its field of scalars $\mathbf{K}$. These spaces have the same dimension, and thus are isomorphic as abstract vector spaces (since algebraically, vector spaces are classified by dimension, just as sets are classified by cardinality), but there is no "natural" choice of isomorphism $V \xrightarrow{\sim} V^*$. If one chooses a basis for V, then this yields an isomorphism: For all $u, v \in V$,

$$v \xrightarrow{\sim} \phi_v \in V^* \text{ that such } \phi_v(u) = v^T u$$

This corresponds to transforming a column vector (element of V) to a row vector (element of V^*) by transpose, but a different choice of basis gives a different isomorphism: the isomorphism "depends on the choice of basis". More subtly, there *is* a map from a vector space V to its double dual $V^{**} = \{\, x\colon V^* \to K \,\}$ that does not depend on the choice of basis: For all $v \in V$ and $\varphi \in V^*$,

$$v \xmapsto{\sim} x_v \in V^{**} \quad \text{that such} \quad x_v(\phi) = \phi(v)$$

This leads to a third notion, that of a natural isomorphism: while V and V^{**} are different sets, there is a "natural" choice of isomorphism between them. This intuitive notion of "an isomorphism that does not depend on an arbitrary choice" is formalized in the notion of a natural transformation; briefly, that one may *consistently* identify, or more generally map from, a vector space to its double dual, $V \xrightarrow{\sim} V^{**}$, for *any* vector space in a consistent way. Formalizing this intuition is a motivation for the development of category theory.

However, there is a case where the distinction between natural isomorphism and equality is usually not made. That is for the objects that may be characterized by a universal property. In fact, there is a unique isomorphism, necessarily natural, between two objects sharing the same universal property. A typical example is the set of real numbers, which may be defined through infinite decimal expansion, infinite binary expansion, Cauchy sequences, Dedekind cuts and many other ways. Formally these constructions define different objects, which all are solutions of the same universal property. As these objects have exactly the same properties, one may forget the method of construction and considering them as equal. This is what everybody does when talking of "*the* set of the real numbers". The same occurs with quotient spaces: they are commonly constructed as sets of equivalence classes. However, talking of set of sets may be counterintuitive, and quotient spaces are commonly considered as a pair of a set of undetermined objects, often called "points", and a surjective map onto this set.

If one wishes to draw a distinction between an arbitrary isomorphism (one that depends on a choice) and a natural isomorphism (one that can be done consistently), one may write $\approx$ for an unnatural isomorphism and $\cong$ for a natural isomorphism, as in $V \approx V^*$ and $V \cong V^{**}$. This convention is not universally followed, and authors who wish to distinguish between unnatural isomorphisms and natural isomorphisms will generally explicitly state the distinction.

Generally, saying that two objects are *equal* is reserved for when there is a notion of a larger (ambient) space that these objects live in. Most often, one speaks of equality of two subsets of a given set (as in the integer set example above), but not of two objects abstractly presented. For example, the 2-dimensional unit sphere in 3-dimensional space

$$S^2 := \{(x, y, z) \in \mathbb{R}^3 \mid x^2 + y^2 + z^2 = 1\} \text{ and the Riemann sphere } \widehat{\mathbb{C}}$$

which can be presented as the one-point compactification of the complex plane $\mathbb{C} \cup \{\infty\}$ *or* as the complex projective line (a quotient space)

$$\mathbf{P}^1_{\mathbb{C}} := (\mathbb{C}^2 \setminus \{(0, 0)\})/(\mathbb{C}^*)$$

are three different descriptions for a mathematical object, all of which are isomorphic, but not *equal* because they are not all subsets of a single space: the first is a subset of $\mathbf{R}^3$, the second is $\mathbf{C} \cong \mathbf{R}^{2\,[note\ 4]}$ plus an additional point, and the third is a subquotient of $\mathbf{C}^2$

In the context of category theory, objects are usually at most isomorphic—indeed, a motivation for the development of category theory was showing that different constructions in homology theory yielded equivalent (isomorphic) groups. Given maps between two objects X and Y, however, one asks if they are equal or not (they are both elements of the set $\mathrm{Hom}(X, Y)$, hence equality is the proper relationship), particularly in commutative diagrams.

11.5   See also

- Bisimulation

- Heap (mathematics)

- Isometry

- Isomorphism class
- Isomorphism theorem
- Universal property

11.6 Notes

[1] For clarity, by *inverse* is meant *inverse homomorphism* or *inverse morphism* respectively, not *inverse function*.

[2] The careful reader may note that A, B, C have a conventional order, namely alphabetical order, and similarly 1, 2, 3 have the order from the integers, and thus one particular isomorphism is "natural", namely

$$A \mapsto 1, B \mapsto 2, C \mapsto 3$$

More formally, as *sets* these are isomorphic, but not naturally isomorphic (there are multiple choices of isomorphism), while as *ordered sets* they are naturally isomorphic (there is a unique isomorphism, given above), since finite total orders are uniquely determined up to unique isomorphism by cardinality. This intuition can be formalized by saying that any two finite totally ordered sets of the same cardinality have a natural isomorphism, the one that sends the least element of the first to the least element of the second, the least element of what remains in the first to the least element of what remains in the second, and so forth, but in general, pairs of sets of a given finite cardinality are not naturally isomorphic because there is more than one choice of map—except if the cardinality is 0 or 1, where there is a unique choice.

[3] In fact, there are precisely $3! = 6$ different isomorphisms between two sets with three elements. This is equal to the number of automorphisms of a given three-element set (which in turn is equal to the order of the symmetric group on three letters), and more generally one has that the set of isomorphisms between two objects, denoted $\text{Iso}(A, B)$, is a torsor for the automorphism group of A, $\text{Aut}(A)$ and also a torsor for the automorphism group of B. In fact, automorphisms of an object are a key reason to be concerned with the distinction between isomorphism and equality, as demonstrated in the effect of change of basis on the identification of a vector space with its dual or with its double dual, as elaborated in the sequel.

[4] Being precise, the identification of the complex numbers with the real plane,

$$\mathbf{C} \cong \mathbf{R} \cdot 1 \oplus \mathbf{R} \cdot i = \mathbf{R}^2$$

depends on a choice of i; one can just as easily choose $(-i)$, , which yields a different identification—formally, complex conjugation is an automorphism—but in practice one often assumes that one has made such an identification.

11.7 References

[1] Awodey, Steve (2006). "Isomorphisms". *Category theory*. Oxford University Press. p. 11. ISBN 9780198568612.

[2] Vinberg, Ėrnest Borisovich (2003). *A Course in Algebra*. American Mathematical Society. p. 3. ISBN 9780821834138.

[3] Mazur 2007

11.8 Further reading

- Mazur, Barry (12 June 2007), *When is one thing equal to some other thing?* (PDF)

11.9 External links

- Hazewinkel, Michiel, ed. (2001), "Isomorphism", *Encyclopedia of Mathematics*, Springer, ISBN 978-1-55608-010-4
- Isomorphism at PlanetMath.org.
- Weisstein, Eric W., "Isomorphism", *MathWorld*.

Chapter 12

Penrose transform

In mathematical physics, the **Penrose transform**, introduced by Roger Penrose (1967, 1968, 1969), is a complex analogue of the Radon transform that relates massless fields on spacetime to cohomology of sheaves on complex projective space. The projective space in question is the twistor space, a geometrical space naturally associated to the original spacetime, and the twistor transform is also geometrically natural in the sense of integral geometry. The Penrose transform is a major component of classical twistor theory.

12.1   Overview

Abstractly, the Penrose transform operates on a double fibration of a space Y, over two spaces X and Z

$$Z \xleftarrow{\eta} Y \xrightarrow{\tau} X.$$

In the classical Penrose transform, Y is the spin bundle, X is a compactified and complexified form of Minkowski space and Z is the twistor space. More generally examples come from double fibrations of the form

$$G/H_1 \xleftarrow{\eta} G/(H_1 \cap H_2) \xrightarrow{\tau} G/H_2$$

where G is a complex semisimple Lie group and H_1 and H_2 are parabolic subgroups.

The Penrose transform operates in two stages. First, one pulls back the sheaf cohomology groups $H^r(Z, \mathbf{F})$ to the sheaf cohomology $H^r(Y, \eta^{-1}\mathbf{F})$ on Y; in many cases where the Penrose transform is of interest, this pullback turns out to be an isomorphism. One then pushes the resulting cohomology classes down to X; that is, one investigates the direct image of a cohomology class by means of the Leray spectral sequence. The resulting direct image is then interpreted in terms of differential equations. In the case of the classical Penrose transform, the resulting differential equations are precisely the massless field equations for a given spin.

12.2   Example

The classical example is given as follows

- The "twistor space" Z is complex projective 3-space $\mathbf{CP}^3$, which is also the Grassmannian $\mathrm{Gr}_1(\mathbf{C}^4)$ of lines in 4-dimensional complex space.

- $X = \mathrm{Gr}_2(\mathbf{C}^4)$, the Grassmannian of 2-planes in 4-dimensional complex space. This is a compactification of complex Minkowski space.

- Y is the flag manifold whose elements correspond to a line in a plane of $\mathbf{C}^4$.

- G is the group $\mathrm{SL}_4(\mathbf{C})$ and H_1 and H_2 are the parabolic subgroups fixing a line or a plane containing this line.

The maps from Y to X and Z are the natural projections.

12.3 Penrose–Ward transform

The Penrose–Ward transform is a non-linear modification of the Penrose transform, introduced by Ward (1977), that (among other things) relates holomorphic vector bundles on 3-dimensional complex projective space $\mathbf{CP}^3$ to solutions of the self-dual Yang–Mills equations on $\mathbf{S}^4$. Atiyah & Ward (1977) used this to describe instantons in terms of algebraic vector bundles on complex projective 3-space. and Atiyah (1979) explained how this could be used to classify instantons on a 4-sphere.

12.4 References

- Atiyah, Michael Francis; Ward, R. S. (1977), "Instantons and algebraic geometry", *Communications in Mathematical Physics* (Springer Berlin / Heidelberg) **55**: 117–124, Bibcode:1977CMaPh..55..117A, doi:10.1007/BF01626514, ISSN 0010-3616, MR 0494098

- Atiyah, Michael Francis (1979), *Geometry of Yang-Mills fields*, Lezioni Fermiane, Scuola Normale Superiore Pisa, Pisa, ISBN 978-88-7642-303-1, MR 554924

- Baston, Robert J.; Eastwood, Michael G. (1989), *The Penrose transform*, Oxford Mathematical Monographs, The Clarendon Press Oxford University Press, ISBN 978-0-19-853565-2, MR 1038279.

- Eastwood, Michael (1993), "Introduction to Penrose transform", in Eastwood, Michael; Wolf, Joseph; Zierau., Roger, *The Penrose transform and analytic cohomology in representation theory (South Hadley, MA, 1992)*, Contemp. Math. **154**, Providence, R.I.: Amer. Math. Soc., pp. 71–75, ISBN 978-0-8218-5176-0, MR 1246377

- Eastwood, M.G. (2001), "P/p120100", in Hazewinkel, Michiel, *Encyclopedia of Mathematics*, Springer, ISBN 978-1-55608-010-4

- David, Liana (2001), *The Penrose transform and its applications* (PDF), University of Edinburgh; Doctor of Philosophy thesis.

- Penrose,Roger(1967),"Twistor algebra",*Journal of Mathematical Physics***8**: 345–366,Bibcode:1967JMP.....8.. doi:10.1063/1.1705200, ISSN 0022-2488, MR 0216828

- Penrose, Roger (1968), "Twistor quantisation and curved space-time", *International Journal of Theoretical Physics* (Springer Netherlands) **1**: 61–99, Bibcode:1968IJTP....1...61P, doi:10.1007/BF00668831, ISSN 0020-7748

- Penrose, Roger (1969), "Solutions of the Zero-Rest-Mass Equations", *Journal of Mathematical Physics* **10** (1): 38–39, Bibcode:1969JMP....10..38P, doi:10.1063/1.1664756, ISSN 0022-2488

- Penrose, Roger; Rindler, Wolfgang (1986), *Spinors and space-time. Vol. 2*, Cambridge Monographs on Mathematical Physics, Cambridge University Press, ISBN 978-0-521-25267-6, MR 838301.

- Ward, R. S. (1977), "On self-dual gauge fields", *Physics Letters A* **61** (2): 81–82, Bibcode:1977PhLA...61...81W, doi:10.1016/0375-9601(77)90842-8, ISSN 0375-9601, MR 0443823

Chapter 13

Sesquilinear form

In mathematics, a **sesquilinear form** is a generalization of a bilinear form that, in turn, is a generalization of the concept of the dot product of Euclidean space. A bilinear form is linear in each of its arguments, but a sesquilinear form allows one of the arguments to be "twisted" in a semilinear manner, thus the name; which originates from the Latin numerical prefix *sesqui-* meaning "one and a half". The basic concept of the dot product – producing a scalar from a pair of vectors – can be generalized by allowing a broader range of scalar values and, perhaps simultaneously, by widening the definition of what a vector is.

A motivating special case is a sesquilinear form on a complex vector space, V. This is a map $V \times V \to \mathbf{C}$ that is linear in one argument and "twists" the linearity of other argument by complex conjugation (referred to as being antilinear in the other argument). This case arises naturally in mathematical physics applications. Another important case allows the scalars to come from any field and the twist is provided by a field automorphism. Many authors assume that this automorphism is an involution (has order two) to stay in analogy with the complex case, but others prove this property when introducing Hermitian forms.

An application in projective geometry requires that the scalars come from a division ring (skewfield), K, and this means that the "vectors" should be replaced by elements of a K-module. In a very general setting, sesquilinear forms can be defined over R-modules for arbitrary rings R.

13.1 Convention

Conventions differ as to which argument should be linear. We shall take the first to be linear, as is common in the mathematical literature, except in the section devoted to sesquilinear forms on complex vector spaces. There we use the other convention and take the first argument to be conjugate-linear (i.e. antilinear) and the second to be linear. This is the convention used mostly by mathematical physicists[1] and originates in Dirac's bra–ket notation in quantum mechanics.

13.2 Complex vector spaces

Over a complex vector space V a map $\varphi : V \times V \to \mathbf{C}$ is sesquilinear if

$$\varphi(x + y, z + w) = \varphi(x, z) + \varphi(x, w) + \varphi(y, z) + \varphi(y, w)$$
$$\varphi(ax, by) = \overline{a}b\,\varphi(x, y)$$

for all $x, y, z, w \in V$ and all $a, b \in \mathbf{C}$. a is the complex conjugate of a.

A complex sesquilinear form can also be viewed as a complex bilinear map

$$\overline{V} \times V \to \mathbf{C}$$

where V is the complex conjugate vector space to V. By the universal property of tensor products these are in one-to-one correspondence with complex linear maps

$$\overline{V} \otimes V \to \mathbf{C}.$$

For a fixed z in V the map $w \mapsto \varphi(z, w)$ is a linear functional on V (i.e. an element of the dual space V^*). Likewise, the $w \mapsto \varphi(w, z)$ is a conjugate-linear functional on V.

Given any complex sesquilinear form φ on V we can define a second complex sesquilinear form ψ via the conjugate transpose:

$$\psi(w, z) = \overline{\varphi(z, w)}.$$

In general, ψ and φ will be different. If they are the same then φ is said to be *Hermitian*. If they are negatives of one another, then φ is said to be *skew-Hermitian*. Every sesquilinear form can be written as a sum of a Hermitian form and a skew-Hermitian form.

13.2.1 Geometric motivation

Bilinear forms are to squaring (z^2), what complex sesquilinear forms are to the squared magnitude ($|z|^2 = z\bar{z}$). Regarding the complex plane geometrically as a two-dimensional real vector space, the latter corresponds with the square of the Euclidean norm.

The norm associated to a complex sesquilinear form is invariant under multiplication by complex numbers of unit norm (elements of the complex unit circle), while the norm associated to a bilinear form is equivariant (with respect to squaring). Bilinear forms are *algebraically* more natural, while sesquilinear forms are *geometrically* more natural.

If B is a bilinear form on a complex vector space and $|x|B := B(x, x)$ is the associated norm, then $|ix|B = B(ix, ix) = i^2 B(x, x) = -|x|B$.

By contrast, if S is a sesquilinear form on a complex vector space and $|x|S := S(x, x)$ is the associated norm, then $|ix|S = S(ix, ix) = i\bar{i}S(x, x) = |x|S$.

13.2.2 Hermitian form

> *The term **Hermitian form** may also refer to a different concept than that explained below: it may refer to a certain differential form on a Hermitian manifold.*

A complex **Hermitian form** (also called a **symmetric sesquilinear form**), is a sesquilinear form $h : V \times V \to \mathbf{C}$ such that

$$h(w, z) = \overline{h(z, w)}.$$

The standard Hermitian form on $\mathbf{C}^n$ is given (again, using the "physics" convention of linearity in the second and conjugate linearity in the first variable) by

$$\langle w, z \rangle = \sum_{i=1}^{n} \overline{w_i} z_i.$$

More generally, the inner product on any complex Hilbert space is a Hermitian form.

A vector space with a Hermitian form (V, h) is called a **Hermitian space**.

If V is a finite-dimensional complex vector space, then relative to any basis $\{\, e_i \,\}$ of V, a complex Hermitian form is represented by a Hermitian matrix $\mathbf{H}$, w by the column vector $\mathbf{w}$, and z by the column vector $\mathbf{z}$:

$$h(w, z) = \overline{\mathbf{w}}^{\mathsf{T}} \mathbf{H} \mathbf{z}.$$

The components of $\mathbf{H}$ are given by $H_{ij} = h(e_i, e_j)$.

The quadratic form associated to a complex Hermitian form

$$Q(z) = h(z, z)$$

is always real. Actually, one can show that a complex sesquilinear form is Hermitian iff the associated quadratic form is real for all $z \in V$.

13.2.3   Skew-Hermitian form

A complex **skew-Hermitian form** (also called an **antisymmetric sesquilinear form**), is a complex sesquilinear form s : $V \times V \to \mathbf{C}$ such that

$$s(w, z) = -\overline{s(z, w)}.$$

Every complex skew-Hermitian form can be written as i times a Hermitian form.

If V is a finite-dimensional complex vector space, then relative to any basis $\{\, e_i \,\}$ of V, a complex skew-Hermitian form s is represented by a skew-Hermitian matrix $\mathbf{S}$, w by the column vector $\mathbf{w}$, and z by the column vector $\mathbf{z}$:

$$s(w, z) = \overline{\mathbf{w}}^{\mathsf{T}} \mathbf{S} \mathbf{z}.$$

The quadratic form associated to a complex skew-Hermitian form

$$Q(z) = s(z, z)$$

is always pure imaginary.

13.3   Over arbitrary fields

On a vector space V defined over an arbitrary field F having a distinguished automorphism σ of order two (an involution known as the *companion automorphism*), a map φ : $V \times V \to F$ is **sesquilinear** if

$$\varphi(x + y, z + w) = \varphi(x, z) + \varphi(x, w) + \varphi(y, z) + \varphi(y, w)$$
$$\varphi(cx, dy) = cd^{\sigma}\varphi(x, y) = c\,\sigma(d)\,\varphi(x, y)$$

for all $x, y, z, w \in V$ and all $c, d \in F$. Recall the convention of having the first argument linear and notice the use of the "transformation exponential notation" $t \mapsto t^{\sigma}$.

If the automorphism $\sigma = $ id then the sesquilinear form is a bilinear form.

A sesquilinear form φ is *reflexive* if for every pair $x, y \in V$, $\varphi(x, y) = 0$ implies $\varphi(y, x) = 0$.

A sesquilinear form φ is said to be σ-*Hermitian* (sometimes referred to as being *conjugate-symmetric*) if

$$\varphi(x, y) = \varphi(y, x)^\sigma$$

for all $x, y \in V$. It follows from this definition that $\varphi(x, x)$ always lies in the fixed field of σ. In the bilinear case ($\sigma = $ id) these forms are called *symmetric*.

Reflexive sesquilinear forms are either bilinear or Hermitian.[2]

Given an ordered basis { ei } of the vector space V, a sesquilinear form φ on V uniquely determines the matrix $M\varphi$ by:

$$\varphi(x, y) = x M_\varphi y^{\sigma T}.$$

A sesquilinear form can also be viewed as an F-bilinear map

$$V \times V^* \to F$$

where V^* is the dual space of V.

13.3.1 Example

Let V be the three dimensional vector space over the finite field $F = \mathrm{GF}(q^2)$, where q is a prime power. With respect to the standard basis we can write $x = (x_1, x_2, x_3)$ and $y = (y_1, y_2, y_3)$ and define the map φ by:

$$\varphi(x, y) = x_1 y_1{}^q + x_2 y_2{}^q + x_3 y_3{}^q.$$

The map $\sigma : t \mapsto t^q$ is an involutory automorphism of F. The map φ is then a σ-sesquilinear form. The matrix $M\varphi$ associated to this form is the identity matrix. This is a Hermitian form.

13.4 In projective geometry

In a projective geometry G a permutation δ of the subspaces which inverts inclusion, i.e.

$$S \subseteq T \Rightarrow T^\delta \subseteq S^\delta \text{ for all subspaces } S, T \text{ of } G,$$

is called a *correlation*. A result of Birkhoff and von Neumann (1936)[3] shows that the correlations of Desarguesian projective geometries correspond exactly to the nondegenerate sesquilinear forms on the underlying vector space.[2] A sesquilinear form φ is *nondegenerate* if $\varphi(x,y) = 0$ for all y in V (if and) only if $x = 0$.

To achieve full generality of this statement Reinhold Baer extended the definition of sesquilinear form to skewfields (division rings) which, in turn, requires replacing vector spaces by R-modules,[4] (in the geometric literature these are still referred to as either left or right vector spaces over skewfields.)[5]

13.5 Over arbitrary rings

The specialization of the above section to skewfields was a consequence of the application to projective geometry, and not intrinsic to the nature of sesquilinear forms. Only the minor modifications needed to take into account the non-commutativity of multiplication are required to generalize the arbitrary field version of the definition to arbitrary rings.

Let R be a ring, V an R-module and σ an antiautomorphism of R of order two. A map $\varphi : V \times V \to R$ is **sesquilinear** if

$$\varphi(x + y, z + w) = \varphi(x, z) + \varphi(x, w) + \varphi(y, z) + \varphi(y, w)$$
$$\varphi(cx, dy) = c\varphi(x, y)d^{\sigma}$$

for all $x, y, z, w \in V$ and all $c, d \in R$.

Since for an antiautomorphism σ we have $\sigma(st) = \sigma(t)\,\sigma(s)$ for all s, t in R, if $\sigma = \mathrm{id}$, then R must be commutative and φ is a bilinear form. In particular, if, in this case, R is a skewfield, then R is a field and V is a vector space with a bilinear form.

An antiautomorphism $\sigma\colon R \to R$ can also be viewed as an isomorphism of $R \to R^{\mathrm{op}}$, the *opposite ring* based on the same set with the same addition, but whose multiplication operation ($*$) is defined by $a * b = ba$, where the product on the right is the product in R. It follows from this that a right (left) R-module V can be turned into a left (right) R^{op}-module, V^*.[6] Thus, the sesquilinear form $\varphi : V \times V \to R$ can be viewed as a bilinear form $\varphi' : V \times V^* \to R$.

13.6   See also

*-ring

13.7   Notes

[1] footnote 1 in Anthony Knapp *Basic Algebra* (2007) pg. 255

[2] Dembowski 1968, p. 42

[3] Birkhoff, G.; von Neumann, J. (1936), "The logic of quantum mechanics", *Annals of Mathematics* **37**: 823–843

[4] Baer, Reinhold (2005) [1952], *Linear Algebra and Projective Geometry*, Dover, ISBN 978-0-486-44565-6

[5] Baer's terminology gives a third way to refer to these ideas, so he must be read with caution.

[6] Jacobson 2009, p. 164

13.8   References

- Dembowski, Peter (1968), *Finite geometries*, Ergebnisse der Mathematik und ihrer Grenzgebiete, Band 44, Berlin, New York: Springer-Verlag, ISBN 3-540-61786-8, MR 0233275

- Gruenberg, K.W.; Weir, A.J. (1977), *Linear Geometry* (2nd ed.), Springer, ISBN 0-387-90227-9

- Jacobson, Nathan J. (2009) [1985], *Basic Algebra I* (2nd ed.), Dover, ISBN 978-0-486-47189-1

13.9   External links

- Hazewinkel, Michiel, ed. (2001), "Sesquilinear form", *Encyclopedia of Mathematics*, Springer, ISBN 978-1-55608-010-4

Chapter 14

Conformal group

In mathematics, the **conformal group** is the group of transformations from a space to itself that preserve all angles within the space. More formally, it is the group of transformations that preserve the conformal geometry of the space. Several specific conformal groups are particularly important:

- The conformal orthogonal group. If V is a vector space with a quadratic form Q, then the conformal orthogonal group $CO(V, Q)$ is the group of linear transformations T of V such that for all x in V there exists a scalar λ such that

$$Q(Tx) = \lambda^2 Q(x)$$

  The conformal orthogonal group is equal to the orthogonal group times the group of dilations.

- The conformal group of the sphere. The group of conformal transformations of the n-sphere is generated by the inversions in circles. This group is also known as the Möbius group. When applied to electrodynamics it is related to the conformal group of spherical wave transformations.

All conformal groups are Lie Groups.

14.1 References

- Kobayashi, S. (1972). *Transformation Groups in Differential Geometry*. Classics in Mathematics. Springer. ISBN 3-540-58659-8. OCLC 31374337.

- Sharpe, R.W. (1997), *Differential Geometry: Cartan's Generalization of Klein's Erlangen Program*, Springer-Verlag, New York, ISBN 0-387-94732-9.

Chapter 15

Complex manifold

In differential geometry, a **complex manifold** is a manifold with an atlas of charts to the open unit disk[1] in $\mathbf{C}^n$, such that the transition maps are holomorphic.

The term **complex manifold** is variously used to mean a complex manifold in the sense above (which can be specified as an **integrable** complex manifold), and an almost complex manifold.

15.1 Implications of complex structure

Since holomorphic functions are much more rigid than smooth functions, the theories of smooth and complex manifolds have very different flavors: compact complex manifolds are much closer to algebraic varieties than to differentiable manifolds.

For example, the Whitney embedding theorem tells us that every smooth n-dimensional manifold can be embedded as a smooth submanifold of $\mathbf{R}^{2n}$, whereas it is "rare" for a complex manifold to have a holomorphic embedding into $\mathbf{C}^n$. Consider for example any compact connected complex manifold M: any holomorphic function on it is locally constant by Liouville's theorem. Now if we had a holomorphic embedding of M into $\mathbf{C}^n$, then the coordinate functions of $\mathbf{C}^n$ would restrict to nonconstant holomorphic functions on M, contradicting compactness, except in the case that M is just a point. Complex manifolds that can be embedded in $\mathbf{C}^n$ are called Stein manifolds and form a very special class of manifolds including, for example, smooth complex affine algebraic varieties.

The classification of complex manifolds is much more subtle than that of differentiable manifolds. For example, while in dimensions other than four, a given topological manifold has at most finitely many smooth structures, a topological manifold supporting a complex structure can and often does support uncountably many complex structures. Riemann surfaces, two dimensional manifolds equipped with a complex structure, which are topologically classified by the genus, are an important example of this phenomenon. The set of complex structures on a given orientable surface, modulo biholomorphic equivalence, itself forms a complex algebraic variety called a moduli space, the structure of which remains an area of active research.

Since the transition maps between charts are biholomorphic, complex manifolds are, in particular, smooth and canonically oriented (not just orientable: a biholomorphic map to (a subset of) $\mathbf{C}^n$ gives an orientation, as biholomorphic maps are orientation-preserving).

15.2 Examples of complex manifolds

- Riemann surfaces.

- The Cartesian product of two complex manifolds.

- The inverse image of any noncritical value of a holomorphic map.

15.2.1 Smooth complex algebraic varieties

Smooth complex algebraic varieties are complex manifolds, including:

- Complex vector spaces.

- Complex projective spaces,[2] $\mathbf{P}^n(\mathbf{C})$.

- Complex Grassmannians.

- Complex Lie groups such as GL(n, $\mathbf{C}$) or Sp(n, $\mathbf{C}$).

Similarly, the quaternionic analogs of these are also complex manifolds.

15.2.2 Simply connected

The simply connected 1-dimensional complex manifolds are isomorphic to either:

- Δ, the unit disk in $\mathbf{C}$

- $\mathbf{C}$, the complex plane

- $\hat{\mathbf{C}}$, the Riemann sphere

Note that there are inclusions between these as $\Delta \subseteq \mathbf{C} \subseteq \hat{\mathbf{C}}$, but that there are no non-constant maps in the other direction, by Liouville's theorem.

15.3   Disk vs. space vs. polydisk

The following spaces are different as complex manifolds, demonstrating the more rigid geometric character of complex manifolds (compared to smooth manifolds):

- complex space $\mathbf{C}^n$.

- the unit disk or open ball

$$\{z \in \mathbf{C}^n \ : \ \|z\| < 1\}.$$

- the polydisk

$$\{z = (z_1, z_2, \ldots, z_n) \in \mathbf{C}^n \ : \ |z_i| < 1, \text{ for all } i = 1, \ldots, n\}.$$

15.4   Almost complex structures

Main article: Almost complex manifold

An almost complex structure on a real manifold is a $GL(n, \mathbf{C})$-structure (in the sense of G-structures) – that is, the tangent bundle is equipped with a linear complex structure.

Concretely, this is an endomorphism of the tangent bundle whose square is $-I$; this endomorphism is analogous to multiplication by the imaginary number i, and is denoted J (to avoid confusion with the identity matrix I). An almost complex manifold is necessarily even-dimensional.

An almost complex structure is *weaker* than a complex structure: any complex manifold has an almost complex structure, but not every almost complex structure comes from a complex structure. Note that every even-dimensional real manifold has an almost complex structure defined locally from the local coordinate chart. The question is whether this complex structure can be defined globally. An almost complex structure that comes from a complex structure is called integrable, and when one wishes to specify a complex structure as opposed to an almost complex structure, one says an *integrable* complex structure. For integrable complex structures the so-called Nijenhuis tensor vanishes. This tensor is defined on pairs of vector fields, X, Y by

$$N_J(X, Y) = [X, Y] + J[JX, Y] + J[X, JY] - [JX, JY] \, .$$

For example, the 6-dimensional sphere $\mathbf{S}^6$ has a natural almost complex structure arising from the fact that it is the orthogonal complement of i in the unit sphere of the octonions, but this is not a complex structure. (It is not currently known whether or not the 6-sphere has a complex structure.) Using an almost complex structure we can make sense of holomorphic maps and ask about the existence of holomorphic coordinates on the manifold. The existence of holomorphic coordinates is equivalent to saying the manifold is complex (which is what the chart definition says).

Tensoring the tangent bundle with the complex numbers we get the *complexified* tangent bundle, on which multiplication by complex numbers makes sense (even if we started with a real manifold). The eigenvalues of an almost complex structure are $\pm i$ and the eigenspaces form sub-bundles denoted by $T^{0.1}M$ and $T^{1.0}M$. The Newlander–Nirenberg theorem shows that an almost complex structure is actually a complex structure precisely when these subbundles are *involutive*, i.e., closed under the Lie bracket of vector fields, and such an almost complex structure is called integrable.

15.5   Kähler and Calabi–Yau manifolds

One can define an analogue of a Riemannian metric for complex manifolds, called a Hermitian metric. Like a Riemannian metric, a Hermitian metric consists of a smoothly varying, positive definite inner product on the tangent bundle, which is Hermitian with respect to the complex structure on the tangent space at each point. As in the Riemannian case, such metrics always exist in abundance on any complex manifold. If the skew symmetric part of such a metric is symplectic, i.e. closed and nondegenerate, then the metric is called Kähler. Kähler structures are much more difficult to come by and are much more rigid.

Examples of Kähler manifolds include smooth projective varieties and more generally any complex submanifold of a Kähler manifold. The Hopf manifolds are examples of complex manifolds that are not Kähler. To construct one, take a complex vector space minus the origin and consider the action of the group of integers on this space by multiplication by $\exp(n)$. The quotient is a complex manifold whose first Betti number is one, so by the Hodge theory, it cannot be Kähler.

A Calabi–Yau manifold can be defined as a compact Ricci-flat Kähler manifold or equivalently one whose first Chern class vanishes.

15.6   See also

- Quaternionic manifold

- Real-complex manifold

15.7   Footnotes

[1] One must use the open unit disk in $\mathbf{C}^n$ as the model space instead of $\mathbf{C}^n$ because these are not isomorphic, unlike for real manifolds.

[2] This means that all complex projective spaces are *orientable*, in contrast to the real case

15.8   References

- Kodaira, Kunihiko. *Complex Manifolds and Deformation of Complex Structures*. Classics in Mathematics. Springer. ISBN 3-540-22614-1.

Chapter 16

Homogeneous space

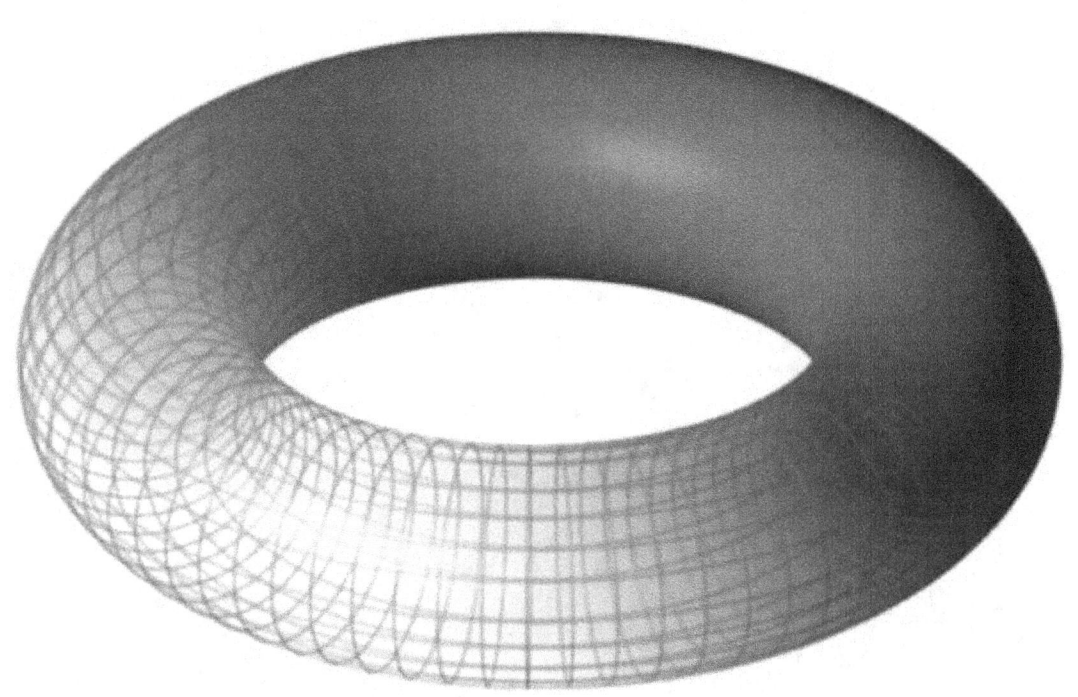

A torus. The standard torus is homogeneous under its diffeomorphism and homeomorphism groups, and the flat torus is homogeneous under its diffeomorphism, homeomorphism, and isometry groups.

In mathematics, particularly in the theories of Lie groups, algebraic groups and topological groups, a **homogeneous space** for a group G is a non-empty manifold or topological space X on which G acts transitively. The elements of G are called the **symmetries** of X. A special case of this is when the group G in question is the automorphism group of the space X – here "automorphism group" can mean isometry group, diffeomorphism group, or homeomorphism group. In this case X is homogeneous if intuitively X looks locally the same at each point, either in the sense of isometry (rigid geometry), diffeomorphism (differential geometry), or homeomorphism (topology). Some authors insist that the action of G be faithful (non-identity elements act non-trivially), although the present article does not. Thus there is a group action of G on X which can be thought of as preserving some "geometric structure" on X, and making X into a single G-orbit.

16.1 Formal definition

Let X be a non-empty set and G a group. Then X is called a G-space if it is equipped with an action of G on X.[1] Note that automatically G acts by automorphisms (bijections) on the set. If X in addition belongs to some category, then the elements of G are assumed to act as automorphisms in the same category. Thus the maps on X effected by G are structure preserving. A homogeneous space is a G-space on which G acts transitively.

Succinctly, if X is an object of the category $\mathbf{C}$, then the structure of a G-space is a homomorphism:

$$\rho : G \to \mathrm{Aut}_{\mathbf{C}}(X)$$

into the group of automorphisms of the object X in the category $\mathbf{C}$. The pair (X, ρ) defines a homogeneous space provided $\rho(G)$ is a transitive group of symmetries of the underlying set of X.

16.1.1 Examples

For example, if X is a topological space, then group elements are assumed to act as homeomorphisms on X. The structure of a G-space is a group homomorphism $\rho : G \to \mathrm{Homeo}(X)$ into the homeomorphism group of X.

Similarly, if X is a differentiable manifold, then the group elements are diffeomorphisms. The structure of a G-space is a group homomorphism $\rho : G \to \mathrm{Diffeo}(X)$ into the diffeomorphism group of X.

Riemannian symmetric spaces are an important class of homogeneous spaces, and include many of the examples listed below.

Concrete examples include:

Isometry groups

- Positive curvature:

1. Sphere (orthogonal group): $S^{n-1} \cong O(n)/O(n-1)$
2. Oriented sphere (special orthogonal group): $S^{n-1} \cong SO(n)/SO(n-1)$
3. Projective space (projective orthogonal group): $P^{n-1} \cong PO(n)/PO(n-1)$

- Flat (zero curvature):

1. Euclidean space (Euclidean group, point stabilizer is orthogonal group): $A^n \cong E(n)/O(n)$

- Negative curvature:

1. Hyperbolic space (orthochronous Lorentz group, point stabilizer orthogonal group, corresponding to hyperboloid model): $H^n \cong O^+(1, n)/O(n)$
2. Oriented hyperbolic space: $SO^+(1, n)/SO(n)$
3. Anti-de Sitter space: $AdS_{n+1} = O(2, n)/O(1, n)$

Others

- Affine space (for affine group, point stabilizer general linear group): $A^n = \mathrm{Aff}(n, K)/GL(n, k)$.
- Grassmannian: $Gr(r, n) = O(n)/(O(r) \times O(n - r))$

16.2   Geometry

From the point of view of the Erlangen program, one may understand that "all points are the same", in the geometry of X. This was true of essentially all geometries proposed before Riemannian geometry, in the middle of the nineteenth century.

Thus, for example, Euclidean space, affine space and projective space are all in natural ways homogeneous spaces for their respective symmetry groups. The same is true of the models found of non-Euclidean geometry of constant curvature, such as hyperbolic space.

A further classical example is the space of lines in projective space of three dimensions (equivalently, the space of two-dimensional subspaces of a four-dimensional vector space). It is simple linear algebra to show that GL_4 acts transitively on those. We can parameterize them by *line co-ordinates*: these are the 2×2 minors of the 4×2 matrix with columns two basis vectors for the subspace. The geometry of the resulting homogeneous space is the line geometry of Julius Plücker.

16.3   Homogeneous spaces as coset spaces

In general, if X is a homogeneous space, and Ho is the stabilizer of some marked point o in X (a choice of origin), the points of X correspond to the left cosets G/Ho, and the marked point o corresponds to the coset of the identity. Conversely, given a coset space G/H, it is a homogeneous space for G with a distinguished point, namely the coset of the identity. Thus a homogeneous space can be thought of as a coset space without a choice of origin.

In general, a different choice of origin o will lead to a quotient of G by a different subgroup Ho' which is related to Ho by an inner automorphism of G. Specifically,

$$H_{o'} = gH_og^{-1} \qquad (1)$$

where g is any element of G for which $go = o'$. Note that the inner automorphism (1) does not depend on which such g is selected; it depends only on g modulo Ho.

If the action of G on X is continuous, then H is a closed subgroup of G. In particular, if G is a Lie group, then H is a Lie subgroup by Cartan's theorem. Hence G/H is a smooth manifold and so X carries a unique smooth structure compatible with the group action.

If H is the identity subgroup $\{e\}$, then X is a principal homogeneous space.

One can go further to *double* coset spaces, notably Clifford–Klein forms $\Gamma\backslash G/H$, where Γ is a discrete subgroup (of G) acting properly discontinuously.

16.4   Example

For example in the line geometry case, we can identify H as a 12-dimensional subgroup of the 16-dimensional general linear group, GL(4), defined by conditions on the matrix entries

$$h_{13} = h_{14} = h_{23} = h_{24} = 0,$$

by looking for the stabilizer of the subspace spanned by the first two standard basis vectors. That shows that X has dimension 4.

Since the homogeneous coordinates given by the minors are 6 in number, this means that the latter are not independent of each other. In fact a single quadratic relation holds between the six minors, as was known to nineteenth-century geometers.

This example was the first known example of a Grassmannian, other than a projective space. There are many further homogeneous spaces of the classical linear groups in common use in mathematics.

16.5 Prehomogeneous vector spaces

The idea of a prehomogeneous vector space was introduced by Mikio Sato.

It is a finite-dimensional vector space V with a group action of an algebraic group G, such that there is an orbit of G that is open for the Zariski topology (and so, dense). An example is $GL(1)$ acting on a one-dimensional space.

The definition is more restrictive than it initially appears: such spaces have remarkable properties, and there is a classification of irreducible prehomogeneous vector spaces, up to a transformation known as "castling".

16.6 Homogeneous spaces in physics

Cosmology using the general theory of relativity makes use of the Bianchi classification system. Homogeneous spaces in relativity represent the space part of background metrics for some cosmological models; for example, the three cases of the Friedmann–Lemaître–Robertson–Walker metric may be represented by subsets of the Bianchi I (flat), V (open), VII (flat or open) and IX (closed) types, while the Mixmaster universe represents an anisotropic example of a Bianchi IX cosmology.[2]

A homogeneous space of N dimensions admits a set of $\frac{1}{2}N(N+1)$ Killing vectors.[3] For three dimensions, this gives a total of six linearly independent Killing vector fields; homogeneous 3-spaces have the property that one may use linear combinations of these to find three everywhere non-vanishing Killing vector fields $\xi_i^{(a)}$.

$$\xi_{[i;k]}^{(a)} = C_{bc}^a \xi_i^{(b)} \xi_k^{(c)}$$

where the object C_{bc}^a , the "structure constants", form a constant order-three tensor antisymmetric in its lower two indices (on the left-hand side, the brackets denote antisymmetrisation and ";" represents the covariant differential operator). In the case of a flat isotropic universe, one possibility is $C_{bc}^a = 0$ (type I), but in the case of a closed FLRW universe, $C_{bc}^a = \varepsilon_{bc}^a$ where ε_{bc}^a is the Levi-Civita symbol.

16.7 See also

- Erlangen program
- Klein geometry
- Heap (mathematics)
- Homogeneous variety

16.8 References

[1] We assume that the action is on the *left*. The distinction is only important in the description of X as a coset space.

[2] Lev Landau and Evgeny Lifshitz (1980), *Course of Theoretical Physics vol. 2: The Classical Theory of Fields*, Butterworth-Heinemann, ISBN 978-0-7506-2768-9

[3] Steven Weinberg (1972), *Gravitation and Cosmology*, John Wiley and Sons

Chapter 17

Conformal map

For other uses, see Conformal (disambiguation).

In mathematics, a **conformal map** is a function that preserves angles locally. In the most common case, the function has a domain and an image in the complex plane.

More formally, a map

$$f : U \to V \text{ with } U, V \subseteq \mathbb{C}^n$$

is called **conformal** (or **angle-preserving**) at a point u_0 if it preserves oriented angles between curves through u_0 with respect to their orientation (i.e. not just the magnitude of the angle). Conformal maps preserve both angles and the shapes of infinitesimally small figures, but not necessarily their size or curvature.

The conformal property may be described in terms of the Jacobian derivative matrix of a coordinate transformation. If the Jacobian matrix of the transformation is everywhere a scalar times a rotation matrix, then the transformation is conformal.

Conformal maps can be defined between domains in higher-dimensional Euclidean spaces, and more generally on a Riemannian or semi-Riemannian manifold.

17.1 Complex analysis

An important family of examples of conformal maps comes from complex analysis. If U is an open subset of the complex plane $\mathbb{C}$, then a function

$$f : U \to \mathbb{C}$$

is conformal if and only if it is holomorphic and its derivative is everywhere non-zero on U. If f is antiholomorphic (that is, the conjugate to a holomorphic function), it still preserves angles, but it reverses their orientation.

In the literature, there is another definition of conformal maps: a map f defined on an open set is said to be conformal if it is one-to-one and holomorphic. Since a one-to-one map defined on a non-empty open set cannot be constant, the open mapping theorem forces the inverse function (defined on the image of f) to be holomorphic. Thus, under this definition, a map is conformal if and only if it is biholomorphic. The two definitions for conformal maps are not equivalent. Being one-to-one and holomorphic implies having a non-zero derivative. However, the exponential function is a holomorphic function with a non-zero derivative, but is not one-to-one since it is periodic. [1]

The Riemann mapping theorem, one of the profound results of complex analysis, states that any non-empty open simply connected proper subset of $\mathbb{C}$ admits a bijective conformal map to the open unit disk in $\mathbb{C}$.

A map of the extended complex plane (which is conformally equivalent to a sphere) onto itself is conformal if and only if it is a Möbius transformation. Again, for the conjugate, angles are preserved, but orientation is reversed.

An example of the latter is taking the reciprocal of the conjugate, which corresponds to **circle inversion** with respect to the unit circle. This can also be expressed as taking the reciprocal of the radial coordinate in circular coordinates, keeping the angle the same. See also inversive geometry.

17.2 Riemannian geometry

See also: Conformal geometry

In Riemannian geometry, two Riemannian metrics g and h on smooth manifold M are called **conformally equivalent** if $g = uh$ for some positive function u on M. The function u is called the **conformal factor**.

A diffeomorphism between two Riemannian manifolds is called a **conformal map** if the pulled back metric is conformally equivalent to the original one. For example, stereographic projection of a sphere onto the plane augmented with a point at infinity is a conformal map.

One can also define a **conformal structure** on a smooth manifold, as a class of conformally equivalent Riemannian metrics.

17.3 Higher-dimensional Euclidean space

A classical theorem of Joseph Liouville called Liouville's theorem shows the higher-dimensions have less varied conformal maps:

Any conformal map on a portion of Euclidean space of dimension greater than 2 can be composed from three types of transformation: a homothetic transformation, an isometry, and a special conformal transformation. (A *special conformal transformation* is the composition of a reflection and an inversion in a sphere.) Thus, the group of conformal transformations in spaces of dimension greater than 2 are much more restricted than the planar case, where the Riemann mapping theorem provides a large group of conformal transformations.

17.4 Uses

If a function is harmonic (that is, it satisfies Laplace's equation $\nabla^2 f = 0$) over a plane domain (which is two-dimensional), and is transformed via a conformal map to another plane domain, the transformation is also harmonic. For this reason, any function which is defined by a potential can be transformed by a conformal map and still remain governed by a potential. Examples in physics of equations defined by a potential include the electromagnetic field, the gravitational field, and, in fluid dynamics, potential flow, which is an approximation to fluid flow assuming constant density, zero viscosity, and irrotational flow. One example of a fluid dynamic application of a conformal map is the Joukowsky transform.

Conformal mappings are invaluable for solving problems in engineering and physics that can be expressed in terms of functions of a complex variable but that exhibit inconvenient geometries. By choosing an appropriate mapping, the analyst can transform the inconvenient geometry into a much more convenient one. For example, one may wish to calculate the electric field, $E(z)$, arising from a point charge located near the corner of two conducting planes separated by a certain angle (where z is the complex coordinate of a point in 2-space). This problem *per se* is quite clumsy to solve in closed form. However, by employing a very simple conformal mapping, the inconvenient angle is mapped to one of precisely pi radians, meaning that the corner of two planes is transformed to a straight line. In this new domain, the problem (that of calculating the electric field impressed by a point charge located near a conducting wall) is quite easy to solve. The solution is obtained in this domain, $E(w)$, and then mapped back to the original domain by noting that w was obtained as a function (viz., the composition of E and w) of z, whence $E(w)$ can be viewed as $E(w(z))$, which is a function of z, the original coordinate basis. Note that this application is not a contradiction to the fact that conformal mappings preserve angles, they do so only for points in the interior of their domain, and not at the boundary.

A large group of conformal maps for relating solutions of Maxwell's equations was identified by Ebenezer Cunningham

(1908) and Harry Bateman (1910) (see spherical wave transformation). Their training at Cambridge University had given them facility with the method of image charges and associated methods of images for spheres and inversion. As recounted by Andrew Warwick (2003) *Masters of Theory*: [2]

> Each four-dimensional solution could be inverted in a four-dimensional hyper-sphere of pseudo-radius K in order to produce a new solution.

Warwick highlights (pages 404 to 424) this "new theorem of relativity" as a Cambridge response to Einstein, and as founded on exercises using the method of inversion, such as found in James Hopwood Jeans textbook *Mathematical Theory of Electricity and Magnetism*.

In cartography, several named map projections (including the Mercator projection) are conformal.

In general relativity, conformal maps are the simplest and thus most common type of causal transformations. Physically, these describe different universes in which all the same events and interactions are still (causally) possible, but a new additional force is necessary to effect this (that is, replication of all the same trajectories would necessitate departures from geodesic motion because the metric is different). It is often used to try to make models amenable to extension beyond curvature singularities, for example to permit description of the universe even before the big bang.

17.5   Alternative angles

A *conformal map* is called that because it preserves the shapes of things (at an infinitesimal scale). The term is based on the Latin prefix *com-* (together, with, near) and the Latin noun *forma* (shape, appearance).[3][4][5] The presumption often is that the shape being preserved is measured by the standard Euclidean angle, say parameterized in degrees or radians. However, in plane mapping there are two other angles to consider: the hyperbolic angle and the slope, which is the analogue of angle for dual numbers.

Suppose $f:U \to V$ is a mapping of surfaces parameterized by (x,y) and (u,v). The Jacobian matrix of f is formed by the four partial derivatives of u and v with respect to x and y.

If the Jacobian g has a non-zero determinant, then f is "conformal in the generalized sense" with respect to one of the three angle types, depending on the real matrix expressed by the Jacobian g.

Indeed, any such g lies in a particular *planar* commutative subring, and g has a polar coordinate form determined by parameters of radial and angular nature. The radial parameter corresponds to a similarity mapping and can be taken as 1 for purposes of conformal examination. The angular parameter of g is one of the three types, shear, hyperbolic, or Euclidean:

- When the subring is isomorphic to the dual number plane, then g acts as a shear mapping and preserves the dual angle.

- When the subring is isomorphic to the split-complex number plane, then g acts as a squeeze mapping and preserves the hyperbolic angle.

- When the subring is isomorphic to the ordinary complex number plane, then g acts as a rotation and preserves the Euclidean angle.

While describing analytic functions of a bireal variable, U. Bencivenga and G. Fox have written about conformal maps that preserve the hyperbolic angle. In general, a linear fractional transformation on any one of the types of complex plane listed provides a conformal map.

17.6   See also

- Conformal pictures

- Schwarz–Christoffel mapping – a conformal transformation of the upper half-plane onto the interior of a simple polygon.

- Penrose diagram

- Carathéodory's theorem – A conformal map extends continuously to the boundary.

17.7 References

[1] http://www.maths.tcd.ie/~{}richardt/414/414-ch7.pdf

[2] Warwick, Andrew (2003). *Masters of theory : Cambridge and the rise of mathematical physics*. Chicago: University of Chicago press. ISBN 978-0226873756.

[3] "conformal – definition and meaning". Wordnik.com. Retrieved 2013-09-05.

[4] "English etymology of conformal". myEtymology.com. Retrieved 2013-09-05.

[5] "conformal – Memidex dictionary/thesaurus". Memidex.com. 2013-06-26. Retrieved 2013-09-05.

- Ahlfors, Lars V. (1973), *Conformal invariants: topics in geometric function theory*, New York: McGraw–Hill Book Co., MR 0357743

- Chanson, H. (2009), *Applied Hydrodynamics: An Introduction to Ideal and Real Fluid Flows*, CRC Press, Taylor & Francis Group, Leiden, The Netherlands, 478 pages, ISBN 978-0-415-49271-3

- E.P. Dolzhenko (2001), "Conformal mapping", in Hazewinkel, Michiel, *Encyclopedia of Mathematics*, Springer, ISBN 978-1-55608-010-4

- Rudin, Walter (1987), *Real and complex analysis* (3rd ed.), New York: McGraw–Hill Book Co., ISBN 978-0-07-054234-1, MR 924157

- Churchill, Ruel V. (1974), *Complex Variables and Applications*, New York: McGraw–Hill Book Co., ISBN 0-07-010855-2

- Weisstein, Eric W., "Conformal Mapping", *MathWorld*.

17.8 External links

- Conformal Mapping Module by John H. Mathews

- interactive visualizations of many conformal maps

- Conformal Maps by Michael Trott, Wolfram Demonstrations Project.

- Java applet by Jürgen Richter-Gebert using Cinderella.

- Java applet by Christian Mercat to deform pictures; MacOSX Java applet that deforms the video flux from the webcam.

- Conformal Mapping images of current flow in different geometries without and with magnetic field by Gerhard Brunthaler.

- Conformal Transformation: from Circle to Square.

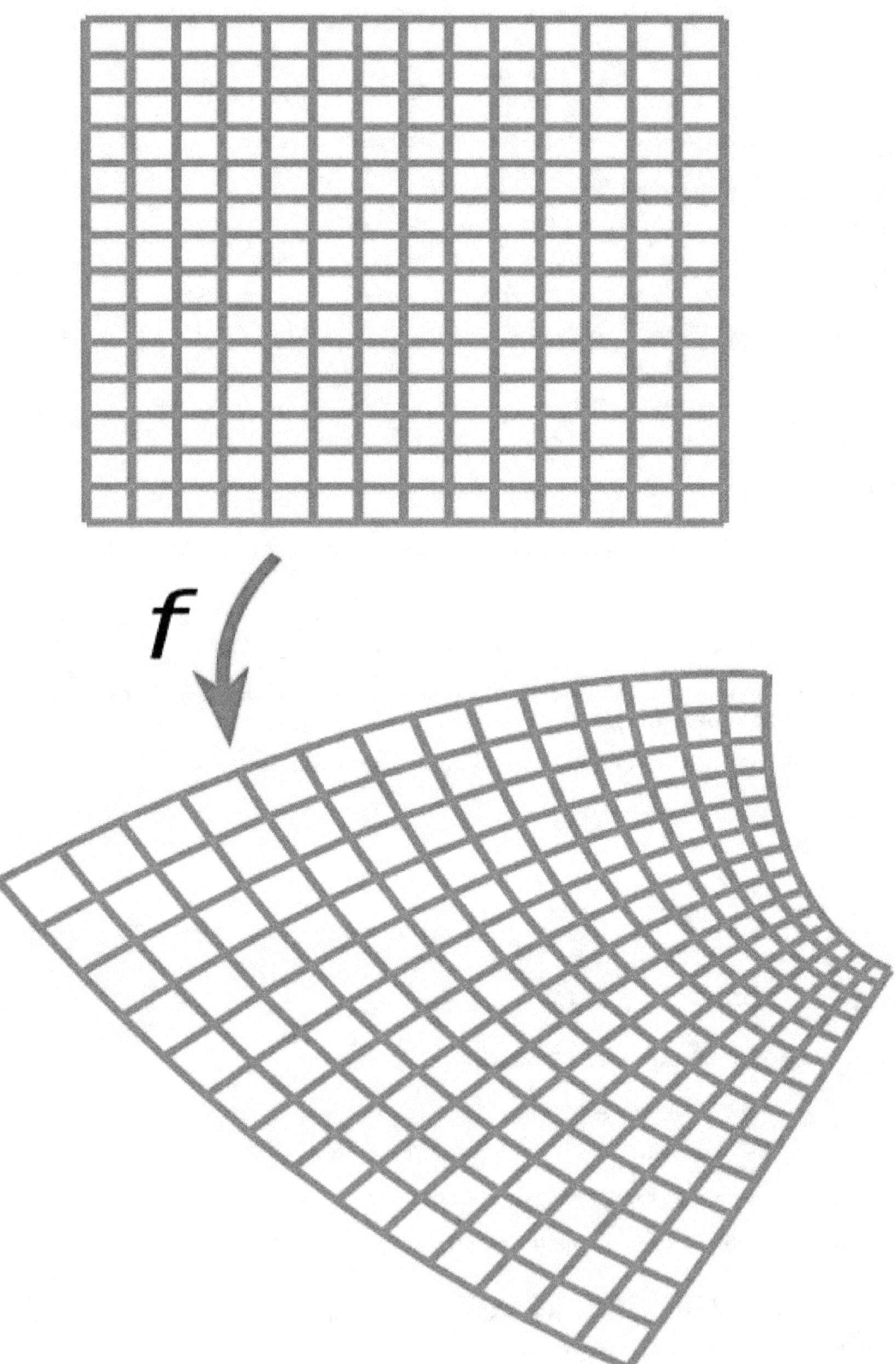

A rectangular grid (top) and its image under a conformal map f (bottom). It is seen that f maps pairs of lines intersecting at 90° to pairs of curves still intersecting at 90°

Chapter 18

Scattering amplitude

In quantum physics, the **scattering amplitude** is the amplitude of the outgoing spherical wave relative to the incoming plane wave in a stationary-state scattering process.[1] The latter is described by the wavefunction

$$\psi(\mathbf{r}) = e^{ikz} + f(\theta)\frac{e^{ikr}}{r} \ ,$$

where $\mathbf{r} \equiv (x, y, z)$ is the position vector; $r \equiv |\mathbf{r}|$; e^{ikz} is the incoming plane wave with the wavenumber k along the z axis; e^{ikr}/r is the outgoing spherical wave; θ is the scattering angle; and $f(\theta)$ is the scattering amplitude. The dimension of the scattering amplitude is length.

The scattering amplitude is a probability amplitude and the differential cross-section as a function of scattering angle is given as its modulus squared

$$\frac{d\sigma}{d\Omega} = |f(\theta)|^2 \ .$$

In the low-energy regime the scattering amplitude is determined by the scattering length.

18.1 Partial wave expansion

Main article: Partial wave analysis

In the partial wave expansion the scattering amplitude is represented as a sum over the partial waves,[2]

$$f = \sum_{\ell=0}^{\infty}(2\ell + 1)f_\ell P_\ell(\cos\theta)$$

where $f\ell$ is the partial scattering amplitude and $P\ell$ are the Legendre polynomials.

The partial amplitude can be expressed via the partial wave S-matrix element $S\ell$ ($= e^{2i\delta_\ell}$) and the scattering phase shift $\delta\ell$ as

$$f_\ell = \frac{S_\ell - 1}{2ik} = \frac{e^{2i\delta_\ell} - 1}{2ik} = \frac{e^{i\delta_\ell}\sin\delta_\ell}{k} = \frac{1}{k\cot\delta_\ell - ik} \ .$$

Then the differential cross section is given by[3]

$$\frac{d\sigma}{d\Omega} = |f(\theta)|^2 = \frac{1}{k^2}\left|\sum_{\ell=0}^{\infty}(2\ell+1)e^{i\delta_\ell}\sin\delta_\ell P_\ell(\cos\theta)\right|^2$$

and the total elastic cross section becomes

$$\sigma = 2\pi\int_0^\pi \frac{d\sigma}{d\Omega}\sin\theta\,d\theta = \frac{4\pi}{k}\operatorname{Im}f(0)$$

where $\operatorname{Im}f(0)$ is the imaginary part of $f(0)$.

18.2   X-rays

The scattering length for X-rays is the Thomson scattering length or classical electron radius, r_0 .

18.3   Neutrons

The nuclear neutron scattering process involves the coherent neutron scattering length, often described by b .

18.4   Quantum mechanical formalism

A quantum mechanical approach is given by the S matrix formalism.

18.5   References

[1] Quantum Mechanics: Concepts and Applications By Nouredine Zettili, 2nd edition, page 623. ISBN 978-0-470-02679-3 Paperback 688 pages January 2009, ©2008

[2] Michael Fowler/ 1/17/08 Plane Waves and Partial Waves

[3] Schiff, Leonard I. (1968). *Quantum Mechanics*. New York: McGraw Hill. pp. 119–120.

Chapter 19

Feynman diagram

For a less technical version, see this article on the Simple English Wikipedia.

In theoretical physics, **Feynman diagrams** are pictorial representations of the mathematical expressions describing the behavior of subatomic particles. The scheme is named for its inventor, American physicist Richard Feynman, and was first introduced in 1948. The interaction of sub-atomic particles can be complex and difficult to understand intuitively. Feynman diagrams give a simple visualization of what would otherwise be a rather arcane and abstract formula. As David Kaiser writes, "since the middle of the 20th century, theoretical physicists have increasingly turned to this tool to help them undertake critical calculations", and as such "Feynman diagrams have revolutionized nearly every aspect of theoretical physics".[1] While the diagrams are applied primarily to quantum field theory, they can also be used in other fields, such as solid-state theory.

Feynman used Ernst Stueckelberg's interpretation of the positron as if it were an electron moving backward in time.[2] Thus, antiparticles are represented as moving backward along the time axis in Feynman diagrams.

The calculation of probability amplitudes in theoretical particle physics requires the use of rather large and complicated integrals over a large number of variables. These integrals do, however, have a regular structure, and may be represented graphically as Feynman diagrams. A Feynman diagram is a contribution of a particular class of particle paths, which join and split as described by the diagram. More precisely, and technically, a Feynman diagram is a graphical representation of a perturbative contribution to the transition amplitude or correlation function of a quantum mechanical or statistical field theory. Within the canonical formulation of quantum field theory, a Feynman diagram represents a term in the Wick's expansion of the perturbative S-matrix. Alternatively, the path integral formulation of quantum field theory represents the transition amplitude as a weighted sum of all possible histories of the system from the initial to the final state, in terms of either particles or fields. The transition amplitude is then given as the matrix element of the S-matrix between the initial and the final states of the quantum system.

19.1 Motivation and history

When calculating scattering cross-sections in particle physics, the interaction between particles can be described by starting from a free field that describes the incoming and outgoing particles, and including an interaction Hamiltonian to describe how the particles deflect one another. The amplitude for scattering is the sum of each possible interaction history over all possible intermediate particle states. The number of times the interaction Hamiltonian acts is the order of the perturbation expansion, and the time-dependent perturbation theory for fields is known as the Dyson series. When the intermediate states at intermediate times are energy eigenstates (collections of particles with a definite momentum) the series is called old-fashioned perturbation theory.

The Dyson series can be alternatively rewritten as a sum over Feynman diagrams, where at each interaction vertex both the energy and momentum are conserved, but where the length of the energy momentum four vector is not equal to the mass. The Feynman diagrams are much easier to keep track of than old-fashioned terms, because the old-fashioned way

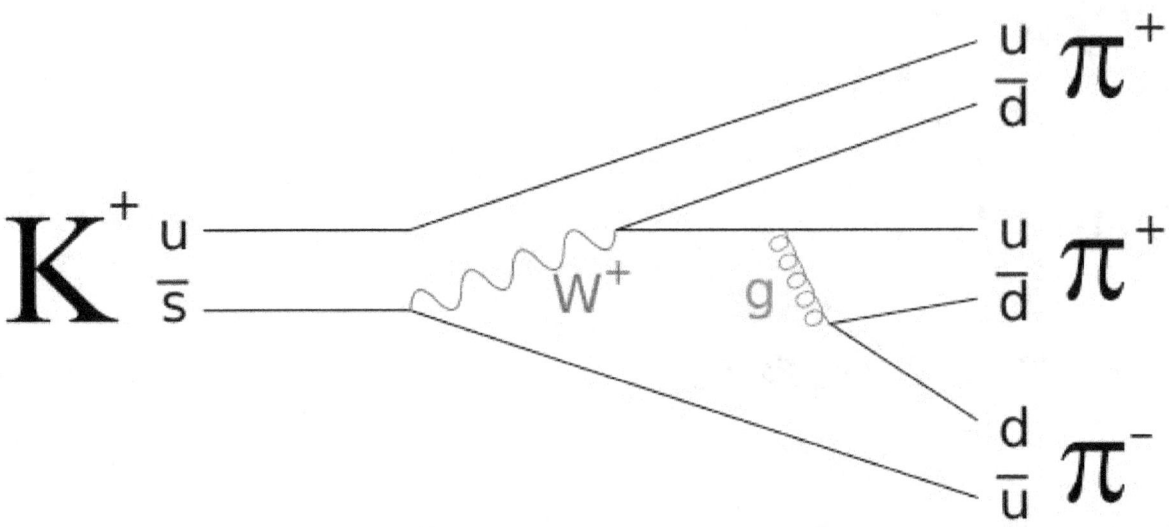

In this diagram, a kaon, made of an up and anti-strange quark, decays both weakly and strongly into three pions, with intermediate steps involving a W boson and a gluon (represented by the blue sine wave and green spiral, respectively).

treats the particle and antiparticle contributions as separate. Each Feynman diagram is the sum of exponentially many old-fashioned terms, because each internal line can separately represent either a particle or an antiparticle. In a non-relativistic theory, there are no antiparticles and there is no doubling, so each Feynman diagram includes only one term.

Feynman gave a prescription for calculating the amplitude for any given diagram from a field theory Lagrangian—the Feynman rules. Each internal line corresponds to a factor of the corresponding virtual particle's propagator; each vertex where lines meet gives a factor derived from an interaction term in the Lagrangian, and incoming and outgoing lines carry an energy, momentum, and spin.

In addition to their value as a mathematical tool, Feynman diagrams provide deep physical insight into the nature of particle interactions. Particles interact in every way available; in fact, intermediate virtual particles are allowed to propagate faster than light. The probability of each final state is then obtained by summing over all such possibilities. This is closely tied to the functional integral formulation of quantum mechanics, also invented by Feynman—see path integral formulation.

The naïve application of such calculations often produces diagrams whose amplitudes are infinite, because the short-distance particle interactions require a careful limiting procedure, to include particle self-interactions. The technique of renormalization, suggested by Ernst Stueckelberg and Hans Bethe and implemented by Dyson, Feynman, Schwinger, and Tomonaga compensates for this effect and eliminates the troublesome infinities. After renormalization, calculations using Feynman diagrams match experimental results with very high accuracy.

Feynman diagram and path integral methods are also used in statistical mechanics and can even be applied to classical mechanics.[3]

19.1.1   Alternative names

Murray Gell-Mann always referred to Feynman diagrams as **Stueckelberg diagrams**, after a Swiss physicist, Ernst Stueckelberg, who devised a similar notation many years earlier. Stueckelberg was motivated by the need for a manifestly covariant formalism for quantum field theory, but did not provide as automated a way to handle symmetry factors and loops, although he was first to find the correct physical interpretation in terms of forward and backward in time particle paths, all without the path-integral.[4] Historically they were sometimes called **Feynman–Dyson diagrams** or **Dyson graphs**,[5] because when they were introduced the path integral was unfamiliar, and Freeman Dyson's derivation from old-fashioned perturbation theory was easier to follow for physicists trained in earlier methods. However, in 2006 Dyson himself stated that the diagrams should be called *Feynman diagrams* because "he taught us how to use them". This reflects historical fact: Feynman had to lobby hard for the diagrams which confused the establishment physicists trained in equations and graphs.[6]

19.2 Representation of physical reality

In their presentations of fundamental interactions,[7][8] written from the particle physics perspective, Gerard 't Hooft and Martinus Veltman gave good arguments for taking the original, non-regularized Feynman diagrams as the most succinct representation of our present knowledge about the physics of quantum scattering of fundamental particles. Their motivations are consistent with the convictions of James Daniel Bjorken and Sidney Drell:[9]

> The Feynman graphs and rules of calculation summarize quantum field theory in a form in close contact with the experimental numbers one wants to understand. Although the statement of the theory in terms of graphs may imply perturbation theory, use of graphical methods in the many-body problem shows that this formalism is flexible enough to deal with phenomena of nonperturbative characters ... Some modification of the Feynman rules of calculation may well outlive the elaborate mathematical structure of local canonical quantum field theory ...

So far there are no opposing opinions. In quantum field theories the Feynman diagrams are obtained from Lagrangian by Feynman rules.

19.3 Particle-path interpretation

A Feynman diagram is a representation of quantum field theory processes in terms of particle paths. The particle trajectories are represented by the lines of the diagram, which can be squiggly or straight, with an arrow or without, depending on the type of particle. A point where lines connect to other lines is an interaction vertex, and this is where the particles meet and interact: by emitting or absorbing new particles, deflecting one another, or changing type.

There are three different types of lines: *internal lines* connect two vertices, *incoming lines* extend from "the past" to a vertex and represent an initial state, and *outgoing lines* extend from a vertex to "the future" and represent the final state. Sometimes, the bottom of the diagram is the past and the top the future; other times, the past is to the left and the future to the right. When calculating correlation functions instead of scattering amplitudes, there is no past and future and all the lines are internal. The particles then begin and end on little x's, which represent the positions of the operators whose correlation is being calculated.

Feynman diagrams are a pictorial representation of a contribution to the total amplitude for a process that can happen in several different ways. When a group of incoming particles are to scatter off each other, the process can be thought of as one where the particles travel over all possible paths, including paths that go backward in time.

Feynman diagrams are often confused with spacetime diagrams and bubble chamber images because they all describe particle scattering. Feynman diagrams are graphs that represent the trajectories of particles in intermediate stages of a scattering process. Unlike a bubble chamber picture, only the sum of all the Feynman diagrams represent any given particle interaction; particles do not choose a particular diagram each time they interact. The law of summation is in accord with the principle of superposition—every diagram contributes to the total amplitude for the process.

19.4 Description

A Feynman diagram represents a perturbative contribution to the amplitude of a quantum transition from some initial quantum state to some final quantum state.

For example, in the process of electron-positron annihilation the initial state is one electron and one positron, the final state: two photons.

The initial state is often assumed to be at the left of the diagram and the final state at the right (although other conventions are also used quite often).

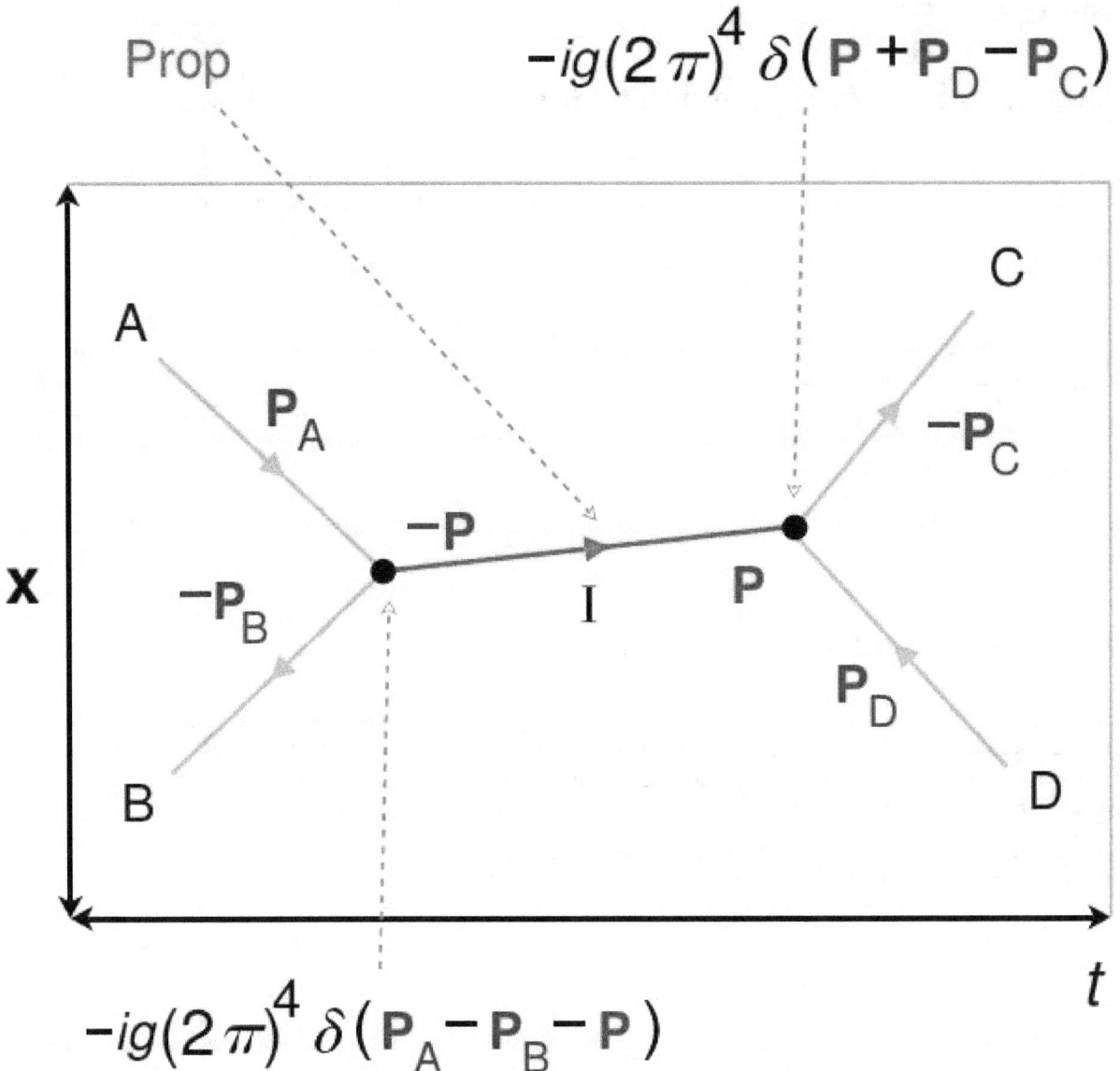

$$\text{Prop} \qquad\qquad -ig(2\pi)^4\,\delta\!\left(P + P_D - P_C\right)$$

$$-ig(2\pi)^4\,\delta\!\left(P_A - P_B - P\right)$$

General features of the scattering process $A + B \to C + D$:

*• internal lines (**red**) for intermediate particles and processes, which has a propagator factor ("prop"), external lines (**orange**) for incoming/outgoing particles to/from vertices (**black**),*

• at each vertex there is 4-momentum conservation using delta functions, 4-momenta entering the vertex are positive while those leaving are negative, the factors at each vertex and internal line are multiplied in the amplitude integral,

• space x and time t axes are not always shown, directions of external lines correspond to passage of time.

A Feynman diagram consists of points, called vertices, and lines attached to the vertices.

The particles in the initial state are depicted by lines sticking out in the direction of the initial state (e.g., to the left), the particles in the final state are represented by lines sticking out in the direction of the final state (e.g., to the right).

In QED there are two types of particles: electrons/positrons (called fermions) and photons (called gauge bosons). They are represented in Feynman diagrams as follows:

1. Electron in the initial state is represented by a solid line with an arrow pointing toward the vertex ($\to\bullet$).

2. Electron in the final state is represented by a line with an arrow pointing away from the vertex: ($\bullet\to$).

3. Positron in the initial state is represented by a solid line with an arrow pointing away from the vertex: ($\leftarrow\bullet$).

4. Positron in the final state is represented by a line with an arrow pointing toward the vertex: ($\bullet\leftarrow$).

5. Photon in the initial and the final state is represented by a wavy line ($\sim\bullet$ and $\bullet\sim$).

In QED a vertex always has three lines attached to it: one bosonic line, one fermionic line with arrow toward the vertex, and one fermionic line with arrow away from the vertex.

The vertices might be connected by a bosonic or fermionic propagator. A bosonic propagator is represented by a wavy line connecting two vertices ($\bullet\sim\bullet$). A fermionic propagator is represented by a solid line (with an arrow in one or another direction) connecting two vertices, ($\bullet\leftarrow\bullet$).

The number of vertices gives the order of the term in the perturbation series expansion of the transition amplitude.

19.4.1 Electron/positron annihilation example

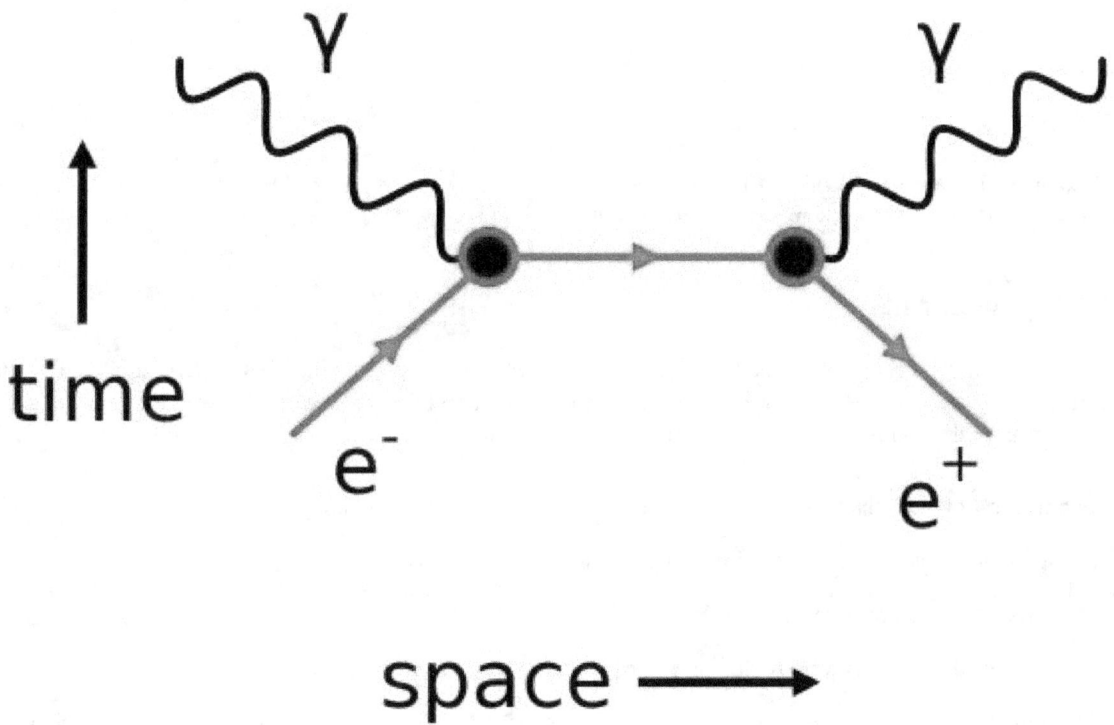

Feynman diagram of electron/positron annihilation

The electron/positron annihilation interaction:

$e^+e^- \rightarrow 2\gamma$

has a contribution from the second order Feynman diagram shown adjacent:

In the initial state (at the bottom; early time) there is one electron (e^-) and one positron (e^+) and in the final state (at the top; late time) there are two photons (γ).

19.5   Canonical quantization formulation

The probability amplitude for a transition of a quantum system from the initial state $|i\rangle$ to the final state $|f\rangle$ is given by the matrix element

$$S_{fi} = \langle f|S|i\rangle ,$$

where S is the S-matrix.

In the canonical quantum field theory the S-matrix is represented within the interaction picture by the perturbation series in the powers of the interaction Lagrangian,

$$S = \sum_{n=0}^{\infty} \frac{i^n}{n!} \int \prod_{j=1}^{n} d^4 x_j\, T \prod_{j=1}^{n} L_v(x_j) \equiv \sum_{n=0}^{\infty} S^{(n)} ,$$

where L_v is the interaction Lagrangian and T signifies the time-ordered product of operators.

A Feynman diagram is a graphical representation of a term in the Wick's expansion of the time-ordered product in the n-th order term $S^{(n)}$ of the S-matrix,

$$T \prod_{j=1}^{n} L_v(x_j) = \sum_{\text{all possible contractions}} (\pm) N \prod_{j=1}^{n} L_v(x_j) ,$$

where N signifies the normal-product of the operators and $(\pm)$ takes care of the possible sign change when commuting the fermionic operators to bring them together for a contraction (a propagator).

19.5.1   Feynman rules

The diagrams are drawn according to the Feynman rules, which depend upon the interaction Lagrangian. For the QED interaction Lagrangian, $L_v = -g\bar{\psi}\gamma^\mu\psi A_\mu$, describing the interaction of a fermionic field ψ with a bosonic gauge field A_μ , the Feynman rules can be formulated in coordinate space as follows:

1. Each integration coordinate x_j is represented by a point (sometimes called a vertex);

2. A bosonic propagator is represented by a wiggly line connecting two points;

3. A fermionic propagator is represented by a solid line connecting two points;

4. A bosonic field $A_\mu(x_i)$ is represented by a wiggly line attached to the point x_i :

5. A fermionic field $\psi(x_i)$ is represented by a solid line attached to the point x_i with an arrow toward the point;

6. A fermionic field $\bar{\psi}(x_i)$ is represented by a solid line attached to the point x_i with an arrow from the point;

19.5.2   Example: second order processes in QED

The second order perturbation term in the S-matrix is

$$S^{(2)} = \frac{(ie)^2}{2!} \int d^4 x\, d^4 x'\, T\bar{\psi}(x)\,\gamma^\mu\,\psi(x)\,A_\mu(x)\,\bar{\psi}(x')\,\gamma^\nu\,\psi(x')\,A_\nu(x').$$

Scattering of fermions

The Wick's expansion of the integrand gives (among others) the following term

$$N\bar{\psi}(x)\gamma^\mu\psi(x)\bar{\psi}(x')\gamma^\nu\psi(x')\underline{A_\mu(x)A_\nu(x')} ,$$

where

$$\underline{A_\mu(x)A_\nu(x')} = \int \frac{d^4k}{(2\pi)^4} \frac{-ig_{\mu\nu}}{k^2+i0} e^{-ik(x-x')}$$

is the electromagnetic contraction (propagator) in the Feynman gauge. This term is represented by the Feynman diagram at the right. This diagram gives contributions to the following processes:

1. e^-e^- scattering (initial state at the right, final state at the left of the diagram);

2. e^+e^+ scattering (initial state at the left, final state at the right of the diagram);

3. e^-e^+ scattering (initial state at the bottom/top, final state at the top/bottom of the diagram).

Compton scattering and annihilation/generation of e^-e^+ pairs

Another interesting term in the expansion is

$$N\bar{\psi}(x)\gamma^\mu \underline{\psi(x)\bar{\psi}(x')}\gamma^\nu \psi(x') A_\mu(x) A_\nu(x') ,$$

where

$$\underline{\psi(x)\bar{\psi}(x')} = \int \frac{d^4p}{(2\pi)^4} \frac{i}{\gamma p - m + i0} e^{-ip(x-x')}$$

is the fermionic contraction (propagator).

19.6 Path integral formulation

In a path-integral, the field Lagrangian, integrated over all possible field histories, defines the probability amplitude to go from one field configuration to another. In order to make sense, the field theory should have a well-defined ground state, and the integral should be performed a little bit rotated into imaginary time, i.e. a Wick Rotation.

19.6.1 Scalar field Lagrangian

A simple example is the free relativistic scalar field in d-dimensions, whose action integral is:

$$S = \int \frac{1}{2}\partial_\mu\phi\partial^\mu\phi\, d^d x .$$

The probability amplitude for a process is:

$$\int_A^B e^{iS} D\phi .$$

where A and B are space-like hypersurfaces that define the boundary conditions. The collection of all the $\phi(A)$ on the starting hypersurface give the initial value of the field, analogous to the starting position for a point particle, and the field values $\phi(B)$ at each point of the final hypersurface defines the final field value, which is allowed to vary, giving a different amplitude to end up at different values. This is the field-to-field transition amplitude.

The path integral gives the expectation value of operators between the initial and final state:

$$\int_A^B e^{iS}\phi(x_1)...\phi(x_n)D\phi = \langle A|\phi(x_1)...\phi(x_n)|B\rangle \, .$$

and in the limit that A and B recede to the infinite past and the infinite future, the only contribution that matters is from the ground state (this is only rigorously true if the path-integral is defined slightly rotated into imaginary time). The path integral should be thought of as analogous to a probability distribution, and it is convenient to define it so that multiplying by a constant doesn't change anything:

$$\frac{\int e^{iS}\phi(x_1)...\phi(x_n)D\phi}{\int e^{iS}D\phi} = \langle 0|\phi(x_1)....\phi(x_n)|0\rangle \, .$$

The normalization factor on the bottom is called the *partition function* for the field, and it coincides with the statistical mechanical partition function at zero temperature when rotated into imaginary time.

The initial-to-final amplitudes are ill-defined if one thinks of the continuum limit right from the beginning, because the fluctuations in the field can become unbounded. So the path-integral should be thought of as on a discrete square lattice, with lattice spacing a and the limit $a \to 0$ should be taken carefully. If the final results do not depend on the shape of the lattice or the value of a, then the continuum limit exists.

19.6.2   On a lattice

On a lattice, (i), the field can be expanded in Fourier modes:

$$\phi(x) = \int \frac{dk}{(2\pi)^d}\phi(k)e^{ik\cdot x} = \int_k \phi(k)e^{ikx} \, .$$

Here the integration domain is over k restricted to a cube of side length $2\pi/a$, so that large values of k are not allowed. It is important to note that the k-measure contains the factors of 2π from Fourier transforms, this is the best standard convention for k-integrals in QFT. The lattice means that fluctuations at large k are not allowed to contribute right away, they only start to contribute in the limit $a \to 0$. Sometimes, instead of a lattice, the field modes are just cut off at high values of k instead.

It is also convenient from time to time to consider the space-time volume to be finite, so that the k modes are also a lattice. This is not strictly as necessary as the space-lattice limit, because interactions in k are not localized, but it is convenient for keeping track of the factors in front of the k-integrals and the momentum-conserving delta functions that will arise.

On a lattice, (ii), the action needs to be discretized:

$$S = \sum_{<x.y>} \frac{1}{2}(\phi(x) - \phi(y))^2 \, ,$$

where $< x, y >$ is a pair of nearest lattice neighbors x and y. The discretization should be thought of as defining what the derivative $\partial_\mu \phi$ means.

In terms of the lattice Fourier modes, the action can be written:

$$S = \int_k ((1 - \cos(k_1)) + (1 - \cos(k_2)) + ... + (1 - \cos(k_d))) \phi_k^* \phi^k .$$

For k near zero this is:

$$S = \int_k \frac{1}{2} k^2 |\phi(k)|^2 .$$

Now we have the continuum Fourier transform of the original action. In finite volume, the quantity $d^d k$ is not infinitesimal, but becomes the volume of a box made by neighboring Fourier modes, or $(2\pi/V)^d$.

The field ϕ is real-valued, so the Fourier transform obeys:

$$\phi(k)^* = \phi(-k) .$$

In terms of real and imaginary parts, the real part of $\phi(k)$ is an even function of k, while the imaginary part is odd. The Fourier transform avoids double-counting, so that it can be written:

$$S = \int_k \frac{1}{2} k^2 \phi(k) \phi(-k)$$

over an integration domain that integrates over each pair (k,−k) exactly once.

For a complex scalar field with action

$$S = \int \frac{1}{2} \partial_\mu \phi^* \partial^\mu \phi \, d^d x$$

the Fourier transform is unconstrained:

$$S = \int_k \frac{1}{2} k^2 |\phi(k)|^2$$

and the integral is over all k.

Integrating over all different values of $\phi(x)$ is equivalent to integrating over all Fourier modes, because taking a Fourier transform is a unitary linear transformation of field coordinates. When you change coordinates in a multidimensional integral by a linear transformation, the value of the new integral is given by the determinant of the transformation matrix. If

$$y_i = A_{ij} x_j ,$$

then

$$\det(A) \int dx_1 dx_2 ... dx_n = \int dy_1 dy_2 ... dy_n .$$

If A is a rotation, then

$$A^T A = I$$

so that $\det A = \pm 1$, and the sign depends on whether the rotation includes a reflection or not.

The matrix that changes coordinates from $\phi(x)$ to $\phi(k)$ can be read off from the definition of a Fourier transform.

$$A_{kx} = e^{ikx}$$

and the Fourier inversion theorem tells you the inverse:

$$A_{kx}^{-1} = e^{-ikx}$$

which is the complex conjugate-transpose, up to factors of 2π . On a finite volume lattice, the determinant is nonzero and independent of the field values.

$$\det A = 1$$

and the path integral is a separate factor at each value of k.

$$\int \exp\left(\frac{i}{2} \sum_k k^2 \phi^*(k)\phi(k) \right) D\phi = \prod_k \int_{\phi_k} e^{\frac{i}{2}k^2 |\phi_k|^2 d^d k}$$

The factor $d^d k$ is the infinitesimal volume of a discrete cell in k-space, in a square lattice box $d^d k = 1/L^d$, where L is the side-length of the box. Each separate factor is an oscillatory Gaussian, and the width of the Gaussian diverges as the volume goes to infinity.

In imaginary time, the *Euclidean action* becomes positive definite, and can be interpreted as a probability distribution. The probability of a field having values ϕ_k is

$$e^{\int_k -\frac{1}{2}k^2 \phi_k^* \phi_k} = \prod_k e^{-k^2 |\phi_k|^2 d^d k}$$

The expectation value of the field is the statistical expectation value of the field when chosen according to the probability distribution:

$$\langle \phi(x_1)...\phi(x_n) \rangle = \frac{\int e^{-S} \phi(x_1)...\phi(x_n) D\phi}{\int e^{-S} D\phi}$$

Since the probability of ϕ_k is a product, the value of $\phi(k)$ at each separate value of k is independently Gaussian distributed. The variance of the Gaussian is 1/ (k^2 d^dk), which is formally infinite, but that just means that the fluctuations are unbounded in infinite volume. In any finite volume, the integral is replaced by a discrete sum, and the variance of the integral is V/k^2 .

19.6.3 Monte Carlo

The path integral defines a probabilistic algorithm to generate a Euclidean scalar field configuration. Randomly pick the real and imaginary parts of each Fourier mode at wavenumber k to be a gaussian random variable with variance $1/k^2$. This generates a configuration $\phi_C(k)$ at random, and the Fourier transform gives $\phi_C(x)$. For real scalar fields, the algorithm must generate only one of each pair $\phi(k), \phi(-k)$, and make the second the complex conjugate of the first.

To find any correlation function, generate a field again and again by this procedure, and find the statistical average:

$$\langle \phi(x_1)...\phi(x_n) \rangle = \lim_{|C| \to \infty} \frac{\sum_C \phi_C(x_1)...\phi_C(x_n)}{|C|}$$

where $|C|$ is the number of configurations, and the sum is of the product of the field values on each configuration. The Euclidean correlation function is just the same as the correlation function in statistics or statistical mechanics. The quantum mechanical correlation functions are an analytic continuation of the Euclidean correlation functions.

For free fields with a quadratic action, the probability distribution is a high-dimensional Gaussian, and the statistical average is given by an explicit formula. But the Monte Carlo method also works well for bosonic interacting field theories where there is no closed form for the correlation functions.

19.6.4 Scalar propagator

Each mode is independently Gaussian distributed. The expectation of field modes is easy to calculate:

$$\langle \phi_k \phi_{k'} \rangle = 0$$

for $k \neq k'$, since then the two Gaussian random variables are independent and both have zero mean.

$$\langle \phi_k \phi_k \rangle = \frac{V}{k^2}$$

in finite volume V, when the two k-values coincide, since this is the variance of the Gaussian. In the infinite volume limit,

$$\langle \phi(k)\phi(k') \rangle = \delta(k - k') \frac{1}{k^2}$$

Strictly speaking, this is an approximation: the lattice propagator is:

$$\langle \phi(k)\phi(k')\rangle = \delta(k - k') \frac{1}{2(d - \cos(k_1) + \cos(k_2)... + \cos(k_d))}$$

But near k=0, for field fluctuations long compared to the lattice spacing, the two forms coincide.

It is important to emphasize that the delta functions contain factors of 2π, so that they cancel out the 2π factors in the measure for k integrals.

$$\delta(k) = (2\pi)^d \delta_D(k_1)\delta_D(k_2)...\delta_D(k_d)$$

where $\delta_D(k)$ is the ordinary one-dimensional Dirac delta function. This convention for delta-functions is not universal—some authors keep the factors of 2π in the delta functions (and in the k-integration) explicit.

19.6.5   Equation of motion

The form of the propagator can be more easily found by using the equation of motion for the field. From the Lagrangian, the equation of motion is:

$$\partial_\mu \partial^\mu \phi = 0$$

and in an expectation value, this says:

$$\partial_\mu \partial^\mu \langle \phi(x)\phi(y)\rangle = 0$$

Where the derivatives act on x, and the identity is true everywhere except when x and y coincide, and the operator order matters. The form of the singularity can be understood from the canonical commutation relations to be a delta-function. Defining the (euclidean) *Feynman propagator* Δ as the Fourier transform of the time-ordered two-point function (the one that comes from the path-integral):

$$\partial^2 \Delta(x) = i\delta(x)$$

So that:

$$\Delta(k) = \frac{i}{k^2}$$

If the equations of motion are linear, the propagator will always be the reciprocal of the quadratic-form matrix that defines the free Lagrangian, since this gives the equations of motion. This is also easy to see directly from the Path integral. The factor of i disappears in the Euclidean theory.

Wick theorem

Main article: Wick's Theorem

Because each field mode is an independent Gaussian, the expectation values for the product of many field modes obeys *Wick's theorem*:

$$\langle \phi(k_1)\phi(k_2)...\phi(k_n)\rangle$$

is zero unless the field modes coincide in pairs. This means that it is zero for an odd number of ϕ's, and for an even number of ϕ's, it is equal to a contribution from each pair separately, with a delta function.

$$\langle \phi(k_1)...\phi(k_{2n})\rangle = \sum \prod_{i,j} \frac{\delta(k_i - k_j)}{k_i^2}$$

where the sum is over each partition of the field modes into pairs, and the product is over the pairs. For example,

$$\langle \phi(k_1)\phi(k_2)\phi(k_3)\phi(k_4)\rangle = \frac{\delta(k_1 - k_2)}{k_1^2}\frac{\delta(k_3 - k_4)}{k_3^2} + \frac{\delta(k_1 - k_3)}{k_3^2}\frac{\delta(k_2 - k_4)}{k_2^2} + \frac{\delta(k_1 - k_4)}{k_1^2}\frac{\delta(k_2 - k_3)}{k_2^2}$$

An interpretation of Wick's theorem is that each field insertion can be thought of as a dangling line, and the expectation value is calculated by linking up the lines in pairs, putting a delta function factor that ensures that the momentum of each partner in the pair is equal, and dividing by the propagator.

Higher Gaussian moments—completing Wick's theorem

There is a subtle point left before Wick's theorem is proved—what if more than two of the phis have the same momentum? If it's an odd number, the integral is zero; negative values cancel with the positive values. But if the number is even, the integral is positive. The previous demonstration assumed that the phis would only match up in pairs.

But the theorem is correct even when arbitrarily many of the phis are equal, and this is a notable property of Gaussian integration:

$$I = \int e^{-ax^2/2}dx = \sqrt{\frac{2\pi}{a}}$$

$$\frac{\partial^n}{\partial a^n}I = \int \frac{x^{2n}}{2^n}e^{-ax^2/2}dx = \frac{1 \cdot 3 \cdot 5... \cdot (2n-1)}{2 \cdot 2 \cdot 2... \cdot 2}\sqrt{2\pi}a^{-\frac{2n+1}{2}}$$

Dividing by I,

$$\langle x^{2n}\rangle = \frac{\int x^{2n}e^{-ax^2/2}}{\int e^{-ax^2/2}} = 1 \cdot 3 \cdot 5... \cdot (2n-1)\frac{1}{a^n}$$

$$\langle x^2\rangle = \frac{1}{a}$$

If Wick's theorem were correct, the higher moments would be given by all possible pairings of a list of 2n x's:

$$\langle x_1 x_2 x_3 ... x_{2n} \rangle$$

where the x's are all the same variable, the index is just to keep track of the number of ways to pair them. The first x can be paired with 2n−1 others, leaving 2n−2. The next unpaired x can be paired with 2n-3 different x's leaving 2n−4, and so on. This means that Wick's theorem, uncorrected, says that the expectation value of x^{2n} should be:

$$\langle x^{2n} \rangle = (2n - 1) \cdot (2n - 3).... \cdot 5 \cdot 3 \cdot 1 (\langle x^2 \rangle)^n$$

and this is in fact the correct answer. So Wick's theorem holds no matter how many of the momenta of the internal variables coincide.

Interaction

Interactions are represented by higher order contributions, since quadratic contributions are always Gaussian. The simplest interaction is the quartic self-interaction, with an action:

$$S = \int \partial^\mu \phi \partial_\mu \phi + \frac{\lambda}{4!} \phi^4.$$

The reason for the combinatorial factor 4! will be clear soon. Writing the action in terms of the lattice (or continuum) Fourier modes:

$$S = \int_k k^2 |\phi(k)|^2 + \int_{k_1 k_2 k_3 k_4} \phi(k_1)\phi(k_2)\phi(k_3)\phi(k_4)\delta(k_1 + k_2 + k_3 + k_4) = S_F + X.$$

Where S_F is the free action, whose correlation functions are given by Wick's theorem. The exponential of S in the path integral can be expanded in powers of λ , giving a series of corrections to the free action.

$$e^{-S} = e^{-S_F}(1 + X + \frac{1}{2!}XX + \frac{1}{3!}XXX + ...)$$

The path integral for the interacting action is then a power series of corrections to the free action. The term represented by X should be thought of as four half-lines, one for each factor of $\phi(k)$. The half-lines meet at a vertex, which contributes a delta-function that ensures that the sum of the momenta are all equal.

To compute a correlation function in the interacting theory, there is a contribution from the X terms now. For example, the path-integral for the four-field correlator:

$$\langle \phi(k_1)\phi(k_2)\phi(k_3)\phi(k_4) \rangle = \frac{\int e^{-S}\phi(k_1)\phi(k_2)\phi(k_3)\phi(k_4)D\phi}{Z}$$

which in the free field was only nonzero when the momenta k were equal in pairs, is now nonzero for all values of the k. The momenta of the insertions $\phi(k_i)$ can now match up with the momenta of the X's in the expansion. The insertions should also be thought of as half-lines, four in this case, which carry a momentum k, but one that is not integrated.

The lowest order contribution comes from the first nontrivial term $e^{-S_F}X$ in the Taylor expansion of the action. Wick's theorem requires that the momenta in the X half-lines, the $\phi(k)$ factors in X, should match up with the momenta of the external half-lines in pairs. The new contribution is equal to:

$$\lambda \frac{1}{k_1^2} \frac{1}{k_2^2} \frac{1}{k_3^2} \frac{1}{k_4^2}.$$

The 4! inside X is canceled because there are exactly 4! ways to match the half-lines in X to the external half-lines. Each of these different ways of matching the half-lines together in pairs contributes exactly once, regardless of the values of the k's, by Wick's theorem.

Feynman diagrams

The expansion of the action in powers of X gives a series of terms with progressively higher number of X's. The contribution from the term with exactly n X's are called n-th order.

The n-th order terms has:

1. 4n internal half-lines, which are the factors of $\phi(k)$ from the X's. These all end on a vertex, and are integrated over all possible k.

2. external half-lines, which are the come from the $\phi(k)$ insertions in the integral.

By Wick's theorem, each pair of half-lines must be paired together to make a *line*, and this line gives a factor of

$$\frac{\delta(k_1 + k_2)}{k_1^2}$$

which multiplies the contribution. This means that the two half-lines that make a line are forced to have equal and opposite momentum. The line itself should be labelled by an arrow, drawn parallel to the line, and labeled by the momentum in the line k. The half-line at the tail end of the arrow carries momentum k, while the half-line at the head-end carries momentum $-k$. If one of the two half-lines is external, this kills the integral over the internal k, since it forces the internal k to be equal to the external k. If both are internal, the integral over k remains.

The diagrams that are formed by linking the half-lines in the X's with the external half-lines, representing insertions, are the Feynman diagrams of this theory. Each line carries a factor of $\frac{1}{k^2}$, the propagator, and either goes from vertex to vertex, or ends at an insertion. If it is internal, it is integrated over. At each vertex, the total incoming k is equal to the total outgoing k.

The number of ways of making a diagram by joining half-lines into lines almost completely cancels the factorial factors coming from the Taylor series of the exponential and the 4! at each vertex.

Loop order

A forest diagram is one where all the internal lines have momentum that is completely determined by the external lines and the condition that the incoming and outgoing momentum are equal at each vertex. The contribution of these diagrams is a product of propagators, without any integration. A tree diagram is a connected forest diagram.

An example of a tree diagram is the one where each of four external lines end on an X. Another is when three external lines end on an X, and the remaining half-line joins up with another X, and the remaining half-lines of this X run off to external lines. These are all also forest diagrams (as every tree is a forest); an example of a forest that is not a tree is when eight external lines end on two X's.

It is easy to verify that in all these cases, the momenta on all the internal lines is determined by the external momenta and the condition of momentum conservation in each vertex.

A diagram that is not a forest diagram is called a *loop* diagram, and an example is one where two lines of an X are joined to external lines, while the remaining two lines are joined to each other. The two lines joined to each other can have any momentum at all, since they both enter and leave the same vertex. A more complicated example is one where two X's are joined to each other by matching the legs one to the other. This diagram has no external lines at all.

The reason loop diagrams are called loop diagrams is because the number of k-integrals that are left undetermined by momentum conservation is equal to the number of independent closed loops in the diagram, where independent loops are counted as in homology theory. The homology is real-valued (actually R^d valued), the value associated with each line is the momentum. The boundary operator takes each line to the sum of the end-vertices with a positive sign at the head and a negative sign at the tail. The condition that the momentum is conserved is exactly the condition that the boundary of the k-valued weighted graph is zero.

A set of k-values can be relabeled whenever there is a closed loop going from vertex to vertex, never revisiting the same vertex. Such a cycle can be thought of as the boundary of a 2-cell. The k-labelings of a graph that conserves momentum (which has zero boundary) up to redefinitions of k (up to boundaries of 2-cells) define the first homology of a graph. The number of independent momenta that are not determined is then equal to the number of independent homology loops. For many graphs, this is equal to the number of loops as counted in the most intuitive way.

Symmetry factors

The number of ways to form a given Feynman diagram by joining together half-lines is large, and by Wick's theorem, each way of pairing up the half-lines contributes equally. Often, this completely cancels the factorials in the denominator of each term, but the cancellation is sometimes incomplete.

The uncancelled denominator is called the *symmetry factor* of the diagram. The contribution of each diagram to the correlation function must be divided by its symmetry factor.

For example, consider the Feynman diagram formed from two external lines joined to one X, and the remaining two half-lines in the X joined to each other. There are 4×3 ways to join the external half-lines to the X, and then there is only one way to join the two remaining lines to each other. The X comes divided by 4!=4×3×2, but the number of ways to link up the X half lines to make the diagram is only 4×3, so the contribution of this diagram is divided by two.

For another example, consider the diagram formed by joining all the half-lines of one X to all the half-lines of another X. This diagram is called a *vacuum bubble*, because it does not link up to any external lines. There are 4! ways to form this diagram, but the denominator includes a 2! (from the expansion of the exponential, there are two X's) and two factors of 4!. The contribution is multiplied by 4!/(2×4!×4!) = 1/48.

Another example is the Feynman diagram formed from two X's where each X links up to two external lines, and the remaining two half-lines of each X are joined to each other. The number of ways to link an X to two external lines is 4×3, and either X could link up to either pair, giving an additional factor of 2. The remaining two half-lines in the two X's can be linked to each other in two ways, so that the total number of ways to form the diagram is 4×3×4×3×2×2, while the denominator is 4!×4!×2!. The total symmetry factor is 2, and the contribution of this diagram is divided by 2.

The symmetry factor theorem gives the symmetry factor for a general diagram: the contribution of each Feynman diagram must be divided by the order of its group of automorphisms, the number of symmetries that it has.

An automorphism of a Feynman graph is a permutation M of the lines and a permutation N of the vertices with the following properties:

1. If a line l goes from vertex v to vertex v', then M(l) goes from N(v) to N(v'). If the line is undirected, as it is for a real scalar field, then M(l) can go from N(v') to N(v) too.

2. If a line l ends on an external line, M(l) ends on the same external line.

3. If there are different types of lines, M(l) should preserve the type.

This theorem has an interpretation in terms of particle-paths: when identical particles are present, the integral over all intermediate particles must not double-count states that differ only by interchanging identical particles.

Proof: To prove this theorem, label all the internal and external lines of a diagram with a unique name. Then form the diagram by linking the a half-line to a name and then to the other half line.

Now count the number of ways to form the named diagram. Each permutation of the X's gives a different pattern of linking names to half-lines, and this is a factor of n!. Each permutation of the half-lines in a single X gives a factor of 4!. So a named diagram can be formed in exactly as many ways as the denominator of the Feynman expansion.

But the number of unnamed diagrams is smaller than the number of named diagram by the order of the automorphism group of the graph.

Connected diagrams: *linked-cluster theorem*

Roughly speaking, a Feynman diagram is called *connected* if all vertices and propagator lines are linked by a sequence of vertices and propagators of the diagram itself. If one views it as a (undirected) graph it is connected. The remarkable relevance of such diagrams in QFTs is due to the fact that they are sufficient to determine the quantum partition function $Z[J]$. More precisely, connected Feynman diagrams determine

$$iW[J] \equiv \ln Z[J].$$

To see this, one should recall that

$$Z[J] \propto \sum_k D_k$$

with D_k constructed from some (arbitrary) Feynman diagram that can be thought to consist of several connected components C_i. If one encounters n_i (identical) copies of a component C_i within the Feynman diagram D_k one has to include a *symmetry factor* $n_i!$. However, in the end each contribution of a Feynman diagram D_k to the partition function has the generic form

$$\prod_i \frac{C_i^{n_i}}{n_i!}$$

where i labels the (infinite) many connected Feynman diagrams possible.

A scheme to successively create such contributions from the D_k to $Z[J]$ is obtained by

$$\left(\frac{1}{0!} + \frac{C_1}{1!} + \frac{C_1^2}{2!} + \dots \right) \left(1 + C_2 + \frac{1}{2}C_2^2 + \dots \right) \dots$$

and therefore yields

$$Z[J] \propto \prod_i \sum_{n_i=0}^{\infty} \frac{C_i^{n_i}}{n_i!} = \exp \sum_i C_i \propto \exp W[J].$$

To establish the *normalization* $Z_0 = \exp W[0] = 1$ one simply calculates all connected *vacuum diagrams*, i.e., the diagrams without any *sources* J (sometimes referred to as *external legs* of a Feynman diagram).

Vacuum bubbles

An immediate consequence of the linked-cluster theorem is that all vacuum bubbles, diagrams without external lines, cancel when calculating correlation functions. A correlation function is given by a ratio of path-integrals:

$$\langle \phi_1(x_1)...\phi_n(x_n) \rangle = \frac{\int e^{-S} \phi_1(x_1)...\phi_n(x_n) D\phi}{\int e^{-S} D\phi}.$$

The top is the sum over all Feynman diagrams, including disconnected diagrams that do not link up to external lines at all. In terms of the connected diagrams, the numerator includes the same contributions of vacuum bubbles as the denominator:

$$\int e^{-S} \phi_1(x_1)...\phi_n(x_n) D\phi = \left(\sum E_i \right)\left(\exp\left(\sum_i C_i \right) \right).$$

Where the sum over E diagrams includes only those diagrams each of whose connected components end on at least one external line. The vacuum bubbles are the same whatever the external lines, and give an overall multiplicative factor. The denominator is the sum over all vacuum bubbles, and dividing gets rid of the second factor.

The vacuum bubbles then are only useful for determining Z itself, which from the definition of the path integral is equal to:

$$Z = \int e^{-S} D\phi = e^{-HT} = e^{-\rho V}$$

where ρ is the energy density in the vacuum. Each vacuum bubble contains a factor of $\delta(k)$ zeroing the total k at each vertex, and when there are no external lines, this contains a factor of $\delta(0)$, because the momentum conservation is over-enforced. In finite volume, this factor can be identified as the total volume of space time. Dividing by the volume, the remaining integral for the vacuum bubble has an interpretation: it is a contribution to the energy density of the vacuum.

Sources

Correlation functions are the sum of the connected Feynman diagrams, but the formalism treats the connected and disconnected diagrams differently. Internal lines end on vertices, while external lines go off to insertions. Introducing *sources* unifies the formalism, by making new vertices where one line can end.

Sources are external fields, fields that contribute to the action, but are not dynamical variables. A scalar field source is another scalar field h that contributes a term to the (Lorentz) Lagrangian:

$$\int h(x)\phi(x) d^d x = \int h(k)\phi(k) d^d k$$

In the Feynman expansion, this contributes H terms with one half-line ending on a vertex. Lines in a Feynman diagram can now end either on an X vertex, or on an H-vertex, and only one line enters an H vertex. The Feynman rule for an H-vertex is that a line from an H with momentum k gets a factor of h(k).

The sum of the connected diagrams in the presence of sources includes a term for each connected diagram in the absence of sources, except now the diagrams can end on the source. Traditionally, a source is represented by a little "x" with one line extending out, exactly as an insertion.

$$\log(Z[h]) = \sum_{n,C} h(k_1)h(k_2)...h(k_n)C(k_1, ..., k_n)$$

where $C(k_1, ..., k_n)$ is the connected diagram with n external lines carrying momentum as indicated. The sum is over all connected diagrams, as before.

The field h is not dynamical, which means that there is no path integral over h: h is just a parameter in the Lagrangian, which varies from point to point. The path integral for the field is:

$$Z[h] = \int e^{iS + i\int h\phi} D\phi$$

and it is a function of the values of h at every point. One way to interpret this expression is that it is taking the Fourier transform in field space. If there is a probability density on R^n, the Fourier transform of the probability density is:

$$\int \rho(y)e^{iky}d^n y = \langle e^{iky} \rangle = \langle \prod_{i=1}^{n} e^{ih_i y_i} \rangle$$

The Fourier transform is the expectation of an oscillatory exponential. The path integral in the presence of a source h(x) is:

$$Z[h] = \int e^{iS} e^{i\int_x h(x)\phi(x)} D\phi = \langle e^{ih\phi} \rangle$$

which, on a lattice, is the product of an oscillatory exponential for each field value:

$$\langle \prod_x e^{ih_x \phi_x} \rangle$$

The fourier transform of a delta-function is a constant, which gives a formal expression for a delta function:

$$\delta(x - y) = \int e^{ik(x-y)} dk$$

This tells you what a field delta function looks like in a path-integral. For two scalar fields ϕ and η,

$$\delta(\phi - \eta) = \int e^{ih(x)(\phi(x)-\eta(x)d^d x} Dh$$

Which integrates over the Fourier transform coordinate, over h. This expression is useful for formally changing field coordinates in the path integral, much as a delta function is used to change coordinates in an ordinary multi-dimensional integral.

The partition function is now a function of the field h, and the physical partition function is the value when h is the zero function:

The correlation functions are derivatives of the path integral with respect to the source:

$$\langle \phi(x) \rangle = \frac{1}{Z} \frac{\partial}{\partial h(x)} Z[h] = \frac{\partial}{\partial h(x)} \log(Z[h]).$$

In Euclidean space, source contributions to the action can still appear with a factor of "i", so that they still do a Fourier transform.

19.6.6 Spin 1/2; "photons" and "ghosts"

Spin 1/2: Grassmann integrals

The field path-integral can be extended to the Fermi case, but only if the notion of integration is expanded. A Grassmann integral of a free Fermi field is a high-dimensional determinant or Pfaffian, which defines the new type of Gaussian integration appropriate for Fermi fields.

The two fundamental formulas of Grassmann integration are:

$$\int e^{M_{ij}\bar{\psi}^i \psi^j} D\bar{\psi} D\psi = \mathrm{Det}(M)$$

where M is an arbitrary matrix and $\psi, \bar{\psi}$ are independent Grassmann variables for each index i, and

$$\int e^{\frac{1}{2} A_{ij} \psi^i \psi^j} D\psi = \mathrm{Pfaff}(A)$$

Where A is an antisymmetric matrix, ψ is a collection of Grassmann variables, and the 1/2 is to prevent double-counting (since $\psi^i \psi^j = -\psi^j \psi^i$). In matrix notation, where $\bar{\psi}$ and $\bar{\eta}$ are Grassmann valued row vectors, η and ψ are Grassmann valued column vectors, and M is a real valued matrix:

$$Z = \int e^{\bar{\psi} M \psi + \bar{\eta}\psi + \bar{\psi}\eta} D\bar{\psi} D\psi = \int e^{(\bar{\psi}+\bar{\eta}M^{-1})M(\psi+M^{-1}\eta) - \bar{\eta}M^{-1}\eta} D\bar{\psi} D\psi = \mathrm{Det}(M)e^{-\bar{\eta}M^{-1}\eta}$$

Where the last equality is a consequence of the translation invariance of the Grassmann integral. The Grassmann variables η are external sources for ψ, and differentiating with respect to η pulls down factors of $\bar{\psi}$.

$$\langle \bar{\psi}\psi \rangle = \frac{1}{Z}\frac{\partial}{\partial \eta}\frac{\partial}{\partial \bar{\eta}}Z|_{\eta=\bar{\eta}=0} = M^{-1}$$

again, in a schematic matrix notation. The meaning of the formula above is that the derivative with respect to the appropriate component of η and $\bar{\eta}$ gives the matrix element of M^{-1} . This is exactly analogous to the Bosonic path integration formula for a Gaussian integral of a complex Bosonic field:

$$\int e^{\phi^* M\phi + h^*\phi + \phi^* h}D\phi^* D\phi = \frac{e^{h^* M^{-1}h}}{\text{Det}(M)}$$

$$\langle \phi^*\phi \rangle = \frac{1}{Z}\frac{\partial}{\partial h}\frac{\partial}{\partial h^*}Z|_{h=h^*=0} = M^{-1}$$

So that the propagator is the inverse of the matrix in the quadratic part of the action in both the Bose and Fermi case.

For real Grassmann fields, for Majorana fermions, the path integral a Pfaffian times a source quadratic form, and the formulas give the square root of the determinant, just as they do for real Bosonic fields. The propagator is still the inverse of the quadratic part.

The free Dirac Lagrangian:

$$\int \bar{\psi}(\gamma^\mu \partial_\mu - m)\psi$$

formally gives the equations of motion and the anticommutation relations of the Dirac field, just as the Klein Gordon Lagrangian in an ordinary path integral gives the equations of motion and commutation relations of the scalar field. By using the spatial Fourier transform of the Dirac field as a new basis for the Grassmann algebra, the quadratic part of the Dirac action becomes simple to invert:

$$S = \int_k \bar{\psi}(i\gamma^\mu k_\mu - m)\psi.$$

The propagator is the inverse of the matrix M linking $\psi(k)$ and $\bar{\psi}(k)$, since different values of k do not mix together.

$$\langle \bar{\psi}(k')\psi(k) \rangle = \delta(k+k')\frac{1}{\gamma \cdot k - m} = \delta(k+k')\frac{\gamma \cdot k + m}{k^2 - m^2}$$

The analog of Wick's theorem matches psi and psi-bars in pairs:

$$\langle \bar{\psi}(k_1)\bar{\psi}(k_2)...\bar{\psi}(k_n)\psi(k_1')...\psi(k_n) \rangle = \sum_{\text{pairings}}(-1)^S \prod_{\text{pairs } i,j} \delta(k_i - k_j)\frac{1}{\gamma \cdot k_i - m}$$

where S is the sign of the permutation that reorders the sequence of psi-bars and psis to put the ones that are paired up to make the delta-functions next to each other, with the psi-bar coming right before the psi. Since a psi-psi-bar pair is a commuting element of the Grassmann algebra, it doesn't matter what order the pairs are in. If more than one psi/psi-bar pair have the same k, the integral is zero, and it is easy to check that the sum over pairings gives zero in this case (there are always an even number of them). This is the Grassmann analog of the higher Gaussian moments that completed the Bosonic Wick's theorem earlier.

The rules for spin-1/2 Dirac particles are as follows: The propagator is the inverse of the Dirac operator, the lines have arrows just as for a complex scalar field, and the diagram acquires an overall factor of −1 for each closed Fermi loop. If there are an odd number of Fermi loops, the diagram changes sign. Historically, the −1 rule was very difficult for Feynman to discover. He discovered it after a long process of trial and error, since he lacked a proper theory of Grassmann integration.

The rule follows from the observation that the number of Fermi lines at a vertex is always even. Each term in the Lagrangian must always be Bosonic. A Fermi loop is counted by following Fermionic lines until one comes back to the starting point, then removing those lines from the diagram. Repeating this process eventually erases all the Fermionic lines: this is the Euler algorithm to 2-color a graph, which works whenever each vertex has even degree. Note that the number of steps in the Euler algorithm is only equal to the number of independent Fermionic homology cycles in the common special case that all terms in the Lagrangian are exactly quadratic in the Fermi fields, so that each vertex has exactly two Fermionic lines. When there are four-Fermi interactions (like in the Fermi effective theory of the Weak interactions) there are more k-integrals than Fermi loops. In this case, the counting rule should apply the Euler algorithm by pairing up the Fermi lines at each vertex into pairs that together form a bosonic factor of the term in the Lagrangian, and when entering a vertex by one line, the algorithm should always leave with the partner line.

To clarify and prove the rule, consider a Feynman diagram formed from vertices, terms in the Lagrangian, with Fermion fields. The full term is Bosonic, it is a commuting element of the Grassmann algebra, so the order in which the vertices appear is not important. The Fermi lines are linked into loops, and when traversing the loop, one can reorder the vertex terms one after the other as one goes around without any sign cost. The exception is when you return to the starting point, and the final half-line must be joined with the unlinked first half-line. This requires one permutation to move the last psi-bar to go in front of the first psi, and this gives the sign.

This rule is the only visible effect of the exclusion principle in internal lines. When there are external lines, the amplitudes are antisymmetric when two Fermi insertions for identical particles are interchanged. This is automatic in the source formalism, because the sources for Fermi fields are themselves Grassmann valued.

Spin 1: photons

The naive propagator for photons is infinite, since the Lagrangian for the A-field is:

$$S = \int \frac{1}{4} F^{\mu\nu} F_{\mu\nu} = \int -\frac{1}{2} (\partial^\mu A_\nu \partial_\mu A^\nu - \partial^\mu A_\mu \partial_\nu A^\nu).$$

The quadratic form defining the propagator is non-invertible. The reason is the gauge invariance of the field, adding a gradient to A does not change the physics.

To fix this problem, one needs to fix a gauge. The most convenient way is to demand that the divergence of A is some function f, whose value is random from point to point. It does no harm to integrate over the values of f, since it only determines the choice of gauge. This procedure inserts the following factor into the path integral for A:

$$\int \delta(\partial_\mu A^\mu - f) e^{-\frac{f^2}{2}} Df.$$

The first factor, the delta function, fixes the gauge. The second factor sums over different values of f that are inequivalent gauge fixings. This is simply

$$e^{-\frac{(\partial_\mu A_\mu)^2}{2}}.$$

The additional contribution from gauge-fixing cancels the second half of the free Lagrangian, giving the Feynman Lagrangian:

$$S = \int \partial^\mu A^\nu \partial_\mu A_\nu$$

which is just like four independent free scalar fields, one for each component of A. The Feynman propagator is:

$$\langle A_\mu(k) A_\nu(k') \rangle = \delta(k + k') \frac{g_{\mu\nu}}{k^2}.$$

The one difference is that the sign of one propagator is wrong in the Lorentz case: the timelike component has an opposite sign propagator. This means that these particle states have negative norm—they are not physical states. In the case of photons, it is easy to show by diagram methods that these states are not physical—their contribution cancels with longitudinal photons to only leave two physical photon polarization contributions for any value of k.

If the averaging over f is done with a coefficient different from 1/2, the two terms don't cancel completely. This gives a covariant Lagrangian with a coefficient λ, which does not affect anything:

$$S = \int \frac{1}{2}(\partial^\mu A^\nu \partial_\mu A_\nu - \lambda(\partial_\mu A^\mu)^2)$$

and the covariant propagator for QED is:

$$\langle A_\mu(k) A_\nu(k') \rangle = \delta(k + k') \frac{g_{\mu\nu} - \lambda \frac{k_\mu k_\nu}{k^2}}{k^2}.$$

Spin 1: nonabelian ghosts

To find the Feynman rules for nonabelian Gauge fields, the procedure that performs the Gauge fixing must be carefully corrected to account for a change of variables in the path-integral.

The gauge fixing factor has an extra determinant from popping the delta function:

$$\delta(\partial_\mu A_\mu - f)e^{-\frac{f^2}{2}} \mathrm{Det} M$$

To find the form of the determinant, consider first a simple two-dimensional integral of a function f that depends only on r, not on the angle θ. Inserting an integral over theta:

$$\int f(r)dxdy = \int f(r) \int d\theta \delta(y) |\frac{dy}{d\theta}| dxdy$$

The derivative-factor ensures that popping the delta function in θ removes the integral. Exchanging the order of integration,

$$\int f(r)dxdy = \int d\theta \int f(r)\delta(y)|\frac{dy}{d\theta}|dxdy$$

but now the delta-function can be popped in y,

$$\int f(r)dxdy = \int d\theta_0 \int f(x)|\frac{dy}{d\theta}|dx .$$

The integral over θ just gives an overall factor of 2π , while the rate of change of y with a change in θ is just x, so this exercise reproduces the standard formula for polar integration of a radial function:

$$\int f(r)dxdy = 2\pi \int f(x)xdx$$

In the path-integral for a nonabelian gauge field, the analogous manipulation is:

$$\int DA \int \delta(F(A))\text{Det}(\frac{\partial F}{\partial G})DGe^{iS} = \int DG \int \delta(F(A))\text{Det}(\frac{\partial F}{\partial G})e^{iS}$$

The factor in front is the volume of the gauge group, and it contributes a constant, which can be discarded. The remaining integral is over the gauge fixed action.

$$\int \text{Det}(\frac{\partial F}{\partial G})e^{iS_{GF}}DA$$

To get a covariant gauge, the gauge fixing condition is the same as in the Abelian case:

$$\partial_\mu A^\mu = f$$

Whose variation under an infinitesimal gauge transformation is given by:

$$\partial_\mu D_\mu \alpha$$

where α is the adjoint valued element of the Lie algebra at every point that performs the infinitesimal gauge transformation. This adds the Faddeev Popov determinant to the action:

$$\text{Det}(\partial_\mu D_\mu)$$

which can be rewritten as a Grassman integral by introducing ghost fields:

$$\int e^{\bar{\eta}\partial_\mu D^\mu \eta} D\bar{\eta} D\eta$$

The determinant is independent of f, so the path-integral over f can give the Feynman propagator (or a covariant propagator) by choosing the measure for f as in the abelian case. The full gauge fixed action is then the Yang Mills action in Feynman gauge with an additional ghost action:

$$S = \int Tr\partial_\mu A_\nu \partial^\mu A^\nu + f^i_{jk}\partial^\nu A^\mu_i A^j_\mu A^k_\nu + f^i_{jr}f^r_{kl}A_iA_jA^kA^l + Tr\partial_\mu\bar{\eta}\partial^\mu\eta + \bar{\eta}A_j\eta$$

The diagrams are derived from this action. The propagator for the spin-1 fields has the usual Feynman form. There are vertices of degree 3 with momentum factors whose couplings are the structure constants, and vertices of degree 4 whose couplings are products of structure constants. There are additional ghost loops, which cancel out timelike and longitudinal states in A loops.

In the Abelian case, the determinant for covariant gauges does not depend on A, so the ghosts do not contribute to the connected diagrams.

19.7 Particle-path representation

Feynman diagrams were originally discovered by Feynman, by trial and error, as a way to represent the contribution to the S-matrix from different classes of particle trajectories.

19.7.1 Schwinger representation

The Euclidean scalar propagator has a suggestive representation:

$$\frac{1}{p^2 + m^2} = \int_0^\infty e^{-\tau(p^2+m^2)}d\tau$$

The meaning of this identity (which is an elementary integration) is made clearer by Fourier transforming to real space.

$$\Delta(x) = \int_0^\infty d\tau e^{-m^2\tau}\frac{1}{(4\pi\tau)^{d/2}}e^{\frac{-x^2}{4\tau}}$$

The contribution at any one value of τ to the propagator is a Gaussian of width $\sqrt{\tau}$. The total propagation function from 0 to x is a weighted sum over all proper times τ of a normalized Gaussian, the probability of ending up at x after a random walk of time τ.

The path-integral representation for the propagator is then:

$$\Delta(x) = \int_0^\infty d\tau \int DX e^{-\int_0^\tau (\dot{x}^2/2+m^2)d\tau'}$$

which is a path-integral rewrite of the Schwinger representation.

The Schwinger representation is both useful for making manifest the particle aspect of the propagator, and for symmetrizing denominators of loop diagrams.

19.7.2   Combining denominators

The Schwinger representation has an immediate practical application to loop diagrams. For example, For the diagram in the phi-4 theory formed by joining two x's together in two half-lines, and making the remaining lines external, the integral over the internal propagators in the loop is:

$$\int_k \frac{1}{(k^2 + m^2)} \frac{1}{((k+p)^2 + m^2)} \; .$$

Here one line carries momentum k and the other k+p. The asymmetry can be fixed by putting everything in the Schwinger representation.

$$\int_{t,t'} e^{-t(k^2+m^2)-t'((k+p)^2+m^2)} dt dt' \; .$$

Now the exponent mostly depends on t+t',

$$\int_{t,t'} e^{-(t+t')(k^2+m^2)-t'2p\cdot k-t'p^2} \; ,$$

except for the asymmetrical little bit. Defining the variable u=(t+t') and $v = t'/u$, the variable u goes from 0 to infinity, while v goes from 0 to 1. The variable u is the total proper time for the loop, while v parametrizes the fraction of the proper time on the top of the loop vs. the bottom.

The Jacobian for this transformation of variables is easy to work out from the identities:

$$d(uv) = dt' \quad du = dt + dt' ,$$

and "wedging" gives

$$udu \wedge dv = dt \wedge dt'$$

This allows the u integral to be evaluated explicitly:

$$\int_{u,v} ue^{-u(k^2+m^2+v2p\cdot k+vp^2)} = \int \frac{1}{(k^2 + m^2 + v2p \cdot k - vp^2)^2} dv$$

leaving only the v -integral. This method, invented by Schwinger but usually attributed to Feynman, is called *combining denominator*. Abstractly, it is the elementary identity:

$$\frac{1}{AB} = \int_0^1 \frac{1}{(vA + (1 - v)B)^2} dv$$

But this form does not provide the physical motivation for introducing v — v is the proportion of proper time on one of the legs of the loop.

Once the denominators are combined, a shift in k to $k' = k + vp$ symmetrizes everything:

$$\int_0^1 \int \frac{1}{(k^2 + m^2 + v2p \cdot k + vp^2)^2} dk dv = \int_0^1 \int \frac{1}{(k'^2 + m^2 + v(1-v)p^2)^2} dk' dv$$

This form shows that the moment that p^2 is more negative than 4 times the mass of the particle in the loop, which happens in a physical region of Lorentz space, the integral has a cut. This is exactly when the external momentum can create physical particles.

When the loop has more vertices, there are more denominators to combine:

$$\int dk \frac{1}{(k^2 + m^2)} \frac{1}{((k+p_1)^2 + m^2)} \cdots \frac{1}{((k+p_n)^2 + m^2)}$$

The general rule follows from the Schwinger prescription for n+1 denominators:

$$\frac{1}{D_0 D_1 ... D_n} = \int_0^\infty ... \int_0^\infty e^{-u_0 D_0 ... - u_n D_n} du_0 ... du_n .$$

The integral over the Schwinger parameters u_i can be split up as before into an integral over the total proper time $u = u_0 + u_1 ... + u_n$ and an integral over the fraction of the proper time in all but the first segment of the loop $v_i = u_i/u$ for $i \in \{1, 2, ..., n\}$. The v's are positive and add up to less than 1, so that the v integral is over an n dimensional simplex.

The Jacobian for the coordinate transformation can be worked out as before:

$$du = du_0 + du_1 ... + du_n$$

$$d(uv_i) = du_i .$$

"Wedging" all these equation together, one obtains

$$u^n du \wedge dv_1 \wedge dv_2 ... \wedge dv_n = du_0 \wedge du_1 ... \wedge du_n .$$

This gives the integral:

$$\int_0^\infty \int_{simplex} u^n e^{-u(v_0 D_0 + v_1 D_1 + v_2 D_2 ... + v_n D_n)} dv_1 ... dv_n du ,$$

where the simplex is the region defined by the conditions $v_i > 0$ and $\sum_{i=1}^n v_i < 1$ as well as $v_0 = 1 - \sum_{i=1}^n v_i$. Performing the u integral gives the general prescription for combining denominators:

$$\frac{1}{D_0 ... D_n} = n! \int_{simplex} \frac{1}{(v_0 D_0 + v_1 D_1 ... + v_n D_n)^{n+1}} dv_1 dv_2 ... dv_n$$

Since the numerator of the integrand is not involved, the same prescription works for any loop, no matter what the spins are carried by the legs. The interpretation of the parameters v_i is that they are the fraction of the total proper time spent on each leg.

19.7.3   Scattering

The correlation functions of a quantum field theory describe the scattering of particles. The definition of "particle" in relativistic field theory is not self-evident, because if you try to determine the position so that the uncertainty is less than the compton wavelength, the uncertainty in energy is large enough to produce more particles and antiparticles of the same type from the vacuum. This means that the notion of a single-particle state is to some extent incompatible with the notion of an object localized in space.

In the 1930s, Wigner gave a mathematical definition for single-particle states: they are a collection of states that form an irreducible representation of the Poincaré group. Single particle states describe an object with a finite mass, a well defined momentum, and a spin. This definition is fine for protons and neutrons, electrons and photons, but it excludes quarks, which are permanently confined, so the modern point of view is more accommodating: a particle is anything whose interaction can be described in terms of Feynman diagrams, which have an interpretation as a sum over particle trajectories.

A field operator can act to produce a one-particle state from the vacuum, which means that the field operator $\phi(x)$ produces a superposition of Wigner particle states. In the free field theory, the field produces one particle states only. But when there are interactions, the field operator can also produce 3-particle, 5-particle (if there is no +/− symmetry also 2, 4, 6 particle) states too. To compute the scattering amplitude for single particle states only requires a careful limit, sending the fields to infinity and integrating over space to get rid of the higher-order corrections.

The relation between scattering and correlation functions is the LSZ-theorem: The scattering amplitude for n particles to go to m-particles in a scattering event is the given by the sum of the Feynman diagrams that go into the correlation function for n+m field insertions, leaving out the propagators for the external legs.

For example, for the $\lambda\phi^4$ interaction of the previous section, the order λ contribution to the (Lorentz) correlation function is:

$$\langle\phi(k_1)\phi(k_2)\phi(k_3)\phi(k_4)\rangle = \frac{i}{k_1^2}\frac{i}{k_2^2}\frac{i}{k_3^2}\frac{i}{k_4^2}i\lambda$$

Stripping off the external propagators, that is, removing the factors of i/k^2, gives the invariant scattering amplitude M:

$$M = i\lambda$$

which is a constant, independent of the incoming and outgoing momentum. The interpretation of the scattering amplitude is that the sum of $|M|^2$ over all possible final states is the probability for the scattering event. The normalization of the single-particle states must be chosen carefully, however, to ensure that M is a relativistic invariant.

Non-relativistic single particle states are labeled by the momentum k, and they are chosen to have the same norm at every value of k. This is because the nonrelativistic unit operator on single particle states is:

$$\int dk|k\rangle\langle k|$$

In relativity, the integral over the k-states for a particle of mass m integrates over a hyperbola in E,k space defined by the energy–momentum relation:

$$E^2 - k^2 = m^2$$

If the integral weighs each k point equally, the measure is not Lorentz invariant. The invariant measure integrates over all values of k and E, restricting to the hyperbola with a Lorentz invariant delta function:

$$\int \delta(E^2 - k^2 - m^2)|E, k\rangle\langle E, k|dEdk = \int \frac{dk}{2E}|k\rangle\langle k|$$

So the normalized k-states are different from the relativistically normalized k-states by a factor of $\sqrt{E} = (k^2 - m^2)^{\frac{1}{4}}$

The invariant amplitude M is then the probability amplitude for relativistically normalized incoming states to become relativistically normalized outgoing states.

For nonrelativistic values of k, the relativistic normalization is the same as the nonrelativistic normalization (up to a constant factor $\sqrt{m}$). In this limit, the ϕ^4 invariant scattering amplitude is still constant. The particles created by the field phi scatter in all directions with equal amplitude.

The nonrelativistic potential, which scatters in all directions with an equal amplitude (in the Born approximation), is one whose Fourier transform is constant—a delta-function potential. The lowest order scattering of the theory reveals the non-relativistic interpretation of this theory—it describes a collection of particles with a delta-function repulsion. Two such particles have an aversion to occupying the same point at the same time.

19.8   Nonperturbative effects

Thinking of Feynman diagrams as a perturbation series, nonperturbative effects like tunneling do not show up, because any effect that goes to zero faster than any polynomial does not affect the Taylor series. Even bound states are absent, since at any finite order particles are only exchanged a finite number of times, and to make a bound state, the binding force must last forever.

But this point of view is misleading, because the diagrams not only describe scattering, but they also are a representation of the short-distance field theory correlations. They encode not only asymptotic processes like particle scattering, they also describe the multiplication rules for fields, the operator product expansion. Nonperturbative tunneling processes involve field configurations that on average get big when the coupling constant gets small, but each configuration is a coherent superposition of particles whose local interactions are described by Feynman diagrams. When the coupling is small, these become collective processes that involve large numbers of particles, but where the interactions between each of the particles is simple.

This means that nonperturbative effects show up asymptotically in resummations of infinite classes of diagrams, and these diagrams can be locally simple. The graphs determine the local equations of motion, while the allowed large-scale configurations describe non-perturbative physics. But because Feynman propagators are nonlocal in time, translating a field process to a coherent particle language is not completely intuitive, and has only been explicitly worked out in certain special cases. In the case of nonrelativistic bound states, the Bethe–Salpeter equation describes the class of diagrams to include to describe a relativistic atom. For quantum chromodynamics, the Shifman Vainshtein Zakharov sum rules describe non-perturbatively excited long-wavelength field modes in particle language, but only in a phenomenological way.

The number of Feynman diagrams at high orders of perturbation theory is very large, because there are as many diagrams as there are graphs with a given number of nodes. Nonperturbative effects leave a signature on the way in which the number of diagrams and resummations diverge at high order. It is only because non-perturbative effects appear in hidden form in diagrams that it was possible to analyze nonperturbative effects in string theory, where in many cases a Feynman description is the only one available.

19.9   In popular culture

- The use of the above diagram of the virtual particle producing a quark–antiquark pair was featured in the television sit-com *The Big Bang Theory*, in the episode "*The Bat Jar Conjecture*".

- *PhD Comics* of January 11. 2012, shows Feynman diagrams that *visualize and describe quantum academic interactions*. i.e. the paths followed by Ph.D. students when interacting with their advisors[10]

19.10   See also

- Julian Schwinger#Schwinger and Feynman
- Stueckelberg–Feynman interpretation
- Invariance mechanics
- Penguin diagram
- Path integral formulation
- Propagator
- List of Feynman diagrams
- Angular momentum diagrams (quantum mechanics)

19.11   Notes

[1] "Physics and Feynman's Diagrams" by David Kaiser. *American Scientist*, Volume 93, p. 156

[2] Feynman,Richard(1949). "The Theory of Positrons". *Physical Review* **76**(76): 749. Bibcode:1949PhRv...76..749F.
In this solution, the "negative energy states" appear in a form which may be pictured (as by Stückelberg) in space-time as waves traveling away from the external potential backwards in time. Experimentally, such a wave corresponds to a positron approaching the potential and annihilating the electron.

[3] R. Penco, D. Mauro (2006). "Perturbation theory via Feynman diagrams in classical mechanics". arXiv:hep-th/0605061v2.

[4] George Johnson (July 2000). "The Jaguar and the Fox". *The Atlantic*. Retrieved February 26, 2013.

[5] Gribbin, John and Mary. *Richard Feynman: A Life in Science*, Penguin-Putnam, 1997 Ch 5.

[6] Leonard Mlodinow. *Feynman's Rainbow*. Vintage, 2011. p. 29

[7] Gerardus 't Hooft, Martinus Veltman, *Diagrammar*, CERN Yellow Report 1973, reprinted in G. 't Hooft, *Under the Spell of Gauge Principle* (World Scientific, Singapore, 1994), Introduction online

[8] Martinus Veltman, *Diagrammatica: The Path to Feynman Diagrams*, Cambridge Lecture Notes in Physics, ISBN 0-521-45692-4

[9] Bjorken, J. D.; Drell, S. D. (1965). "Relativistic Quantum Fields". New York: McGraw-Hill. p. viii.

[10] Jorge Cham, Academic Interaction - Feynman Diagrams, January 11, 2012.

19.12   References

- Gerardus 't Hooft, Martinus Veltman, *Diagrammar*, CERN Yellow Report 1973, online
- David Kaiser, *Drawing Theories Apart: The Dispersion of Feynman Diagrams in Postwar Physics*, Chicago: University of Chicago Press, 2005. ISBN 0-226-42266-6
- Martinus Veltman, *Diagrammatica: The Path to Feynman Diagrams*, Cambridge Lecture Notes in Physics, ISBN 0-521-45692-4 (expanded, updated version of above)
- Mark Srednicki, *Quantum Field Theory*, online Script (2006)

19.13 External links

- AMS article: "What's New in Mathematics: Finite-dimensional Feynman Diagrams"

- Draw Feynman diagrams explained by Flip Tanedo at Quantumdiaries.com

- Drawing Feynman diagrams with FeynDiagram C++ library that produces PostScript output.

- Feynman Diagram Examples using Thorsten Ohl's Feynmf LaTeX package.

- Online Diagram Tool A graphical application for creating publication ready diagrams.

- JaxoDraw A Java program for drawing Feynman diagrams.

- SCaViS – a Java program that can be used for drawing Feynman diagrams using Python scripts

- Bowley, Roger; Copeland, Ed (2010). "Feynman Diagrams". *Sixty Symbols*. Brady Haran for the University of Nottingham.

Chapter 20

Amplituhedron

An **amplituhedron** is a geometric structure that enables simplified calculation of particle interactions in some quantum field theories. In planar $N = 4$ supersymmetric Yang–Mills theory, an amplituhedron is defined as a mathematical space known as the positive Grassmannian.[1]

Amplituhedron theory challenges the notion that space-time locality and unitarity are necessary components of a model of particle interactions. Instead, they are treated as properties that emerge from an underlying phenomenon.[2][3]

The connection between the amplituhedron and scattering amplitudes is at present a conjecture that has passed many non-trivial checks, including an understanding of how locality and unitarity arise as consequences of positivity.[1]

Research has been led by Nima Arkani-Hamed. Edward Witten described the work as "very unexpected" and said that "it is difficult to guess what will happen or what the lessons will turn out to be."[4]

20.1   Description

In the approach, the on-shell scattering process "tree" is described by a positive Grassmannian, a structure in algebraic geometry analogous to a convex polytope, that generalizes the idea of a simplex in projective space.[2] A polytope is the n-dimensional analogue of 3-dimensional polyhedrons, and the values being calculated in this case are scattering amplitudes, and so the object is called an *amplituhedron*.[5][1]

Using twistor theory, BCFW recursion relations involved in the scattering process may be represented as a small number of twistor diagrams. These diagrams effectively provide the recipe for constructing the positive Grassmannian, i.e. the amplituhedron, which may be captured in a single equation.[2] The scattering amplitude can thus be thought of as the volume of a certain polytope, the positive Grassmannian, in momentum twistor space.[1]

When the volume of the amplituhedron is calculated in the planar limit of $N = 4$ $D = 4$ supersymmetric Yang–Mills theory, it describes the scattering amplitudes of subatomic particles.[5] The amplituhedron thus provides a more intuitive geometric model for calculations whose underlying principles were until then highly abstract.[6]

The twistor-based representation provides a recipe for constructing specific cells in the Grassmannian which assemble to form a positive Grassmannian, i.e. the representation describes a specific cell decomposition of the positive Grassmannian.

The recursion relations can be resolved in many different ways, each giving rise to a different representation, with the final amplitude expressed as a sum of on-shell processes in different ways as well. Therefore any given on-shell representation of scattering amplitudes is not unique, but all such representations of a given interaction yield the same amplituhedron.[1]

20.2 Implications

The twistor approach simplifies calculations of particle interactions. In a perturbative approach to quantum field theory, such interactions may require the calculation of hundreds of Feynman diagrams. In contrast, twistor theory provides an approach in which scattering amplitudes can be computed in a way that yields much simpler expressions.[7]

The twistor approach was relatively abstract. The amplituhedron provides an underlying model. Its geometric nature suggests the possibility that the nature of the universe, both classical relativistic spacetime and quantum mechanics, can be described with geometry. Calculations can be done without assuming the quantum mechanical properties of locality and unitarity. In amplituhedron theory, locality and unitarity arise as a direct consequence of positivity. They are encoded in the positive geometry of the amplituhedron, via the singularity structure of the integrand for scattering amplitudes.[1]

Since the planar limit of the $N = 4$ supersymmetric Yang–Mills theory is a toy theory that does not describe the real world, the relevance of this technique for more realistic quantum field theories is currently unknown, but it provides promising directions for research into theories about the real world.

20.3 See also

- Grassmannian
- Invariance mechanics
- Twistor space
- Wilson loop

20.4 References

20.4.1 Notes

[1] Arkani-Hamed & Trnka 2013.

[2] Arkani-Hamed et al. 2012.

[3] Ryan O'Hanlon (September 19, 2013). "How to Feel About Space and Time Maybe Not Existing". Pacific Standard.

[4] Natalie Wolchover (September 17, 2013). "A Jewel at the Heart of Quantum Physics". Quanta Magazine.

[5] Trnka, Jaroslav. "The Amplituhedron" (PDF). Retrieved 19 September 2013.

[6] 4 gravitons and a grad student 2013.

[7] Kevin Drum (September 18, 2013). "Maybe Space-Time Is Just an Illusion". Mother Jones.

20.4.2 Bibliography

- Arkani-Hamed, Nima; Bourjaily, Jacob L.; Cachazo, Freddy; Goncharov, Alexander B.; Postnikov, Alexander; Trnka, Jaroslav (2012). "Scattering Amplitudes and the Positive Grassmannian". arXiv:1212.5605 [hep-th].

- Arkani-Hamed, Nima; Trnka, Jaroslav (2013). "The Amplituhedron". arXiv:1312.2007.

- Nima Arkani-Hamed (2013-08-30). "The Amplituhedron" (video). SUSY 2013 Conference Video Archive.

- Scattering Without Space-Time Subrahmanyan Chandrasekhar Lecture, 25 September 2012 on YouTube

- $N = 4$ $D = 4$ super Yang–Mills theory from nLab

- Arxiv paper on Total positivity, Grassmannians, and networks (Sept 2006)

- 4 gravitons and a grad student (2013). "The Amplituhedron and Other Excellently Silly Words".

20.5 External links

- New Discovery Simplifies Quantum Physics: Introducing the Amplituhedron

- The Amplituhedron – Joel Werner

- Grassmannian Geometry of Scattering Amplitudes Workshop, December 8–12, 2014

Chapter 21

Supersymmetry

"SUSY" redirects here. For other uses, see Susy (disambiguation).
For the episode of the American TV series *Angel*, see Supersymmetry (Angel).

Supersymmetry (SUSY), a theory of particle physics, is a proposed type of spacetime symmetry that relates two basic classes of elementary particles: bosons, which have an integer-valued spin, and fermions, which have a half-integer spin.[1] Each particle from one group is associated with a particle from the other, known as its superpartner, the spin of which differs by a half-integer. In a theory with perfectly "unbroken" supersymmetry, each pair of superpartners would share the same mass and internal quantum numbers besides spin. For example, there would be a "selectron" (superpartner electron), a bosonic version of the electron with the same mass as the electron, that would be easy to find in a laboratory. Thus, since no superpartners have been observed, if supersymmetry exists it must be a spontaneously broken symmetry so that superpartners may differ in mass.[2][3] Spontaneously-broken supersymmetry could solve many mysterious problems in particle physics including the hierarchy problem. The simplest realization of spontaneously-broken supersymmetry, the so-called Minimal Supersymmetric Standard Model, is one of the best studied candidates for physics beyond the Standard Model.

There is only indirect evidence and motivation for the existence of supersymmetry. Direct confirmation would entail production of superpartners in collider experiments, such as the Large Hadron Collider (LHC). The first run of the LHC found no evidence for supersymmetry (all results were consistent with the Standard Model), and thus set limits on superpartner masses in supersymmetric theories. Whilst many remain enthusiastic about supersymmetry,[4] this first run at the LHC led some physicists to explore other ideas.[5] In any case, in 2015 the LHC resumed its search for supersymmetry and other new physics in its second run.

21.1 Motivations

There are numerous phenomenological motivations for supersymmetry close to the electroweak scale, as well as technical motivations for supersymmetry at any scale.

21.1.1 The hierarchy problem

Supersymmetry close to the electroweak scale ameliorates the hierarchy problem that afflicts the Standard Model. In the Standard Model, the electroweak scale receives enormous Planck-scale quantum corrections. The observed hierarchy between the electroweak scale and the Planck scale must be achieved with extraordinary fine tuning. In a supersymmetric theory, on the other hand, Planck-scale quantum corrections cancel between partners and superpartners (owing to a minus sign associated with fermionic loops). The hierarchy between the electroweak scale and the Planck scale is achieved in a natural manner, without miraculous fine-tuning.

21.1.2   Gauge coupling unification

The idea that the gauge symmetry groups unify at high-energy is called Grand unification theory. In the Standard Model, however, the weak, strong and electromagnetic couplings fail to unify at high energy. In a supersymmetry theory, the running of the gauge couplings are modified, and precise high-energy unification of the gauge couplings is achieved. The modified running also provides a natural mechanism for radiative electroweak symmetry breaking.

21.1.3   Dark matter

TeV-scale supersymmetry (augmented with a discrete symmetry) typically provides a candidate dark matter particle at a mass scale consistent with thermal relic abundance calculations.[6][7]

21.1.4   Other technical motivations

Supersymmetry is also motivated by solutions to several theoretical problems, for generally providing many desirable mathematical properties, and for ensuring sensible behavior at high energies. Supersymmetric quantum field theory is often much easier to analyze, as many more problems become exactly solvable. When supersymmetry is imposed as a *local* symmetry, Einstein's theory of general relativity is included automatically, and the result is said to be a theory of supergravity. It is also a necessary feature of the most popular candidate for a theory of everything, superstring theory.

Another theoretically appealing property of supersymmetry is that it offers the only "loophole" to the Coleman–Mandula theorem, which prohibits spacetime and internal symmetries from being combined in any nontrivial way, for quantum field theories like the Standard Model with very general assumptions. The Haag-Lopuszanski-Sohnius theorem demonstrates that supersymmetry is the only way spacetime and internal symmetries can be combined consistently.[8]

21.2   History

A supersymmetry relating mesons and baryons was first proposed, in the context of hadronic physics, by Hironari Miyazawa during 1966. This supersymmetry did not involve spacetime, that is, it concerned internal symmetry, and was broken badly. Miyazawa's work was largely ignored at the time.[9][10][11][12]

J. L. Gervais and B. Sakita (during 1971),[13] Yu. A. Golfand and E. P. Likhtman (also during 1971), and D.V. Volkov and V.P. Akulov (1972),[14] independently rediscovered supersymmetry in the context of quantum field theory, a radically new type of symmetry of spacetime and fundamental fields, which establishes a relationship between elementary particles of different quantum nature, bosons and fermions, and unifies spacetime and internal symmetries of microscopic phenomena. Supersymmetry with a consistent Lie-algebraic graded structure on which the Gervais–Sakita rediscovery was based directly first arose during 1971[15] in the context of an early version of string theory by Pierre Ramond, John H. Schwarz and André Neveu.

Finally, Julius Wess and Bruno Zumino (during 1974)[16] identified the characteristic renormalization features of four-dimensional supersymmetric field theories, which identified them as remarkable QFTs, and they and Abdus Salam and their fellow researchers introduced early particle physics applications. The mathematical structure of supersymmetry (Graded Lie superalgebras) has subsequently been applied successfully to other topics of physics, ranging from nuclear physics,[17][18] critical phenomena,[19] quantum mechanics to statistical physics. It remains a vital part of many proposed theories of physics.

The first realistic supersymmetric version of the Standard Model was proposed during 1977 by Pierre Fayet and is known as the Minimal Supersymmetric Standard Model or MSSM for short. It was proposed to solve, amongst other things, the hierarchy problem.

21.3   Applications

21.3.1 Extension of possible symmetry groups

One reason that physicists explored supersymmetry is because it offers an extension to the more familiar symmetries of quantum field theory. These symmetries are grouped into the Poincaré group and internal symmetries and the Coleman–Mandula theorem showed that under certain assumptions, the symmetries of the S-matrix must be a direct product of the Poincaré group with a compact internal symmetry group or if there is not any mass gap, the conformal group with a compact internal symmetry group. During 1971 Golfand and Likhtman were the first to show that the Poincaré algebra can be extended through introduction of four anticommuting spinor generators (in four dimensions), which later became known as supercharges. During 1975 the Haag-Lopuszanski-Sohnius theorem analyzed all possible superalgebras in the general form, including those with an extended number of the supergenerators and central charges. This extended super-Poincaré algebra paved the way for obtaining a very large and important class of supersymmetric field theories.

The supersymmetry algebra

Main article: Supersymmetry algebra

Traditional symmetries of physics are generated by objects that transform by the tensor representations of the Poincaré group and internal symmetries. Supersymmetries, however, are generated by objects that transform by the spinor representations. According to the spin-statistics theorem, bosonic fields commute while fermionic fields anticommute. Combining the two kinds of fields into a single algebra requires the introduction of a $\mathbb{Z}_2$-grading under which the bosons are the even elements and the fermions are the odd elements. Such an algebra is called a Lie superalgebra.

The simplest supersymmetric extension of the Poincaré algebra is the Super-Poincaré algebra. Expressed in terms of two Weyl spinors, has the following anti-commutation relation:

$$\{Q_\alpha, \bar{Q}_{\dot\beta}\} = 2(\sigma^\mu)_{\alpha\dot\beta} P_\mu$$

and all other anti-commutation relations between the Qs and commutation relations between the Qs and Ps vanish. In the above expression $P_\mu = -i\partial_\mu$ are the generators of translation and σ^μ are the Pauli matrices.

There are representations of a Lie superalgebra that are analogous to representations of a Lie algebra. Each Lie algebra has an associated Lie group and a Lie superalgebra can sometimes be extended into representations of a Lie supergroup.

21.3.2 The Supersymmetric Standard Model

Main article: Minimal Supersymmetric Standard Model

Incorporating supersymmetry into the Standard Model requires doubling the number of particles since there is no way that any of the particles in the Standard Model can be superpartners of each other. With the addition of new particles, there are many possible new interactions. The simplest possible supersymmetric model consistent with the Standard Model is the Minimal Supersymmetric Standard Model (MSSM) which can include the necessary additional new particles that are able to be superpartners of those in the Standard Model.

One of the main motivations for SUSY comes from the quadratically divergent contributions to the Higgs mass squared. The quantum mechanical interactions of the Higgs boson causes a large renormalization of the Higgs mass and unless there is an accidental cancellation, the natural size of the Higgs mass is the greatest scale possible. This problem is known as the hierarchy problem. Supersymmetry reduces the size of the quantum corrections by having automatic cancellations between fermionic and bosonic Higgs interactions. If supersymmetry is restored at the weak scale, then the Higgs mass is related to supersymmetry breaking which can be induced from small non-perturbative effects explaining the vastly different scales in the weak interactions and gravitational interactions.

In many supersymmetric Standard Models there is a heavy stable particle (such as neutralino) which could serve as a weakly interacting massive particle (WIMP) dark matter candidate. The existence of a supersymmetric dark matter candidate is related closely to R-parity.

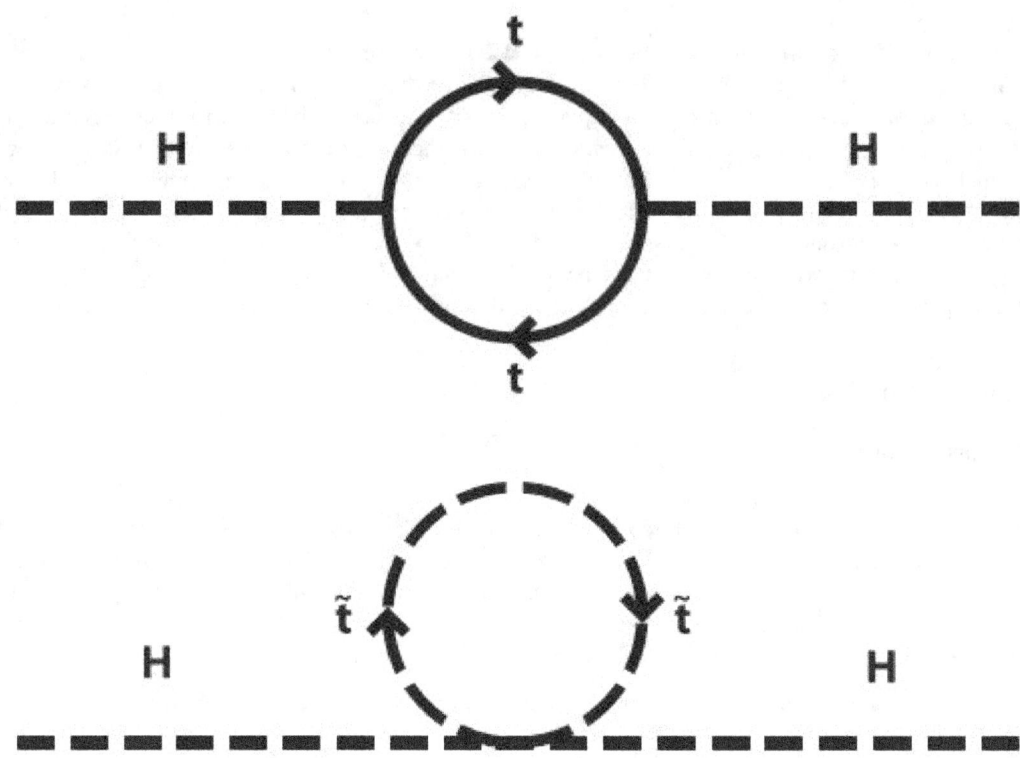

Cancellation of the Higgs boson quadratic mass renormalization between fermionic top quark loop and scalar stop squark tadpole Feynman diagrams in a supersymmetric extension of the Standard Model

The standard paradigm for incorporating supersymmetry into a realistic theory is to have the underlying dynamics of the theory be supersymmetric, but the ground state of the theory does not respect the symmetry and supersymmetry is broken spontaneously. The supersymmetry break can not be done permanently by the particles of the MSSM as they currently appear. This means that there is a new sector of the theory that is responsible for the breaking. The only constraint on this new sector is that it must break supersymmetry permanently and must give superparticles TeV scale masses. There are many models that can do this and most of their details do not matter. In order to parameterize the relevant features of supersymmetry breaking, arbitrary soft SUSY breaking terms are added to the theory which temporarily break SUSY explicitly but could never arise from a complete theory of supersymmetry breaking.

Gauge-coupling unification

Main article: Minimal Supersymmetric Standard Model § Gauge-coupling unification

One piece of evidence for supersymmetry existing is gauge coupling unification. The renormalization group evolution of the three gauge coupling constants of the Standard Model is somewhat sensitive to the present particle content of the theory. These coupling constants do not quite meet together at a common energy scale if we run the renormalization group using the Standard Model.[20] With the addition of minimal SUSY joint convergence of the coupling constants is projected at approximately 10^{16} GeV.[20]

21.3.3 Supersymmetric quantum mechanics

Main article: Supersymmetric quantum mechanics

Supersymmetric quantum mechanics adds the SUSY superalgebra to quantum mechanics as opposed to quantum field theory. Supersymmetric quantum mechanics often becomes relevant when studying the dynamics of supersymmetric solitons, and due to the simplified nature of having fields which are only functions of time (rather than space-time), a great deal of progress has been made in this subject and it is now studied in its own right.

SUSY quantum mechanics involves pairs of Hamiltonians which share a particular mathematical relationship, which are called *partner Hamiltonians*. (The potential energy terms which occur in the Hamiltonians are then known as *partner potentials*.) An introductory theorem shows that for every eigenstate of one Hamiltonian, its partner Hamiltonian has a corresponding eigenstate with the same energy. This fact can be exploited to deduce many properties of the eigenstate spectrum. It is analogous to the original description of SUSY, which referred to bosons and fermions. We can imagine a "bosonic Hamiltonian", whose eigenstates are the various bosons of our theory. The SUSY partner of this Hamiltonian would be "fermionic", and its eigenstates would be the theory's fermions. Each boson would have a fermionic partner of equal energy.

21.3.4 Supersymmetry: Applications to condensed matter physics

SUSY concepts have provided useful extensions to the WKB approximation. Additionally, SUSY has been applied to disorder averaged systems both quantum and non-quantum (through statistical mechanics), the Fokker-Planck equation being an example of a non-quantum theory. The 'supersymmetry' in all these systems arises from the fact that one is modelling one particle and as such the 'statistics' don't matter. The use of the supersymmetry method provides a mathematical rigorous alternative to the replica trick, but only in non-interacting systems, which attempts to address the so-called 'problem of the denominator' under disorder averaging. For more on the applications of supersymmetry in condensed matter physics see the book[21]

21.3.5 Supersymmetry in optics

Integrated optics was recently found[22] to provide a fertile ground on which certain ramifications of SUSY can be explored in readily-accessible laboratory settings. Making use of the analogous mathematical structure of the quantum-mechanical Schrödinger equation and the wave equation governing the evolution of light in one-dimensional settings, one may interpret the refractive index distribution of a structure as a potential landscape in which optical wave packets propagate. In this manner, a new class of functional optical structures with possible applications in phase matching, mode conversion[23] and space-division multiplexing becomes possible. SUSY transformations have been also proposed as a way to address inverse scattering problems in optics and as a one-dimensional transformation optics [24]

21.3.6 Mathematics

SUSY is also sometimes studied mathematically for its intrinsic properties. This is because it describes complex fields satisfying a property known as holomorphy, which allows holomorphic quantities to be exactly computed. This makes supersymmetric models useful "toy models" of more realistic theories. A prime example of this has been the demonstration of S-duality in four-dimensional gauge theories[25] that interchanges particles and monopoles.

The proof of the Atiyah-Singer index theorem is much simplified by the use of supersymmetric quantum mechanics.

21.4 General supersymmetry

Supersymmetry appears in many related contexts of theoretical physics. It is possible to have multiple supersymmetries and also have supersymmetric extra dimensions.

21.4.1   Extended supersymmetry

Main article: Extended supersymmetry

It is possible to have more than one kind of supersymmetry transformation. Theories with more than one supersymmetry transformation are known as extended supersymmetric theories. The more supersymmetry a theory has, the more constrained are the field content and interactions. Typically the number of copies of a supersymmetry is a power of 2, i.e. 1, 2, 4, 8. In four dimensions, a spinor has four degrees of freedom and thus the minimal number of supersymmetry generators is four in four dimensions and having eight copies of supersymmetry means that there are 32 supersymmetry generators.

The maximal number of supersymmetry generators possible is 32. Theories with more than 32 supersymmetry generators automatically have massless fields with spin greater than 2. It is not known how to make massless fields with spin greater than two interact, so the maximal number of supersymmetry generators considered is 32. This corresponds to an $N = 8$ supersymmetry theory. Theories with 32 supersymmetries automatically have a graviton.

For four dimensions there are the following theories, with the corresponding multiplets[26](CPT adds a copy, whenever they are not invariant under such symmetry)

- $N = 1$

Chiral multiplet: $(0, \frac{1}{2})$ Vector multiplet: $(\frac{1}{2}, 1)$ Gravitino multiplet: $(1, \frac{3}{2})$ Graviton multiplet: $(\frac{3}{2}, 2)$

- $N = 2$

hypermultiplet: $(-\frac{1}{2}, 0^2, \frac{1}{2})$ vector multiplet: $(0, \frac{1}{2}^2, 1)$ supergravity multiplet: $(1, \frac{3}{2}^2, 2)$

- $N = 4$

Vector multiplet: $(-1, -\frac{1}{2}^4, 0^6, \frac{1}{2}^4, 1)$ Supergravity multiplet: $(0, \frac{1}{2}^4, 1^6, \frac{3}{2}^4, 2)$

- $N = 8$

Supergravity multiplet: $(-2, -\frac{3}{2}^8, -1^{28}, -\frac{1}{2}^{56}, 0^{70}, \frac{1}{2}^{56}, 1^{28}, \frac{3}{2}^8, 2)$

21.4.2   Supersymmetry in alternate numbers of dimensions

It is possible to have supersymmetry in dimensions other than four. Because the properties of spinors change drastically between different dimensions, each dimension has its characteristic. In d dimensions, the size of spinors is approximately $2^{d/2}$ or $2^{(d-1)/2}$. Since the maximum number of supersymmetries is 32, the greatest number of dimensions in which a supersymmetric theory can exist is eleven.

21.5   Supersymmetry as a quantum group

Main article: Supersymmetry as a quantum group

Supersymmetry can be reinterpreted in the language of noncommutative geometry and quantum groups. In particular, it involves a mild form of noncommutativity, namely supercommutativity. See the main article for more details.

21.6 Supersymmetry in quantum gravity

Supersymmetry is part of a larger enterprise of theoretical physics to unify everything we know about the universe into a single consistent set of physical principles, known as the quest for a Theory of Everything (TOE). A significant part of this larger enterprise is the quest for a theory of quantum gravity, which would unify the classical theory of general relativity and the Standard Model, which explains the other three basic forces in physics (electromagnetism, the strong interaction, and the weak interaction), and provides a palette of fundamental particles upon which all four forces act. Two of the most active methods of forming a theory of quantum gravity are string theory and loop quantum gravity (LQG), although in theory, supersymmetry could be a component of other theories as well.

For string theory to be consistent, supersymmetry seems to be required at some level (although it may be a strongly broken symmetry). In particle theory, supersymmetry is recognized as a way to stabilize the hierarchy between the unification scale and the electroweak scale (or the Higgs boson mass), and can also provide a natural dark matter candidate. String theory also requires extra spatial dimensions which have to be compactified as in Kaluza–Klein theory.

Loop quantum gravity (LQG) predicts no additional spatial dimensions, nor anything else about particle physics. These theories can be formulated in three spatial dimensions and one dimension of time, although in some LQG theories dimensionality is an emergent property of the theory, rather than a fundamental assumption of the theory. Also, LQG is a theory of quantum gravity which does not require supersymmetry. Lee Smolin, one of the originators of LQG, has proposed that a loop quantum gravity theory incorporating either supersymmetry or extra dimensions, or both, be called "loop quantum gravity II".

If experimental evidence confirms supersymmetry in the form of supersymmetric particles such as the neutralino that is often believed to be the lightest superpartner, some people believe this would be a major boost to string theory. Since supersymmetry is a required component of string theory, any discovered supersymmetry would be consistent with string theory. If the Large Hadron Collider and other major particle physics experiments fail to detect supersymmetric partners or evidence of extra dimensions, many versions of string theory which had predicted certain low mass superpartners to existing particles may need to be significantly revised. The failure of experiments to discover either supersymmetric partners or extra spatial dimensions, as of 2013, has encouraged loop quantum gravity researchers.

21.7 Current status

Supersymmetric models are constrained by a variety of experiments, including measurements of low-energy observables – for example, the anomalous magnetic moment of the muon at Brookhaven; the WMAP dark matter density measurement and direct detection experiments – for example, XENON–100 and LUX; and by particle collider experiments, including B-physics, Higgs phenomenology and direct searches for superpartners (sparticles), at the Large Electron–Positron Collider, Tevatron and the LHC.

Historically, the tightest limits were from direct production at colliders. The first mass limits for squarks and gluinos were made at CERN by the UA1 experiment and the UA2 experiment at the Super Proton Synchrotron. LEP later set very strong limits,[27] which in 2006 were extended by the D0 experiment at the Tevatron.[28][29] From 2003, WMAP's and Planck's dark matter density measurements have strongly constrained supersymmetry models, which, if they explain dark matter, have to be tuned to invoke a particular mechanism to sufficiently reduce the neutralino density.

Prior to the beginning of the LHC, in 2009 fits of available data to CMSSM and NUHM1 indicated that squarks and gluinos were most likely to have masses in the 500 to 800 GeV range, though values as high as 2.5 TeV were allowed with low probabilities. Neutralinos and sleptons were expected to be quite light, with the lightest neutralino and the lightest stau most likely to be found between 100 to 150 GeV.[30]

The first run of the LHC found no evidence for supersymmetry, and, as a result, surpassed existing experimental limits from the Large Electron–Positron Collider and Tevatron and partially excluded the aforementioned expected ranges.[31]

During 2011 and 2012, the LHC discovered a Higgs boson with a mass of about 125 GeV, and with couplings to fermions and bosons which are consistent with the Standard Model. The MSSM predicts that the mass of the lightest Higgs boson should not be much higher than the mass of the Z boson, and, in the absence of fine tuning (with the supersymmetry breaking scale on the order of 1 TeV), should not exceed 130 GeV. Furthermore, for values of the MSSM parameter *tan* β ≤ 3, it predicts a Higgs mass below 114 GeV over most of the parameter space.[32] This region of Higgs mass was excluded by LEP by 2000. The LHC result is somewhat problematic for the minimal supersymmetric model, as the value of 125 GeV is relatively large for the model and can only be achieved with large radiative loop corrections from top squarks, which many theorists consider to be "unnatural" (see naturalness and fine tuning).[33] On the other hand, the lightest Higgs boson in the MSSM is Standard Model-like, which is consistent with measurements of the Higgs boson couplings at the LHC.

In spite of the null searches and the heavy Higgs, a recent analysis of the constrained minimal supersymmetric Standard Model, the CMSSM, suggests that the model is still compatible with all present experimental constraints.[34][35] The preferred masses for squarks and gluinos is about 2 TeV. The resulting fine-tuning of the electroweak scale, however, is considered "unnatural" (see little hierarchy problem), and some theorists now favor extended supersymmetry models – for example, the NMSSM.

21.8   See also

- Supersymmetric gauge theory

- Wess–Zumino model

- Minimal Supersymmetric Standard Model

- Supersymmetry as a quantum group

- Quantum group

- Supercharge

- Superfield

- Supergeometry

- Supergravity

- Supergroup

- Superspace

21.9   References

[1] Haber, Howie. "SUPERSYMMETRY, PART I (THEORY)" (PDF). *Reviews, Tables and Plots*. Particle Data Group (PDG). Retrieved 8 July 2015.

[2] Martin, Stephen P. (1997). "A Supersymmetry Primer". arXiv:hep-ph/9709356.

[3] Dine, Michael (2007). *Supersymmetry and String Theory: Beyond the Standard Model*. p. 169.

[4] Ellis, John. "The Physics Landscape after the Higgs Discovery at the LHC". *arXiv*. Invited plenary talk at SILAFAE 2014. Retrieved 8 July 2015.

[5] Wolchover, Natalie (November 20, 2012). "Supersymmetry Fails Test, Forcing Physics to Seek New Ideas". *Quanta Magazine*.

[6] Jonathan Feng: Supersymmetric Dark Matter *(pdf)*, University of California, Irvine, 11 May 2007

[7] Torsten Bringmann: The WIMP "Miracle" *(pdf)* University of Hamburg

[8] R. Haag, J. T. Lopuszanski and M. Sohnius, "All Possible Generators Of Supersymmetries Of The S Matrix", Nucl. Phys. B 88 (1975) 257

[9] H.Miyazawa(1966). "Baryon Number Changing Currents". *Prog. Theor. Phys.* **36**(6): 1266–1276. Bibcode:1966PThPh.. doi:10.1143/PTP.36.1266.

[10] H.Miyazawa(1968). "Spinor Currents and Symmetries of Baryons and Mesons". *Phys. Rev.* **170**(5): 1586–1590. doi:10.1103/PhysRev.170.1586.

[11] Michio Kaku, *Quantum Field Theory*, ISBN 0-19-509158-2, pg 663.

[12] Peter Freund, *Introduction to Supersymmetry*, ISBN 0-521-35675-X, pages 26-27, 138.

[13] Gervais, J. -L.; Sakita, B. (1971). "Field theory interpretation of supergauges in dual models". *Nuclear Physics B* **34** (2): 632. Bibcode:1971NuPhB..34..632G. doi:10.1016/0550-3213(71)90351-8.

[14] D.V. Volkov, V.P. Akulov, Pisma Zh.Eksp.Teor.Fiz. 16 (1972) 621; Phys.Lett. B46 (1973) 109; V.P. Akulov, D.V. Volkov, Teor.Mat.Fiz. 18 (1974) 39

[15] Ramond, P. (1971). "Dual Theory for Free Fermions". *Physical Review D* **3** (10): 2415. Bibcode:1971PhRvD...3.2415R. doi:10.1103/PhysRevD.3.2415.

[16] Wess,J.;Zumino,B. (1974). "Supergauge transformations in four dimensions". *Nuclear Physics B***70**: 39. Bibcode:1974NuPh doi:10.1016/0550-3213(74)90355-1.

[17] http://users.physik.fu-berlin.de/~{}kleinert/kleinert/?p=supersym suggested here

[18] Iachello,F. (1980). "Dynamical Supersymmetries in Nuclei". *Physical Review Letters***44**(12): 772. Bibcode:1980PhRvL..44. doi:10.1103/PhysRevLett.44.772.

[19] Friedan, D.; Qiu, Z.; Shenker, S. (1984). "Conformal Invariance, Unitarity, and Critical Exponents in Two Dimensions". *Physical Review Letters* **52** (18): 1575. Bibcode:1984PhRvL..52.1575F. doi:10.1103/PhysRevLett.52.1575.

[20] Gordon L. Kane, *The Dawn of Physics Beyond the Standard Model*, Scientific American, June 2003, page 60 and *The frontiers of physics*, special edition, Vol 15, #3, page 8

[21] *Supersymmetry in Disorder and Chaos*, Konstantin Efetov, Cambridge university press, 1997.

[22] Miri, M.-A.; Heinrich, M.; El-Ganainy, R.; Christodoulides, D. N. (2013). "Superymmetric optical structures". *Physical Review Letters* (APS) **110** (23): 233902. arXiv:1304.6646. Bibcode:2013PhRvL.110w3902M. doi:10.1103/PhysRevLett.110.233902. PMID 25167493. Retrieved April 2014.

[23] Heinrich, M.; Miri, M.-A.; Stützer, S.; El-Ganainy, R.; Nolte, S.; Szameit, A.; Christodoulides, D. N. (2014). "Superymmetric mode converters". *Nature Communications*(NPG)**5**: 3698. arXiv:1401.5734. Bibcode:2014NatCo...5E3698H.doi:10.1038/n PMID 24739256. Retrieved April 2014.

[24] Miri, M.-A.; Heinrich, Matthias; Christodoulides, D. N. (2014). "SUSY-inspired one-dimensional transformation optics". *Optica* (OSA) **1** (2): 89. arXiv:1408.0832. doi:10.1364/OPTICA.1.000089. Retrieved August 2014.

[25] Krasnitz, Michael (2002). *Correlation functions in supersymmetric gauge theories from supergravity fluctuafluctuations hHKtions* (PDF). Princeton University Department of Physics: Princeton University Department of Physics. p. 91.

[26] Polchinski,J. *String theory. Vol. 2: Superstring theory and beyond*, Appendix B

[27] LEPSUSYWG, ALEPH, DELPHI, L3 and OPAL experiments, charginos, large m0 LEPSUSYWG/01-03.1

[28] The D0-Collaboration (2009). "Search for associated production of charginos and neutralinos in the trilepton final state using 2.3 fb^{-1} of data". arXiv:0901.0646. Bibcode:2009PhLB..680...34D. doi:10.1016/j.physletb.2009.08.011.

[29] The D0 Collaboration (2006). "Search for squarks and gluinos in events with jets and missing transverse energy using 2.1 fb-1 of pp¯ collision data at s=1.96 TeV". arXiv:0712.3805. Bibcode:2008PhLB..660..449D. doi:10.1016/j.physletb.2008.01.042.

[30] O. Buchmueller; et al. (2009). "Likelihood Functions for Supersymmetric Observables in Frequentist Analyses of the CMSSM and NUHM1". *The European Physical Journal C* **64**(3): 391–415. arXiv:0907.5568. Bibcode:2009EPJC...64..391B.doi:10. 009-1159-z.

[31] Roszkowski, Leszek; Sessolo, Enrico Maria; Williams, Andrew J. (11 August 2014). "What next for the CMSSM and the NUHM:improved prospects for superpartner and dark matter detection". *Journal of High Energy Physics***2014**(8). doi:10.1007

[32] Marcela Carena and Howard E. Haber; Haber (1970). "Higgs Boson Theory and Phenomenology". *Progress in Particle and Nuclear Physics* **50**: 63. arXiv:hep-ph/0208209v3. Bibcode:2003PrPNP..50...63C. doi:10.1016/S0146-6410(02)00177-1.

[33] Patrick Draper; et al. (December 2011). "Implications of a 125 GeV Higgs for the MSSM and Low-Scale SUSY Breaking". *Physical Review D* **85** (9): 095007. arXiv:1112.3068. Bibcode:2012PhRvD..85i5007D. doi:10.1103/PhysRevD.85.095007.

[34] Bechtle, Philip. "How alive is constrained SUSY really?". *arXiv*. Retrieved 8 July 2015.

[35] Jan de Vries, Kees. "SUSY fits with full LHC Run I data". *arXiv*. Retrieved 8 July 2015.

21.10   Further reading

- Supersymmetry and Supergravity page in String Theory Wiki lists more books and reviews.

21.10.1   Theoretical introductions, free and online

- S. Martin (2011). "A Supersymmetry Primer". arXiv:hep-ph/9709356.

- Joseph D. Lykken (1996). "Introduction to Supersymmetry". arXiv:hep-th/9612114.

- Manuel Drees (1996). "An Introduction to Supersymmetry". arXiv:hep-ph/9611409.

- Adel Bilal (2001). "Introduction to Supersymmetry". arXiv:hep-th/0101055.

- An Introduction to Global Supersymmetry by Philip Arygres, 2001

21.10.2   Monographs

- Weak Scale Supersymmetry by Howard Baer and Xerxes Tata, 2006.

- Cooper, F.; Khare, A.; Sukhatme, U. (1995). "Supersymmetry and quantum mechanics". *Physics Reports* **251** (5–6): 267. doi:10.1016/0370-1573(94)00080-M. (arXiv:hep-th/9405029).

- Junker, G. (1996). "Supersymmetric Methods in Quantum and Statistical Physics". doi:10.1007/978-3-642-61194-0. ISBN 978-3-540-61591-0..

- Gordon L. Kane.*Supersymmetry: Unveiling the Ultimate Laws of Nature* Basic Books, New York (2001). ISBN 0-7382-0489-7.

- Gordon L. Kane and Shifman, M., eds. *The Supersymmetric World: The Beginnings of the Theory*, World Scientific, Singapore (2000). ISBN 981-02-4522-X.

- Weinberg, Steven, *The Quantum Theory of Fields, Volume 3: Supersymmetry*, Cambridge University Press, Cambridge, (1999). ISBN 0-521-66000-9.

- Wess, Julius, and Jonathan Bagger, *Supersymmetry and Supergravity*, Princeton University Press, Princeton, (1992). ISBN 0-691-02530-4.

- "Concise Encyclopedia of Supersymmetry". 2003. doi:10.1007/1-4020-4522-0. ISBN 978-1-4020-1338-6.

21.10.3   On experiments

- Bennett GW; Muon (g–2) Collaboration; Bousquet; Brown; Bunce; Carey; Cushman; Danby; Debevec; Deile; Deng; Dhawan; Druzhinin; Duong; Farley; Fedotovich; Gray; Grigoriev; Grosse-Perdekamp; Grossmann; Hare; Hertzog; Huang; Hughes; Iwasaki; Jungmann; Kawall; Khazin; Krienen; Kronkvist; et al. (2004). "Measurement of the negative muon anomalous magnetic moment to 0.7 ppm". *Physical Review Letters* **92** (16): 161802. arXiv:hep-ex/0401008. Bibcode:2004PhRvL..92p1802B. doi:10.1103/PhysRevLett.92.161802. PMID 15169217.

- Brookhaven National Laboratory (Jan. 8, 2004). *New g−2 measurement deviates further from Standard Model.* Press Release.

- Fermi National Accelerator Laboratory (Sept 25, 2006). *Fermilab's CDF scientists have discovered the quick-change behavior of the B-sub-s meson.* Press Release.

21.11   External links

- Supersymmetry (physics) at *Encyclopædia Britannica*

- What do current LHC results (mid-August 2011) imply about supersymmetry? Matt Strassler

- ATLAS Experiment Supersymmetry search documents

- CMS Experiment Supersymmetry search documents

- "Particle wobble shakes up supersymmetry", *Cosmos* magazine, September 2006

- LHC results put supersymmetry theory 'on the spot' BBC news 27/8/2011

- SUSY running out of hiding places BBC news 12/11/2012

- Supersymmetry in optics? "Skulls in the Stars" blog 22/08/2013

Chapter 22

Supersymmetry algebra

In theoretical physics, a **supersymmetry algebra** (or **SUSY algebra**) is a mathematical formalism for describing the relation between bosons and fermions. The supersymmetry algebra contains not only the Poincaré algebra and a compact subalgebra of internal symmetries, but also contains some fermionic supercharges, transforming as a sum of N real spinor representations of the Poincaré group. Such symmetries are allowed by the Haag–Lopuszanski–Sohnius theorem. When $N>1$ the algebra is said to have **extended supersymmetry**. The supersymmetry algebra is a semidirect product of a central extension of the super-Poincaré algebra by a compact Lie algebra B of internal symmetries.

Bosonic fields commute while fermionic fields anticommute. In order to have a transformation that relates the two kinds of fields, the introduction of a $\mathbf{Z}_2$-grading under which the even elements are bosonic and the odd elements are fermionic is required. Such an algebra is called a Lie superalgebra.

Just as one can have representations of a Lie algebra, one can also have representations of a Lie superalgebra, called supermultiplets. For each Lie algebra, there exists an associated Lie group which is connected and simply connected, unique up to isomorphism, and the representations of the algebra can be extended to create group representations. In the same way, representations of a Lie superalgebra can sometimes be extended into representations of a Lie supergroup.

22.1 Structure of a supersymmetry algebra

The general supersymmetry algebra for spacetime dimension d, and with the fermionic piece consisting of a sum of N irreducible real spinor representations, has a structure of the form

$$(P \times Z).Q.(L \times B)$$

where

- P is a bosonic abelian vector normal subalgebra of dimension d, normally identified with translations of spacetime. It is a vector representation of L.

- Z is a scalar bosonic algebra in the center whose elements are called central charges.

- Q is an abelian fermionic spinor subquotient algebra, and is a sum of N real spinor representations of L. (When the signature of spacetime is divisible by 4 there are two different spinor representations of L, so there is some ambiguity about the structure of Q as a representation of L.) The elements of Q, or rather their inverse images in the supersymmetry algebra, are called supercharges. The subalgebra $(P \times Z).Q$ is sometimes also called the supersymmetry algebra and is nilpotent of length at most 2, with the Lie bracket of two supercharges lying in $P \times Z$.

- L is a bosonic subalgebra, isomorphic to the Lorentz algebra in d dimensions, of dimension $d(d-1)/2$

- *B* is a scalar bosonic subalgebra, given by the Lie algebra of some compact group, called the group of internal symmetries. It commutes with *P,Z,* and *L,* but may act non-trivially on the supercharges *Q.*

The terms "bosonic" and "fermionic" refer to even and odd subspaces of the superalgebra.

The terms "scalar", "spinor", "vector", refer to the behavior of subalgebras under the action of the Lorentz algebra *L.*

The number *N* is the number of irreducible real spin representations. When the signature of spacetime is divisible by 4 this is ambiguous as in this case there are two different irreducible real spinor representations, and the number *N* is sometimes replaced by a pair of integers (N_1, N_2).

The supersymmetry algebra is sometimes regarded as a real super algebra, and sometimes as a complex algebra with a hermitian conjugation. These two views are essentially equivalent, as the real algebra can be constructed from the complex algebra by taking the skew-Hermitian elements, and the complex algebra can be constructed from the real one by taking tensor product with the complex numbers.

The bosonic part of the superalgebra is isomorphic to the product of the Poincaré algebra *P.L* with the algebra *Z×B* of internal symmetries.

When *N>1* the algebra is said to have **extended supersymmetry**.

When *Z* is trivial, the subalgebra *P.Q.L* is the Super-Poincaré algebra.

22.2 See also

- adinkra symbols

- super-Poincaré algebra

- superconformal algebra

- *N* = 1 supersymmetry algebra in 1 + 1 dimensions

- *N* = 2 superconformal algebra

22.3 References

- Bagger, Jonathan; Wess, Julius (1992), *Supersymmetry and supergravity,* Princeton Series in Physics (2nd ed.), Princeton University Press, ISBN 0-691-02530-4, MR 1152804

- Haag, Rudolf; Sohnius, Martin; Łopuszański, Jan T. (1975), "All possible generators of supersymmetries of the S-matrix", *Nuclear Physics B* **88**: 257–274, Bibcode:1975NuPhB..88..257H, doi:10.1016/0550-3213(75)90279-5, MR 0411396

22.4 Text and image sources, contributors, and licenses

22.4.1 Text

sus22, Introductory adverb clause, MelonBot, SoxBot III, Egmontaz, Notpayingthepsychiatrist, DumZiBoT, BahTab, TimothyRias, Aj00200, Reaperfromhell, Dunkaroo207, XLinkBot, AlexGWU, Impshum, Saeed.Veradi, Little Mountain 5, Guy392, David424, Truthnlove, Qweeveen, Tayste, Addbot, Steven66s, Denali134, Elemented9, Varrey280303, Eric Drexler, Some jerk on the Internet, Fizzycyst, Uruk2008, DOI bot, Jojhutton, AngryBacon, Captain-tucker, Auspex1729, Kongr43gpen, Fgnievinski, Rhetoric Of A Sophist, Ronhjones, CanadianLinuxUser, Cst17, Download, Glane23, Bassbonerocks, Chzz, Favonian, Kronix35, LinkFA-Bot, Udugunit, Aktsu, Tassedethe, Numbo3-bot, Anpecota, Tide rolls, HerpesVirus, SDJ, OlEnglish, Scourge of God, Davidmedlar, Couldbenoway66, Yobot, Maxdamantus, Terrisknickers, Kartano, TaBOT-zerem, Julia W, Unique and proud of it, FireMouseHQ, Terrifictriffid, ArchonMagnus, CinchBug, Synchronism, AnomieBOT, Cleese-heb, 1exec1, Charlesvi, Bigdaddy4x4, Gitman4, Jim1138, IRP, Mintrick, Drweetmola, Ornamentalone, M00npirate, Gautam10, Csigabi, Poli-Psy, Materialscientist, 90 Auto, Citation bot, Teleprinter Sleuth, Vuerqex, Twri, Frankenpuppy, Fuzzy Bob Saget, DirlBot, Georgepowell2008, Heidisql, Cureden, Ekwos, Capricorn42, Gensanders, NFD9001, Anna Frodesiak, Tomwsulcer, A23649, Pra1998, Coretheapple, Ruy Pugliesi, Jagbag2, Vandalism destroyer, Ab1, Omnipaedista, Bandit5005, Shirik, RibotBOT, Waleswatcher, Saalstin, Amaury, Aaron35510, Caz34, Doulos Christos, Sewblon, Born Gay, Capricorn24, SchnitzelMannGreek, A. di M., SpacePyjamas, Kierkkadon, A.amitkumar, Dougofborg, StringLove, Nobelprizewinner, Astiburg, FrescoBot, Fortdj33, Paine Ellsworth, Goodbye Galaxy, HJ Mitchell, Steve Quinn, Vhann, Kwiki, Xhaoz, Citation bot 1, Batong, Gil987, Pinethicket, I dream of horses, Tallboyhoops1991, Three887, Steveo27five, RedBot, Sardinita, Serols, Vhsatheeshkumar, Swisstingle, DeletionUK, Reconsider the static, IVAN3MAN, Remingtonhill1, Orenburg1, Coltonhs, Willy Weazley, Sma-maret, Bethovenn, Dinamik-bot, Dc987, Oswaldo Zapata, Egemont, Syebo, Alaithiran, Reaper Eternal, Seahorseruler, Ybungalobill, Quaker phil, Specs112, Dr. Aakash Patel, Tbhotch, StormbringerUK, Minimac, Mathgenius3141592, Keegscee, Omgwaffels, Mick le pick, Solancel, Aznhero3793, Dwielark, Afteread, Enauspeaker, EmausBot, MaooaM, Immunize, Az29, Milkocookie, Faolin42, Fotoni, RA0808, Rename-dUser01302013, 8digits, Yukiseaside, Slightsmile, Tommy2010, Winner 42, Wikipelli, JonezyKiDx, Joe Gazz84, ZéroBot, Timeitsways, John Cline, Cogiati, Quaqa, Chrispaps2413, Nasulikid, Vollrath2323, Benjamin1414141414141414, Arbnos, Green Lane, A930913, Bamy-ers99, Azeraphale, H3llBot, Encyclopadia, Danga1988, Ollainen, PoisonGM, Wayne Slam, OnePt618, Knome335, L Kensington, Lulzpro-tuns, Kranix, Rpcappello, Maschen, Vastly~enwiki, Donner60, CatFiggy, CountMacula, Orange Suede Sofa, Etov, M1k3 101, Bill william compton, Wakabaloola, TERBAFAN, Nickslspride34, NeuralLotus, Isocliff, Brechbill123, Xanchester, ClueBot NG, Martti Muukkonen, KagakuKyouju, Jeff Song, This lousy T-shirt, Satellizer, Name Omitted, Marcdean123, Wiki incorp, Frietjes, O.Koslowski, Alexdamaino9, Dream of Nyx, Blackhall616, Widr, Sashhere, WikiPuppies, Stu181, T00g00d96, Pluma, Storm.sarup, Helpful Pixie Bot, Manzeet, Waffle-boy36, HMSSolent, Mikeshelton1, Bibcode Bot, 2001:db8, Phillip.phillipson, Hoaxinator, Lowercase sigmabot, Thor cherubim, Mrshabam, Nischt, Flowerhat15, AvocatoBot, Housegeek224, MahRanch, Benzband, Altaïr, Benhenchdickthomas, Shreyakstring, Sweaty maori sphinc-ter, DaFalk, Dsabo74, Ratanmaitra, MM4EVAH, Steven.w.kowalski, Minsbot, JGallardo2600, Dylanlatham, Myfriendganesha, OCCullens, Likeaboss189, Sean271293, LinusE8, BattyBot, Several Pending, Aldrich2122, CommanderMoka, The Illusive Man, ChrisGualtieri, Koala-maN2, Trevorkid45, Catsloveit07, Alex Modzz, Rustyjamsen, Goh ryangoh, Dexbot, Exolius, Hilander316, Alman1234321, SuperCalzer, LightandDark2000, MeekMelange, BQND, Cdarrai1, Kephir, TheMonkeyboy524, Michael Anon, Mattfat8, Lugia2453, Anruy, Rachel weld, Jamesx12345, AHusain314, BossEditors, Hillbillyholiday, Joeinwiki, Mattninja, Theshadow444, Asaa82, Jakemarz197, Kzhang1025, Epic-genius, Spongbob456789, ⚛, TestMaster, Ianreisterariola, GrapperJ, Makeitnasty, Moemajdi, I am One of Many, NualaIvy, BAZINGASS, St3fanPC, Eyesnore, Isaac grozd, Jordanissexyaf1999, Baruch6525, Mosbruckercj, Ihatedirac2k13, Jonamithy121314, 123physicsquantum, Jt198, RaphaelQS, HeyJude70, AParker628, DimReg, A.k.blaze1, Joshuk, Zenibus, Nianoobasik, Ihelpapplen, Gamo To Apoel, Sacred-Labyrinth, Ginsuloft, Vampre1122, Dimension10, Howard Wolowitz, AddWittyNameHere, Polytope24, Elysion, Tutun12S, Longerboats5, SimonWombat8, Konveyor Belt, Vtank54, Micheal545, Hck24, Caliae19, Hexafish, Simpick, TheRealTheKoi, Bballbro62, Monkbot, Army-Path, TheQ Editor, Jtsmith098, Joshmiller1, Hanseer360, XXvPIEvXx, Dbennett 24, Ghikpenos, Nick65633, Saundra03, Thehippothatknows, Sewwgers, Teelaskeletor, Cirksena, Balockaye1234, PloppyDoo, Yesufu29, Lumpy2k14, Podayeruma, Abstract92, Sbenfiel, Monkman2k4, Swegwegdgfyetkfotfkkfkfkv, John95541234, Poopman224, ScrapIronIV, Tetra quark, GeneralizationsAreBad, Shivansh2014n, KasparBot, SHUCKYLUCKY, Fabiotheoto, FartGoblin, Joca potato, Joshcool246, Theoretical Physisist4444 and Anonymous: 1546

• **Topological string theory** *Source:* https://en.wikipedia.org/wiki/Topological_string_theory?oldid=646032537 *Contributors:* Charles Matthews, Giftlite, LeYaYa, Lumidek, Woohookitty, Rjwilmsi, John Baez, Mgnbar, SmackBot, Colonies Chris, Dreadstar, Harryboyles, JarahE, Twyder, R'n'B, Natsirtguy, Henry Delforn (old), ClueBot, Mild Bill Hiccup, SchreiberBike, Rror, Addbot, Ozob, Wireader, Raven1977, Omnipaedista, Charvest, FrescoBot, EmausBot, Timothyklustnuh, Quondum, Enyokoyama, Makecat-bot, Jamesmcmahon0, Polytope24, Broido and Anonymous: 10

• **Quantum field theory** *Source:* https://en.wikipedia.org/wiki/Quantum_field_theory?oldid=682800089 *Contributors:* AxelBoldt, CYD, Mav, The Anome, XJaM, Roadrunner, Stevertigo, Michael Hardy, Tim Starling, IZAK, TakuyaMurata, SebastianHelm, Looxix~enwiki, Ahoerste-meier, Cyp, Glenn, Rotem Dan, Stupidmoron, Charles Matthews, Timwi, Jitse Niesen, Kbk, Rudminjd, Wik, Phys, Bevo, BenRG, Northgrove, Robbot, Bkalafut, Gandalf61, Rursus, Fuelbottle, Tobias Bergemann, Ancheta Wis, Giftlite, Lethe, Dratman, Alison, St3vo, Mboverload, DefLog~enwiki, ConradPino, Amarvc, Pearbonn, Karol Langner, APH, AmarChandra, D6, CALR, Urvabara, Discospinster, Guanabot, Ig-orivanov~enwiki, Masudr, Pjacobi, Vsmith, Nvj, MuDavid, Bender235, Pt, El C, Shanes, Sietse Snel, Physicistjedi, KarlHallowell, PWilkinson, Helix84, Thialfi, Varuna, Gcbirzan, Docboat, Count Iblis, Egg, Mpatel, Marudubshinki, Graham87, Opie, Vanderdecken, Rjwilmsi, MarSch, Earin, R.e.b., RE, Strobilomyces, Arnero, Itinerant1, Alfred Centauri, Srleffler, Chobot, UkPaolo, Wavelength, Bambaiah, Hairy Dude, Russ-Bot, TimNelson, Archelon, CambridgeBayWeather, SCZenz, Odddmonster, E2mb0t~enwiki, Semperf, Tetracube, Garion96, Erik J, Robert L. Banus, RG2, SmackBot, Stephan Schneider, Tom Lougheed, Melchoir, KocjoBot~enwiki, Mcld, Dauto, Chris the speller, Complexica, Threepounds, RuudVisser, QFT, Jmnbatista, Cybercobra, Rebooted, Victor Eremita, DJIndica, Lambiam, Mgiganteus1, Zarniwoot, Jim.belk, Stwalkerster, SirFozzie, Hu12, Dan Gluck, Iridescent, Joseph Solis in Australia, Albertod4, Van helsing, BeenAroundAWhile, Witten Is God, Cydebot, Jamie Lokier, Meno25, Michael C Price, The 80s chick, Mendicus~enwiki, AstroPig7, Msebast~enwiki, Mbell, Headbomb, Nick Number, Mentifisto, AntiVandalBot, Bt414, Bananan~enwiki, Martin Kostner, Moltrix, Kasimann, Kromatol, Puksik, Lerman, LLHolm, RogueNinja, Tlabshier, JEH, Nikolas Karalis, Storkk, JAnDbot, Igodard, Four Dog Night, N shaji, Bongwarrior, Andrea Allais, Soulbot, Etale, Maliz, Custos0, HEL, J.delanoy, Acalamari, Jeepday, Policron, Blckavnger, Juliancolton, Skou, Telecomtom, GrahamHardy, Sheliak, Cuzkatz-imhut, VolkovBot, Bktennis2006, Marksr, HowardFrampton, The Original Wildbear, Dj thegreat, Markisgreen, TBond, Lejarrag, Moose-32, Raphtee, Sue Rangell, Neparis, Drschawrz, YohanN7, SieBot, TCO, Yintan, Likebox, Paolo.dL, Tugjob, Henry Delforn (old), Jecht (Final Fantasy X), OKBot, StewartMH, ClueBot, EoGuy, Wwheaton, The Wild West guy, Shvav~enwiki, Bob108, Brews ohare, Thingg, Count Truthstein, XLinkBot, PSimeon, SilvonenBot, Truthnlove, HexaChord, Addbot, ConCompS, Pinkgoanna, Leapold~enwiki, Dmhowarth26, Glane23, Hanish.polavarapu, Lightbot, Scientryst, R.ductor, Ettrig, Yndurain, Legobot, Luckas-bot, Yobot, Ht686rg90, Niout, Tamtamar,

norge, ElPeste, Afteread, EmausBot, Detogain, John of Reading, Racerx11, GoingBatty, XinaNicole, Ensabah6, Uploadvirus, ZéroBot, Arbnos, Zueignung, WaterCrane, Crown Prince, LaurentRDC, Isocliff, Vodkacannon, Raidr, Helpful Pixie Bot, Titodutta, Bibcode Bot, BG19bot, Spaligo, KateWishing, PhnomPencil, Sylvain.maurin, Kecchina, Halfb1t, Brad7777, Fylbecatulous, Jimw338, MyTuppence, Mogism, LT-Woods, Andyhowlett, Jawa0, &reasNink, SomeFreakOnTheInternet, Tentinator, EvergreenFir, DimReg, Pedarkwa, Db9199 24, Anrnusna, Notspelly, Ntomlin1996, Monkbot, Isbromberg, Dsprc, YeOldeGentleman, Tetra quark and Anonymous: 333

- **Isomorphism**Source: https://en.wikipedia.org/wiki/Isomorphism?oldid=682549504Contributors: AxelBoldt,Zundark,Andre Engels, san, Ghakko, Edemaine, Ryguasu, Youandme, Stevertigo, Patrick, Michael Hardy, Isomorphic, TakuyaMurata, Glenn, Netsnipe, Mxn, Revolver, Charles Matthews, Reddi, Dysprosia, Andrewman327, Zero0000, Phys, Robbot, Bkell, Marc Venot, Tosha, Giftlite, MathKnight, Peruvianllama, MarkSweep, PhotoBox, Wrp103, Paul August, Bender235, Elwikipedista~enwiki, Nabla, Rgdboer, EmilJ, Army1987, Msh210, Philip Cross, Oleg Alexandrov, LOL, BD2412, Yurik, Zbxgscqf, Mattmacf, Mathbot, Chobot, YurikBot, Wavelength, Spacepotato, Jlittlet, Michael Slone, KSmrq, Mathwiz777, Grubber, Vanished user 1029384756, Stuhacking, Banus, SmackBot, Rljacobson, The Rhymesmith, Kmarinas86, Lubos, Nbarth, Chlewbot, Maksim-bot, Bryanmcdonald, Spinality, Nick Green, Fantomdrives, Cronholm144, Jim.belk, 16@r, Mets501, Rschwieb, Yuide, CRGreathouse, Krauss, Sam Staton, Rlupsa, QuiteUnusual, Hannes Eder, TK-925, JAnDbot, Avaya1, Bahar, Coolhandscot, Magioladitis, Koberozendaal, JamesBWatson, Albmont, Uncle Dick, Smite-Meister, Cpiral, Trumpet marietta 45750, Policron, Bigdumbdinosaur, STBotD, Mcole13, Michael Angelkovich, Borhan0, LokiClock, Thaddeus Slamp, Anonymous Dissident, JhsBot, PaulTanenbaum, Spinningspark, Mike4ty4, SieBot, Ivan Štambuk, Ssavelan, Iamthedeus, Taemyr, Thehotelambush, Richard Molnár-Szipai, Anchor Link Bot, DixonD, Superbatfish, Martarius, Alexbot, James.Demetriou, Brews ohare, Subversive.sound, Addbot, Omnipedian, Debresser, Ozob, PV=nRT, Jarble, Luckas-bot, Ezequiels.90, Nallimbot, AnomieBOT, Xqbot, XZeroBot, Omnipaedista, AllCluesKey, Sławomir Biały, Ebony Jackson, RedBot, Lars Washington, Pokus9999, Gryllida, عقيل كاشف, EmausBot, Fly by Night, Racerx11, GoingBatty, Slawekb, Zell08v, Quondum, D.Lazard, SporkBot, YnnusOiramo, ClueBot NG, Frietjes, Mesoderm, HMSSolent, BG19bot, Канеюку, M hariprasad, Brad7777, Pratyya Ghosh, Pariefracture, Mark viking, Eyesnore and Anonymous: 88

- **Penrose transform** Source: https://en.wikipedia.org/wiki/Penrose_transform?oldid=601794084 Contributors: Michael Hardy, MarSch, R.e.b., Wikid77, Headbomb, Sławomir Biały, Citation bot 1, Helpful Pixie Bot, Bibcode Bot, Dimension10, Anrnusna and Anonymous: 1

- **Sesquilinear form** Source: https://en.wikipedia.org/wiki/Sesquilinear_form?oldid=661453323 Contributors: Charles Matthews, Phys, Sverdrup, Centrx, Giftlite, Fropuff, Matt Crypto, Almit39, Rgdboer, Kusma, Uncle G, FlaBot, Mathbot, Don Gosiewski, Bgwhite, YurikBot, Michael Slone, Markus Schmaus, Evilbu, Maksim-e~enwiki, Mhss, Nbarth, Colonies Chris, Hongooi, Summentier, Jaksmata, KarolS, Albmont, Plclark, Yoda of Borg, Akiry, Addbot, Мышка, LaaknorBot, EmausBot, Quondum, Wcherowi, Nbrothers, RudolfRed, Aszilagyi and Anonymous: 23

- **Conformal group** Source: https://en.wikipedia.org/wiki/Conformal_group?oldid=648705164 Contributors: Phys, Nbarth, D.H. Antic-Hay, Mr. Stradivarius, Addbot, Yobot, Sławomir Biały, Quondum, Rezabot, Jmiki, Qetuth, Khazar2 and Anonymous: 4

- **Complex manifold**Source: https://en.wikipedia.org/wiki/Complex_manifold?oldid=667946938Contributors: Zundark,Michael Hardy, Matthews, Dysprosia, MathMartin, Tosha, Giftlite, Lethe, Fropuff, Waltpohl, Pjacobi, Oleg Alexandrov, Joriki, Linas, MarSch, R.e.b., Wavelength, Michael Slone, Tong~enwiki, Larsobrien, Nicholas Jackson, SmackBot, Nbarth, Michael Kinyon, Harej bot, Xantharius, Palbin, Sophie means wisdom, Strange but untrue, Red Act, Kesseki, AnonyScientist, Addbot, Legobot, Yobot, 9258fahsflkh917fas, Howard McCay, LucienBOT, EmausBot, ZéroBot, Quondum, Taladris, Helpful Pixie Bot, NotWith, Brad7777, Qetuth, Enyokoyama, AHusain314, KasparBot and Anonymous: 22

- **Homogeneous space** Source: https://en.wikipedia.org/wiki/Homogeneous_space?oldid=660895840 Contributors: Michael Hardy, TakuyaMurata, Charles Matthews, Dcoetzee, Dysprosia, Phys, Choni, Tobias Bergemann, Giftlite, Fropuff, Fleminra, Tomruen, Paul August, Gauge, Killing Vector, Oleg Alexandrov, Joriki, Linas, MarSch, Mathbot, Chobot, Eienmaru, Siddhant, YurikBot, Archelon, Silly rabbit, Nbarth, YK Times, Apon, Ixionid, Lantonov, Squids and Chips, Trigamma, YoungFrog, LokiClock, TXiKiBoT, Mr. Stradivarius, Alexbot, Nilradical, SilvonenBot, Addbot, Topology Expert, Fluffernutter, Point-set topologist, Jschnur, Fly by Night, Dewritech, Quondum, D.Lazard, Helpful Pixie Bot, Brad7777, Qetuth, Brirush, Vskrin and Anonymous: 25

- **Conformal map** Source: https://en.wikipedia.org/wiki/Conformal_map?oldid=680445730 Contributors: AxelBoldt, PierreAbbat, Patrick, Michael Hardy, Meekohi, Smack, Charles Matthews, Phys, Topbanana, Rogper~enwiki, Robbot, Rvollmert, Moink, Paul Murray, Cyrius, Tosha, Giftlite, MSGJ, DefLog~enwiki, PhotoBox, Zowie, Pjacobi, Billlion, El C, Rgdboer, Reinyday, Mattpickman, Oleg Alexandrov, OdedSchramm, Mpatel, Kri, Adoniscik, Wavelength, Tong~enwiki, Grafen, Daniel Mietchen, Morpheios Melas, Paul D. Anderson, Mebden, That Guy, From That Show!, Adam majewski, RowBean, Silly rabbit, Vanished User 0001, Just plain Bill, SingCal, Sbmehta, Ulner, JorisvS, George100, Cydebot, WISo, Michael C Price, Thijs!bot, Headbomb, Paquitotrek, D.H, BigJohnHenry, Widefox, Magioladitis, Christian.Mercat, Lantonov, Policron, Pleasantville, Rei-bot, Marcosaedro, Ishboyfay, SieBot, Steven Crossin, Chansonh, Huku-chan, ClueBot, JP.Martin-Flatin, Ledkar, MisterCircle, Awickert, Dthomsen8, Addbot, Fgnievinski, AkhtaBot, Blmichel, Cesiumfrog, Legobot, Luckas-bot, Timeroot, Erel Segal, Xqbot, Nickkid5, FrescoBot, Citation bot 1, Didactik, Miracle Pen, David.c.stone, EmausBot, ZéroBot, ClueBot NG, Helpful Pixie Bot, Giuseppe Negro, IkamusumeFan, Mark viking, ReconditeRodent, WillemienH, KasparBot and Anonymous: 40

- **Scattering amplitude** Source: https://en.wikipedia.org/wiki/Scattering_amplitude?oldid=661582111 Contributors: Timwi, Lumidek, Drw25, Tpikonen, Modify, Teply, SmackBot, Kdliss, Myasuda, Thijs!bot, Headbomb, Bakken, Fylwind, Mutlay, Bmtran, Hugh16, RedBlade7, Polyamorph, Addbot, Luckas-bot, Yobot, ^musaz, Unara, Mnmngb, Mforesto, Tom.Reding, Affen15, GoingBatty, ZéroBot, Mjbmrbot, Mark viking and Anonymous: 7

- **Feynman diagram** Source: https://en.wikipedia.org/wiki/Feynman_diagram?oldid=682793503 Contributors: CYD, Bryan Derksen, XJaM, Nate Silva, Shii, Stevertigo, Edward, Michael Hardy, Looxix~enwiki, Ahoerstemeier, Ryan Cable, Nikai, Kaihsu, Charles Matthews, Ww, Doradus, Furrykef, Phys, Omegatron, Bevo, Bhiggs, BenRG, Peak, Rasnus Faber, Intangir, SC, Wikibot, Anthony, Tobias Bergemann, Danenberg, Ancheta Wis, Giftlite, Sj, Harp, Alison, Remy B, Vivektewary, Sigfpe, Beland, Pmanderson, Icairns, Sam Hocevar, Robin klein, AlexChurchill, Urvabara, Jkl, Pyrop, Guanabot, Pjacobi, Vsmith, Rspeer, Mal~enwiki, Bender235, Ben Standeven, Robert P. O'Shea, AnyFile, Jjk, Matt McIrvin, Scentoni, Tritium6, Mdd, Neonumbers, Rgclegg, Ferrierd, Mac Davis, Tony Sidaway, RJFJR, Drat, Dominic, Markko, Linas, LoopZilla, Daira Hopwood, Ketiltrout, Koavf, Zbxgscqf, Commander, Strait, Miserlou, Ligulem, RE, FlaBot, Moskvax, RobertG, Gnostic804, Acyso, DoomBringer, Chobot, YurikBot, JabberWok, Gaius Cornelius, Rodier, Anomalocaris, SEWilcoBot, Welsh, SCZenz, Ragesoss, Nubby, Larsobrien, Ejl, DRosenbach, Dna-webmaster, Tomj, Caco de vidro, GrinBot~enwiki, KasugaHuang, SmackBot, Tom Lougheed,

GaeusOctavius, Chris the speller, Jjalexand, JustThisGuy, Pieter Kuiper, Complexica, Colonies Chris, Salmar, Wikipedia brown, Xiner, Huon, Tesseran, DJIndica, Eliyak, SilverStar, Xiphoris, JanBielawski, Bitwise, Beefyt, Paul venter, Joseph Solis in Australia, Wikifarzin, Patrickwooldridge, 8754865, CmdrObot, Van helsing, Jsmaye, Joelholdsworth, Myasuda, Cydebot, Xxanthippe, Michael C Price, Thijs!bot, Epbr123, Barticus88, Headbomb, Davidhorman, Shlomi Hillel, Leevclarke, Sluzzelin, AniRaptor2001, CosineKitty, Jameskeates, Bakken, SHCarter, Swpb, Warchef, Connor Behan, R'n'B, Choihei, Sefog, Chiswick Chap, Fylwind, Joshmt, Vyn, Brvman, Sheliak, Mulanhua, Quilbert, Rei-bot, Kevin Steinhardt, Mbusux, Richwil, Ptrslv72, Drschawrz, SieBot, Likebox, Taemyr, Staylor71, Martarius, WurmWoode, ChandlerMapBot, Sjdunn9, DragonBot, Chutsu, GlasGhost, DumZiBoT, TimothyRias, Mchaddock, PSimeon, SilvonenBot, Truthnlove, Out of Phase User, Metsavend, Download, Chamal N, Favonian, ChenzwBot, Barak Sh, Mikkim64, AgadaUrbanit, Conroy23, Alfie66, Legobot, Luckas-bot, Yobot, Amirobot, AnomieBOT, Piano non troppo, Xqbot, Srich32977, PhysicsR, Noamz, Seeleschneider, Createangelos, Kismalac, Ysyoon, Jonesey95, Nurefsan, Fizzotter, Meier99, Hickorybark, Earthandmoon, Aaivazis, EmausBot, John of Reading, Bookalign, WikitanvirBot, Bornerdogge, Maschen, Zueignung, Xanchester, Anagogist, Antiqueight, Smack the donkey, Pfeiferwalter, OCCullens, Bakkedal, ChrisGualtieri, Equatorbit, Mrmagikpants, Jamesx12345, Pjpeters, Mark viking, MutluMan, Someone not using his real name, UltraBird, Airwoz, Monkbot, Quogle, Cheweblaze, FivePillarPurist and Anonymous: 127

- **Amplituhedron** *Source:* https://en.wikipedia.org/wiki/Amplituhedron?oldid=678112546 *Contributors:* The Anome, Michael Hardy, Julesd, Timwi, Ancheta Wis, DocWatson42, Tomruen, Hydrox, Koavf, Miserlou, FayssalF, Sus scrofa, Bhny, Melchoir, Rrburke, Camilo Sanchez, Lentower, Myasuda, Headbomb, Yellowdesk, Lfstevens, Steelpillow, Planeta~enwiki, Tonyfaull, David Eppstein, GermanX, OttoMäkelä, Camrn86, Satani, Rodney1h, Topquark22, AnomieBOT, Dr. Günter Bechly, Joshuafilmer, Omnipaedista, RockMagnetist, ClueBot NG, Cliff12345, EuroCarGT, AHusain314, Mark viking, MikeRosseel, Mvonhipp and Anonymous: 15

- **Supersymmetry** *Source:* https://en.wikipedia.org/wiki/Supersymmetry?oldid=681009502 *Contributors:* Bryan Derksen, Taw, Andre Engels, Roadrunner, Maury Markowitz, Ewen, Stevertigo, Edward, Michael Hardy, Arpingstone, Theresa knott, IMSoP, Jeandré du Toit, Samw, Smack, Charles Matthews, Maximus Rex, Phys, Raul654, BenRG, Rursus, Mor~enwiki, Ancheta Wis, Giftlite, Mporter, Ferkelparade, Monedula, Fropuff, Xerxes314, Anville, Gus Polly, Moyogo, Unconcerned, DO'Neil, Maarten van Vliet, Pharotic, LiDaobing, Sam Hocevar, Lumidek, Deglr6328, Arivero, Rich Farmbrough, Roybb95~enwiki, Bender235, El C, Nornagon~enwiki, Duk, Tweet Tweet, Russ3Z, LostLeviathan, Pearle, Gary, Francescog~enwiki, Wtmitchell, RJFJR, Reaverdrop, Blaxthos, Killing Vector, Jordan14, Ted BJ, MONGO, Mpatel, MFH, SeventyThree, Bodera, VermillionBird, Drbogdan, Rjwilmsi, Josiah Rowe, R.e.b., Bubba73, Maxim Razin, Drrngrvy, FlaBot, Cless Alvein, Nowhither, Itinerant1, Gparker, KFP, Lmatt, Chobot, Vyroglyph, YurikBot, Wavelength, RussBot, Ohwilleke, Bhny, Epolk, Sasuke Sarutobi, Maxim Leyenson, Chaos, Romanc19s, Bota47, Mgnbar, Closedmouth, Arthur Rubin, RG2, That Guy, From That Show!, A bit iffy, SmackBot, Mira, Kurochka, Wangjiaji, Gilliam, Bluebot, Cadmasteradam, Complexica, Bazonka, Colonies Chris, Can't sleep, clown will eat me, QFT, Ruff ilb, Wen D House, Solarapex, Radagast83, Jgwacker, TheMaster42, Martijn Hoekstra, Ligulembot, Acjohnson55, Yevgeny Kats, Charleswestbrook, TriTertButoxy, Lambiam, Tktktk, Xiaphias, JarahE, Mdanziger, Dan Gluck, Newone, Marysunshine, Tawkerbot2, Cydebot, Hydraton31, Bazzargh, David edwards, Michael C Price, Crum375, Koeplinger, Headbomb, J.christianson, Escarbot, Salgueiro~enwiki, Kborland, Jpod2, Cgingold, Maliz, TimidGuy, C9, Kostisl, R'n'B, Zentropa77, Natsirtguy, Maurice Carbonaro, Kevin Hickerson, Shawn in Montreal, Idioma-bot, Sheliak, Cuzkatzimhut, Nxavar, Kawakameha, Cuboidal, Ptrslv72, PhysPhD, Kbrose, SieBot, Nn123645, ClueBot, Jcpilman, Chessmaster7m, Kitsunegami, Rhododendrites, Mastertek, Mishas42, Scrabby~enwiki, TimothyRias, WikHead, MystBot, Addbot, DOI bot, Zahd, Barak Sh, F Notebook, Lightbot, Windward1, Luckas-bot, Yobot, Ibayn, TaBOT-zerem, Amirobot, Nonnormalizable, AnomieBOT, Girl Scout cookie, Materialscientist, Citation bot, ArthurBot, Plumpurple, Tomwsulcer, Omnipaedista, Gsard, CES1596, FrescoBot, HaloStereo1, Paine Ellsworth, Xmikywayx, Citation bot 1, Gil987, Kikeku, Jonesey95, Eddie Nixon, MondalorBot, Aknochel, Tom1661, Gagoga ju, TobeBot, Puzl bustr, Andraas, EmausBot, Djloststylez, Ddimensões, Arbnos, Susy is it, ChuispastonBot, Isocliff, ClueBot NG, KagakuKyouju, IJVin, Frietjes, Helpful Pixie Bot, Bibcode Bot, BG19bot, Teika kazura, JayBeeEye, Ninmacer20, ChrisGualtieri, Dexbot, Logosun, AHusain314, NA48, Rfassbind, Katherine Pendleton, Lioinnisfree, Laplacemat, Liquidityinsta, TaiSakuma, Stamptrader, Kdmeaney, Qxxxxxq, Almaionescu, Monkbot, Janhaithabu, Mammoth2011, Jwill530, Stacie Croquet, Cuttlas1 and Anonymous: 173

- **Supersymmetry algebra** *Source:* https://en.wikipedia.org/wiki/Supersymmetry_algebra?oldid=670730267 *Contributors:* Michael Hardy, Dod1, R.e.b., Pred, Papa November, Mathsci, Myasuda, Michael C Price, Thijs!bot, Headbomb, Nilradical, MystBot, Addbot, Barak Sh, AnomieBOT, Omnipaedista, Charvest, Jonesey95, ClueBot NG, Bibcode Bot, Colbert Sesanker, Mark viking, TuxLibNit, AHusain3141 and Anonymous: 5

22.4.2 Images

- **File:AdS3.svg** *Source:* https://upload.wikimedia.org/wikipedia/commons/4/47/AdS3.svg *License:* CC BY-SA 3.0 *Contributors:* This file was derived from: AdS3 (new).png
  Original artist:

- derivative work: Alex Dunkel (Maky)

- **File:Affine_subspace.svg** *Source:* https://upload.wikimedia.org/wikipedia/commons/8/8c/Affine_subspace.svg *License:* CC BY-SA 3.0 *Contributors:* Own work *Original artist:* Jakob.scholbach

- **File:Ambox_important.svg** *Source:* https://upload.wikimedia.org/wikipedia/commons/b/b4/Ambox_important.svg *License:* Public domain *Contributors:* Own work, based off of Image:Ambox scales.svg *Original artist:* Dsmurat (talk · contribs)

- **File:Black_Hole_Merger.jpg** *Source:* https://upload.wikimedia.org/wikipedia/commons/d/d1/Black_Hole_Merger.jpg *License:* Public domain *Contributors:* Taken from http://www.space.com/imageoftheday/image_of_day_060203.html credit is listed to NASA. *Original artist:* NASA

- **File:Calabi-Yau.png** *Source:* https://upload.wikimedia.org/wikipedia/commons/d/d4/Calabi-Yau.png *License:* CC BY-SA 2.5 *Contributors:* own work by Lunch
  http://en.wikipedia.org/wiki/Image:Calabi-Yau.png (english Wikipedia) *Original artist:* Lunch

- **File:Calabi_yau.jpg** *Source:* https://upload.wikimedia.org/wikipedia/commons/f/f3/Calabi_yau.jpg *License:* Public domain *Contributors:* Mathematica output, created by author *Original artist:* Jbourjai

22.4.3 Content license

- Creative Commons Attribution-Share Alike 3.0